DIGITAL TRANSMISSION LINES

DIGITAL
TRANSMISSION LINES
COMPUTER MODELLING AND ANALYSIS

Kenneth D. Granzow

Logicon R&D Associates

New York   Oxford
OXFORD UNIVERSITY PRESS
1998

OXFORD UNIVERSITY PRESS

Oxford   New York
Athens   Auckland   Bangkok   Bogota   Bombay   Buenos Aires
Calcutta   Cape Town   Dar es Salaam   Delhi   Florence   Hong Kong
Istanbul   Karachi   Kuala Lumpur   Madras   Madrid   Melbourne
Mexico City   Nairobi   Paris   Singapore   Taipei   Tokyo   Toronto   Warsaw

and associated companies in
Berlin   Ibadan

Published by Oxford University Press, Inc.,
198 Madison Avenue, New York, New York, 10016
http://www.oup-usa.org
1-800-334-4249

Library of Congress Cataloging-in-Publication Data
Granzow, Kenneth D.
Digital transmission lines : computer modelling and analysis /
Kenneth D. Granzow.
p.   cm.
Includes bibliographical references and index.
ISBN 0-19-511292-X (cloth)
1. Multiconductor transmission lines—Computer simulation.
2. Digital circuits—Computer simulation.
TK7872.T74G7 1998
621.3815—dc21   97-9513
                                                CIP

2 4 6 8 9 7 5 3 1
Printed in the United States of America
on acid-free paper

To the One:
WHO CONCEIVED OF THE COSMOS, ITS PHYSICAL LAW;
FLUNG IT IN SPACE, TURNED IT ON, LET IT BRILLIANTLY RUN.
WHO GAVE MAN A MIND, FILLED IT WITH AWE
AND INQUISITIVE THINKING. . . BUT HE'LL NOT KNOW ALL. . .
UNTIL HE MEETS THE ONE.

"THE FEAR OF THE LORD IS THE BEGINNING OF WISDOM, AND
KNOWLEDGE OF THE HOLY ONE IS UNDERSTANDING."
—PROVERBS 9:10 (NIV)

CONTENTS

FOREWORD

Almost every facet of modern life depends in some degree on man's mastery of electromagnetism; as the governing laws, Maxwell's equations thus command a central importance in technological culture. While few technologically literate people would deny the truth of this assertion, many would be surprised by the limited part that is played in electrical engineering by complete solutions of the 3D Maxwell's equations. Rather, an assumption of symmetry is often justified that yields reduced equations which allow for a tractable analysis. Such symmetry considerations lead in many circumstances to one of the most useful forms of the equations — the transmission line approximation of Maxwell's equations.

As a new graduate, I first encountered transmission line theory and Ken Granzow, the author of this book, in my initial professional situation. Under his guidance my academic, and somewhat uncertain, knowledge of Maxwell's equations was solidified and supplemented with a practical knowledge of circuit theory and transmission lines. Over the ensuing years, I benefited greatly from the techniques that Ken taught me as numerous problems were confronted — EMP coupling to missile bodies and silos, railroad tracks and cars, radio towers, command facilities and to communication and power lines as well — which quickly yielded to a transmission line analysis. In such applied problems, a basic mathematical viewpoint of the underlying physics, and a software implementation unencumbered by special case exceptions, lends great flexibility to Granzow's approach.

This book lays out tools of remarkable applicability; networks of multi-conductor transmission lines, where line junctions can involve arbitrary circuits of passive or reactive or nonlinear elements, where the networks can be embedded in homogeneous or inhomogeneous media, and where skin-depth effects in conductors can be taken into account, are easily modeled. Emphasizing the particular context of circuit boards, the book provides engineers with practical methods for the elimination of crosstalk, a problem of much contemporary interest. Overall, Granzow has employed his long experience in physics and applied mathematics to give a unique and insightful delineation of transmission line theory; the necessary background material, mathematical algorithms, numerical techniques and software implementation are all presented in a most accessible manner. The skills represented in this book should be in the repertoire of anyone engaged with electromagnetic phenomena.

Tom Larkin,
Senior Scientist
Metatech Corporation

PREFACE

This book focuses on practical mathematical algorithms to simulate the propagation of signals on circuit boards. Computer codes are included to demonstrate the algorithms and give the reader "hands-on" experience with them. It is not a review of state-of-the-art methods, but rather a collection of algorithms that the author has successfully used for over 20 years to simulate multi-conductor transmission lines in the time domain. The author began transmission line simulation work as it relates to the coupling and propagation of nuclear burst generated electromagnetic pulses (NEMP) to military systems. Recently, these methods have been applied to a code to simulate propagation and crosstalk on circuit boards. The circuit board simulator was developed for HyperLynx, Inc., Redmond, WA and is to be used in their CAD program BoardSim and other simulation codes.

The methods of the book provide a new tool for designing digital transmission lines that are physically compact, but with low crosstalk. All simulation algorithms that are developed are time-stepping methods that are based on time domain solutions of the (various) appropriate differential equations. The time domain approach to the skin effect provides a way to include skin effect in the propagation of signals with an accuracy not found in former methods.

The book is in five parts:

I. **Transmission Line Fundamentals**. In this part, time domain methods are developed to simulate the propagation of signals on single- and multi-wire transmission lines. The simulation of crosstalk is included. The basic mathematical tool is the transport form of the transmission line equations. The methods presented in this part of the book apply to lines with a single propagation speed. A review of network solution methods for resistive networks and a code that performs general resistive network solutions is included (in Chapter 3). The notion of termination crosstalk and its control is introduced. General lines on a layered circuit board (with multiple propagation speeds) are treated in Part III.

II. **Circuit Solutions at Line Terminations**. Algorithms for time domain solution of general networks using the node-equation approach are presented. Emphasis is placed on solving simultaneously line propagation and the networks resulting from transmission line terminations and interconnection. A general network solving computer code that will treat reactive and nonlinear circuit elements simultaneously with propagation on transmission lines is discussed.

III. Propagation in Layered Media. A method of simulating the propagation of signals on a multi-wire line with multiple propagation speeds is presented. A method is given for separating a signal into "modes" of discreet propagation speed and then recombining the modes to obtain trace voltages and currents at the line terminals. This method is applicable to layered circuit boards with layers of different dielectric constants.

IV. Transmission Line Parameter Determination. Methods to calculate the parameters L (inductance per length) and C (capacitance per length) for a given line cross section on a layered circuit board are given. For multi-wire lines, these parameters are matrices.

V. Simulation of Skin Effect. The physics of the skin effect and the classical frequency domain analysis of it is reviewed. Then, time domain methods are derived for simulating the skin effect for a planar geometry and for a cylindrical geometry.

Propagation is treated by the quasi-static (QS) approximation. That is, the signal is assumed to obey the transmission line equations with parameters L and C determined by a two-dimensional solution of the Laplace equation in the plane perpendicular to the direction of propagation. For lines in a homogeneous (nonlayered) infinite medium, these solutions correspond to the transverse electric and magnetic (TEM) approximation.

Reader (student) involvement is solicited in the book in several ways. *Exercises* are given that request the reader to perform or verify derivations of formulas given. *Hints* are often provided to suggest how to approach the derivation or problem. *Problems* are provided that require numerical solution for a given geometry and parameter values. *Computer codes* are included that allow hands-on use of the algorithms presented for complex problems.

Computer codes are contained on an accompanying disk. The codes are of two types: 1. The source files for codes that are printed and described in the text are included so that the reader may modify them, incorporate them into his own codes or do assigned problems using them on his own computer, text editor and C compiler, 2. Executable demo versions of commercial CAD codes are included that illustrate use of the principles in the book. The executable code is suitable for use on PC compatible computers running Windows 3.1 (using win32s, included) or higher, Windows 95 or Windows NT. The source code in C can be compiled and run on any computer compatible with and running an ANSI standard C compiler. The computer codes were developed in the C language written to the ANSI standard. They were developed on a 486 PC using a Windows NT operating system and Microsoft Visual C++ version 2.0 development environment.

This book is intended for use as a graduate text in electrical engineering or for the use of professionals interested in digital signal transmission. However, the material in Parts I and II could be the basis of a course that included seniors in electrical engineering who have had introductory level courses in circuit analysis, transmission line theory, and matrix analysis.

Colorado Springs, Colorado   K. D. G.
September 1997

ACKNOWLEDGMENTS

Special thanks is extended to my friend and (former) colleague Dr. Tom Larkin who read every word of the manuscript and gave many helpful suggestions that resulted in significant improvements to it. Tom's training is in applied mathematics so that his evaluation was from a perspective somewhat different than my own. His suggestions to add tutorial material in several parts of the book resulted in a work that will benefit a wider audience.

Dr. John Norgard of the University of Colorado, EE and CS Department, Colorado Springs, read the first parts of the manuscript and offered many suggestions for its improvement. He was also very encouraging that I continue the project to completion. Thanks, John.

Milt Radant, my brother-in-law, an electrical engineer and Vice President, Technology, Hughes Aircraft Corporation (retired), read part of the early manuscript. He offered encouragement and several suggestions to strengthen the presentation. He also helped identify appropriate reviewers for the manuscript. Milt, thank you much.

It was my contact with Kellee Crisafulli and Steve Kaufer that made me realize that the approaches that I had developed for EMP interaction simulation had a useful and contemporary application in circuit board simulation and integrity software design. Talking with them about their needs in these areas stimulated the concept of this book.

When I suggested, sort of facetiously, to my wife Elinor that I thought I'd write a book, she encouraged me from the start (without that encouragement, it probably would not have happened). Thanks dear, it's cost us some round dancing time together, but it's been very satisfying for me to write down this approach for all to see. I hope you share in the sense of satisfaction.

I have had many colleagues who have served as sounding boards for the approaches in this book and others who have contributed to the coding of the time domain simulation of transmission lines. They, too, have contributed in an indirect way to this work. Among those (in alphabetical order) are Dave Babb, John Bombardt, Terry Brown, Chris Jones, Don Jones, Dick Maguire, Bill Page, Jim Riker and Mark Stephens; my apology to anyone omitted that should be included. I'm grateful for their help and friendship throughout my professional career.

CD-ROM USER'S GUIDE

The CD-ROM that accompanies this book can be a help to professional and student alike. Both can gain insight into the behavior of closely coupled parallel transmission line paths and networks by studying the theory and then observing its implications by experimenting with the illustrating code that is included. From that point, readers will undoubtedly go in different directions. Researchers may want to generalize the theory and the numerical algorithms, there's plenty of room for that. Others may want to use the methods to build simulation codes to meet their own special needs. The more practical design engineer may simply want to use the knowledge gained to be able to use commercial simulation codes more intelligently. The demo code HyperSuite is one that uses the methods in this book to perform signal simulations on "real" circuit boards that are still "on the drawing board."

The software on the CD-ROM is in five directories, as shown below. There are files named `readme.txt` in each of the higher level directories. In Windows, using the Windows Explorer (Windows 95) or the File Manager (Windows 3.x or Windows NT), double clicking on any of these will load them in Notepad and you can read their contents. It's suggested that you do that when reviewing the CD-ROM. The directories on the CD-ROM are:

```
F:CLAEIGEN      <DIR>       Feb 12, 1997       5:40 PM
F:EIGNLAPK      <DIR>       Feb 12, 1997       5:40 PM
F:EXAMPLES      <DIR>       Feb 12, 1997       5:40 PM
F:HYPRSUIT      <DIR>       Feb 12, 1997       5:40 PM
F:LINPACK       <DIR>       Feb 12, 1997       5:40 PM
```

DIRECTORY CLAEIGEN.

This directory contains routines that are translations to C of a selection of the FORTRAN math routines in LAPACK (the C translations are referred to as CLAPACK — a search on the

The demonstration software in the directory HYPRSUIT is owned and copyrighted by HyperLynx, Inc.; it is included by permission. The utility software win32s (directory HYPRSUIT/WIN32S) is owned and copyrighted by Microsoft Corporation. The software in the EXAMPLES directory is copyrighted by Oxford University Press. Unauthorized distribution of copyrighted software is prohibited by law.

internet will yield many available CLAPACK programs). This module is not used in the illustrating programs but is recommended for the eigenvalue problem of Chapter 10, Section 4 (Diagonalization of a Matrix). It will calculate the eigenvectors and eigenvalues of the product matrix **LC** using these matrices individually as input and utilizing mathematically the fact that they are real symmetric matrices. It is superior to the EISPACK routines used in the illustrating code that use general eigenvalue methods on the matrix **LC** directly. See the readme.txt file in this directory.

DIRECTORY EIGNLAPK.

This directory contains source code in the FORTRAN 77 language that performs the same functions as the C translations in directory CLAEIGEN. Hence, for the user who is developing FORTRAN code to perform the multi-wire line eigenvalue problem, this software is recommended.

DIRECTORY EXAMPLES.

The directory EXAMPLES contains the source code for all of the example programs discussed in the book. The subdirectory names make it easy to find any particular code. For instance, subdirectory ch02_1 contains the first code discussed in Chapter 2. Subdirectories under these contain executable code. The subdirectories named EXE16 contain 16-bit executable code that can be run from any DOS command line. The subdirectories named EXE32 contain 32-bit executable code; it must be run from a DOS window under Windows 95 or Windows NT (i.e., 32-bit operating systems).

DIRECTORY HYPRSUIT.

This directory contains demo software from a commercial producer of Computer Aided Design (CAD) software. This software uses several of the methods developed in this book and is included to illustrate one type of application of these methods. The software in HYPR-SUIT performs an electrical simulation of traces and devices on a circuit board. It is a windows program that will run under Windows 95, Windows NT, or Windows 3.1x (if win32s is installed). See the file named info.txt in this directory, for setup information.

DIRECTORY LINPACK.

LINPACK is a collection of programs that perform the linear algebra operations of solving sets of simultaneous linear equations and inverting matrices. These programs were originally written in FORTRAN. This directory contains translations to C of several LINPACK programs that are useful for problems in digital transmission lines. The program Simple-Gauss is used in the examples to solve these kinds of problems. It is recommended that LIN-PACK programs be used to replace the program SimpleGauss for production codes. The LINPACK programs use pivoting and will provide higher accuracy in some problems.

DIGITAL TRANSMISSION LINES

PART I

TRANSMISSION LINE FUNDAMENTALS

Chapter 1
INTRODUCTION

Miniaturization of digital electronic components and their interconnections continues at an incredible pace. Small separation distances imply unwanted coupling due to both electric and magnetic field effects, i.e., parasitic capacitance and inductance. The limit of such micro-miniaturization may very well depend on how well the designer can cope with these parasitic effects. A major contribution of this book is that it supplies the design engineer with a new method of reducing the effect of parasitic coupling. If parallel conductors are regarded as (multi-conductor) transmission lines and are terminated in their *matrix* characteristic imped-ance, rather than in their *single-conductor* characteristic impedance, crosstalk can *theoreti-cally* be almost entirely eliminated.[1] Using this approach, termination loads would be placed both from each conductor to ground and from conductor to conductor. From a practical point of view, this theory supplies a design approach to allow conductors that carry independent signals to be placed closer together and avoid intolerable crosstalk by clever termination design. The principles of such a design are developed in this book as well as simulation algo-rithms to directly simulate the response of digital transmission lines on circuit boards with any assumed termination.

Another contribution of the book is a new approach to understanding and simulating skin effect. Traditionally, skin effect is evaluated from the perspective of frequency domain analysis. Approximations are attempted in the time domain by estimating the "important fre-quency" and applying the skin effect formula at that frequency. The new approach given in Part 5 of this book, treats the skin effect entirely as a time domain phenomena. It is shown that the skin effect not only results in unwanted signal losses, but that it can cause an "induc-tive kick" and hence, signal distortion. The inductive kick can be accurately simulated using the methods developed in this book.

The book presents a toolbox of mathematical methods and implementing computer codes that can be used to understand the behavior of and to simulate the propagation of sig-nals on multi-layered circuit boards. The methods can be generalized as new designs emerge for transporting digital signals. Multi-conductor-line models are used as a means of repre-

---

1. Line terminations based on this principle are the subject of a patent application.

senting the interaction between traces on a circuit board. Though most traces on a board are driven as single-conductor paths (driven with respect to a ground plane), the multi-conductor-line approach allows one to calculate the signals induced on adjacent traces when a single trace is driven. Hence, crosstalk between traces can be evaluated and designs to reduce the crosstalk can be devised. Because of the interaction between traces, the inductance and capacitance of a single trace cannot be accurately calculated without including the effect of adjacent traces. Hence, this is another motivation for a multi-conductor approach.

1.1 FUNDAMENTAL APPROACH

The basic notions that are developed in this book are the transport or propagation of digital signals along multi-conductor transmission lines and the solution of the circuit equations that arise at the interconnections and end terminations of these lines. The solutions are developed using time-stepping methods. In a multi-conductor transmission line, the coupling between the lines is described by the line's matrix parameters: capacitance, inductance, resistance, and conductance. Crosstalk between the traces and reflections at the terminations are determined by solutions of the multi-conductor transmission line equations and (simultaneously) the junction/termination circuit equations. Methods to simulate the response of arbitrary configurations of such lines, interconnections and terminations are developed in this book. A simple illustration of two interconnected three-conductor transmission lines and a single-wire line is shown below in Figure 1.1.

Much insight is gained by understanding the behavior of the propagation along multi-conductor lines and the effect of various loads on the lines. The circuit board designer can benefit greatly by an in-depth understanding of the principles involved. However, the numerical evaluation of the resulting equations can be done by hand only for the simplest of cases. The evaluations for meaningful problems must be performed by computer. Therefore, each math method of the book is developed into a numerical algorithm; and each algorithm that is developed is accompanied by a computer program written in the C language. Complete listings of short programs are included in the text with descriptions of "how they work." All computer programs discussed are included in their entirety on a disk accompanying the book.

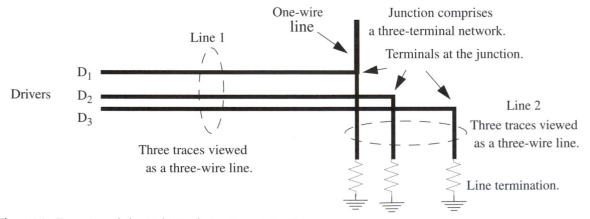

Figure 1.1   Traces Regarded as Multi-Conductor Transmission Lines.

The reader is invited to experiment with these programs, run them, and include the algorithms in their own computer-code inventions. No attempt is made to include derivations of all the methods that are presented, especially those that are well known. References are given where more complete information can be found. New methods are derived in detail. Units are (essentially) all metric (MKS); and are shown in brackets, e.g., [volts] = [V].

The propagation of digital signals along a transmission line is treated with increasing fidelity as the book progresses. The mathematical tool is the transport form of the transmission line equations. The transport equation is derived in Part 1, first for single-conductor lossless transport, then for lossy (resistive-conductor) propagation. Then, the transport equation is developed for multi-conductor lines that have a single transport speed. First for lossless lines, then for resistive lines. In Part 3, transport equations are derived for multi-conductor lines with multiple propagation speeds; first lossless lines are treated, then resistive lines. Lines consisting of traces embedded in a layered circuit board are of this type. Finally, in Part 5, transport is formulated to include diffusion (skin effect) distributed along a general, multi-speed line.

Similarly, the network solutions at multi-wire line junctions and terminations are progressively developed. In Part 1, basic circuit theory is presented for resistive networks. In Part 2, it is shown that resistive network analysis can be applied to networks with reactive or even nonlinear elements if a time-stepping method is used.

1.2 OVERVIEW

The book is composed of five parts. Each part may depend on previously developed methods, but has its own focus and develops methods suitable for part of the overall simulation problem. The parts are entitled

- Transmission Line Fundamentals
- Circuit Solutions at Line Terminations
- Propagation in Layered Media
- Transmission Line Parameters Determination
- Simulation of Skin Effect

The first part of the book, Transmission Line Fundamentals (Chapters 1–5), lays the foundation of the approach. The foundation has two distinct thrusts: (1) the transport of signals along transmission lines, and (2) the circuit solutions at line terminations and interconnections. The transport form of the transmission line equations is derived from the classical form for lines with a single propagation speed. Basic network solution concepts are developed starting with Kirchhoff's laws. The node equation approach is emphasized and used through the remainder of the book.

The second part of the book, Circuit Solutions at Line Terminations (Chapters 6–9), presents the methods used to solve the network equations at line junctions and terminations. Whereas the initial circuit solution concepts of Part 1 deal only with resistive networks, the generalization to time-stepping network solutions that include reactive circuit elements (inductors and capacitors) and nonlinear circuit elements is developed in this part.

The third part of the book, Propagation in Layered Media (Chapters 10–14), generalizes the transmission line solutions to treat lines with more than one speed of propagation. In a circuit board with layers of different permittivity, signals can propagate at different speeds. This causes a time dispersion of the signals arriving at the "far" end of a line. Methods are developed to separate signals into components or "modes" with discreet propagation speeds, transport them along the line at their appropriate speeds, and recombine them at the "far" end of the line.

The fourth part of the book, Transmission Line Parameters Determination (Chapters 15–19), develops methods to simulate the capacitance and inductance matrices of groups of traces in a circuit board. The propagation characteristics of a transmission line depend on these parameters: capacitance, inductance, resistance (or, skin effect), and conductance matrices. In part four, methods are developed to simulate the capacitance and inductance matrices of traces in layered circuit boards.

The fifth part of the book, Simulation of Skin Effect (Chapters 20–23), develops methods to simulate the skin effect for pulsed signals. Traditionally, the skin effect is estimated using the classical formulas based on sinusoidal signals. Digital signals have no simple relationship to sinusoidal signals (they contain a very wide band of sinusoidal frequencies). In Part 5, methods are developed to determine the "skin effect" using a rigorous time-stepping approach appropriate to digital signals.

1.3 THE TRANSMISSION LINE PARTIAL DIFFERENTIAL EQUATIONS (PDEs)

The transmission line equations are a set of two partial differential equations that can be written

$$v_x(x,t) = -Li_t(x,t) - Ri(x,t) \tag{1-1}$$

$$i_x(x,t) = -Cv_t(x,t) - Gv(x,t) \tag{1-2}$$

where

$v(x,t)$ is the potential of a trace with respect to ground (the "reference") [V],
$i(x,t)$ is the current (positive in the x direction, i.e., "to the right") [A],
L is the inductance per unit length [H/m],
R is the resistance per unit length [Ω/m],
C is the capacitance per unit length [C/m],
G is the conductance per unit length [mhos/m].

The constants R, L, C, and G will be referred to as "the parameters of the transmission line." Subscripts are used to denote partial differentiation as follows:

$$v_x(x,t) \equiv \frac{\partial v}{\partial x} \qquad i_x(x,t) \equiv \frac{\partial i}{\partial x}$$

and

$$v_t(x,t) \equiv \frac{\partial v}{\partial t} \qquad i_t(x,t) \equiv \frac{\partial i}{\partial t}$$

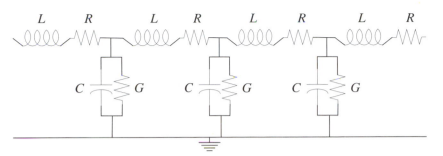

Figure 1.2   Parameters L, R, C, and G along a Transmission Line.

The parameters R, L, C, and G are distributed uniformly along a transmission line. Their effect can be visualized using discrete elements as shown in Figure 1.2.

Physically, the first transmission line equation, Equation (1-1), is a mathematical statement that the potential drops as one moves in the positive direction along the line (i.e., to the right, the assumed positive direction of x) due to two causes, (1) the time derivative of the magnetic flux linking the conductor and ground (Faraday's law), and, (2) the resistive voltage drop due to the current in the resistive conductor (Ohm's law). The second Equation (1-2) can be visualized as the conservation of current. That is, the current on the wire decreases as one moves to right due to two causes, (1) the displacement current per length flowing from the wire (capacitive current), and, (2) the conduction current per length flowing from the wire (through G).

Initially these will be regarded as applying to a single conductor. However, Equations (1-1) and (1-2) are also applicable to multiple conductors. No change in form is required, only a change in interpretation. In the case of multiple conductors, the voltage and current, v and i, are column vectors whose elements are the voltage and current of the individual conductors. The line parameters L, C, R, and G are matrices. It can be shown that L and C are symmetric matrices. The matrix R is diagonal if the resistance of the reference conductor is neglected; the diagonal elements are the resistance per length of the conductors. If the resistance per unit length of the reference conductor is to be included, it must be added to each element of the (otherwise diagonal) matrix (see Appendix B), hence, R is always symmetric. Note that the resistance of the reference conductor causes an additional rate-of-change in the x direction of all of the voltages, proportional to the sum of all the conductor currents. It can contribute to crosstalk between the conductors. The G matrix is a symmetric matrix; if the medium of the transmission line is uniform and conducting, the G matrix can be written

$$G = \frac{\sigma}{\varepsilon}C \qquad (1\text{-}3)$$

where σ is the conductivity of the medium and ε is the permittivity of the medium.

1.3.1 The Quasi-Static Approximation

If a transmission line is in a homogeneous medium, its primary propagation mode (as opposed to other wave-guide modes) is the transverse electric and magnetic (TEM) mode. That is, both the electric and magnetic fields are perpendicular to the axis of propagation (see

Jackson, 1975, Chapter 8 or Stratton, 1941, Chapter 6). The field configuration can be determined by solving a two-dimensional Laplace-equation problem in a plane perpendicular to the axis of the transmission line. However, if a transmission line is in a nonhomogeneous medium, such as a layered dielectric with different dielectric constants in the layers, it can be shown that, for a time-varying signal, a field component tangent to the direction of propagation must exist. If the time variation is small (i.e., wavelengths are long), then the field component parallel to the axis of propagation is small and the propagation approaches TEM though it is not strictly TEM. Hence, for frequencies low enough that wavelengths are long compared to the separation of the conductors, approximate TEM propagation exists. For these signals on a nonhomogeneous transmission line, the description quasi-static (QS) is more appropriate than TEM. The analysis given in this book is TEM when the medium is homogeneous and is QS when the medium is layered. There has been significant research performed on modifying TEM/QS theory to make it accurate for short wavelengths. A recent work on this topic is Faché, Olyslager, and De Zutter, 1993.

1.3.2 Solution Methods for the Transmission Line Equations

The transmission line equations can be solved in several ways. Some of these are given below with a short description of them and their applicability.

Fourier Transform Methods.   The Equations (1-1) and (1-2) can be formally Fourier transformed with respect to t to obtain ordinary differential equations with independent variable x. This is often done in engineering books simply by assuming that all time behavior is sinusoidal and can be represented by the exponential function $\exp(j\omega t)$. Partial derivatives with respect to time then become multiplications by $j\omega$, where, $j = \sqrt{-1}$ and ω is the angular frequency $\omega = 2\pi f$ [radians/second]. This approach is especially suitable if the line is to carry narrow band sinusoidal signals. The Smith chart is an easy-to-use graphical tool for designing lines for narrow band signals. See Everitt and Anner, 1956, Chapters 8 through 12 for methods using this approach. Dworsky, 1988, Chapter 3 also describes this approach.

Laplace Transform Methods.   Laplace transforms provide a powerful analytical tool for solving many kinds of transient problems. Application of Laplace transforms to the transmission line equations can be found in Scott, 1955, Sections 7.5 and 7.6, also in Dworsky, 1988, Section 2.3.

Finite Difference Solutions.   Finite difference approximations provide a powerful numerical tool for computer solutions of many kinds of differential equations. Finite difference solutions applied directly to the transmission line equations are especially appropriate if the parameters of the line change in time or if there are distributed sources (incident radiation or electromagnetic (EM) fields) that are space and time dependent along the line. Such is the case for a line in the hostile environment of a nuclear explosion. Finite difference methods have been used extensively to solve for the response of transmission lines in the presence of EM fields and radiation from a nuclear burst. Such finite difference solutions are the subject of Richtmyer and Morton, 1967 and Smith, 1965.

Equivalent Circuit Methods.   Short sections of a transmission line can be represented by finite circuit elements. Hence, a line of any length can be simulated by connecting such sections together. This method allows one to use circuit solving computer codes to solve for transmission line responses providing a quick way to solve transmission line problems to anyone who has a circuit solving code. This approximation of a transmission line is accurate only for wavelengths long compared to the length of line chosen for each section. For higher frequencies, the model behaves as a low-pass filter. For long lines, this type of model has many circuit elements making it impractical. See Dworsky, 1988, Sections 1.3 and 1.4, also Everitt and Anner, 1956, Chapter 8.

Method of Characteristics.   This is the approach used exclusively in this book. It is a very effective method for simulating transmission lines with low loss; i.e., lines that propagate signals with little or no distortion. When using it, signals are represented by forward and backward traveling components. For lossless lines it yields an exact solution, since signals are unchanged as they propagate, the line is represented simply by a time delay. At the ends of the line, "arriving" signals are forcing functions for the circuit elements (or network) attached as a load or driver to the line. The network solution then provides the "departing" signals that propagate from the line ends back the other direction. This method was developed by the author in Granzow and Jones, 1974.

Chapter 2
SINGLE-WIRE LINES

2.1 THE WAVE EQUATION

The transmission line equations can be combined to form wave equations for the voltage and current. Substituting the time derivative of Equation (1-2) into the x derivative of Equation (1-1) one obtains the wave equation for the voltage

$$v_{xx}(x,t) - LCv_{tt}(x,t) = (LG + RC)v_t(x,t) + RGv(x,t) \qquad (2\text{-}1)$$

And substituting the time derivative of Equation (1-1) into the x derivative of Equation (1-2) one obtains the wave equation for the current

$$i_{xx}(x,t) - CLi_{tt}(x,t) = (CR + GL)i_t(x,t) + GRi(x,t) \qquad (2\text{-}2)$$

Note that, for the general case of multi-wire lines in layered media, the matrices L, C, R, and G do not necessarily commute. Hence, care should be taken to preserve the order of multiplication of the matrices to use these equations for more general analysis later. In particular, $LC \neq CL$ in the general multi-conductor, layered media case. Therefore, the wave equations for the voltage and current are not the same in the most general case, even for lossless lines.

For a single-wire line, all of the quantities in Equations (2-1) and (2-2) are scalars and, of course, commute. For a single-wire line, define c as[1]

$$c = \frac{1}{\sqrt{LC}} \qquad (2\text{-}3)$$

This is the speed of propagation of the lossless line.

---

1. The symbol c is the standard symbol for the speed of light in vacuum, 299,792,458 [m/s]. It will be used in this book for the speed of propagation along a transmission line. If the medium of the transmission line has the permeability and permittivity of free space, its propagation speed will be that of light in a vacuum.

The "characteristic impedance" of the line is defined by[2]

$$z_o = \sqrt{\frac{L}{C}} \tag{2-4}$$

2.2 THE LOSSLESS LINE

For a lossless line, $R = 0$ and $G = 0$ and Equations (2-1) and (2-2) reduce to

$$v_{xx}(x,t) - \frac{1}{c^2}v_{tt}(x,t) = 0 \tag{2-5}$$

and

$$i_{xx}(x,t) - \frac{1}{c^2}i_{tt}(x,t) = 0 \tag{2-6}$$

By direct substitution, it is easy to show that any arbitrary function of $(t - x/c)$ is a solution of Equations (2-5) or (2-6). Such a function represents a forward-moving signal because it remains constant along the path $(t - x/c) = $ a constant; i.e., if the observer is moving in the x direction with speed c, he sees a constant value of i or v. Similarly, any arbitrary function of $(t + x/c)$ is also a solution of Equations (2-5) or (2-6). It represents a backward-moving signal since an observer moving in the $-x$ direction with speed c sees a constant signal value. This is illustrated in Figure 2.1 for a forward signal.

EXERCISE 2-1.   Show by substitution that if $v(x,t) = f(t - x/c)$ for any function $f(\tau)$, then $v(x,t)$ satisfies Equation (2-5).

EXERCISE 2-2.   Assume that $f(t - x/c) = C$, where C is a constant. Show that for any point where $\dfrac{df(\tau)}{d\tau} \neq 0$, the motion is such that $\dfrac{dx}{dt} = c$.

Figure 2.1 Illustration of Forward-Moving Signal. Snapshot at Two Times.

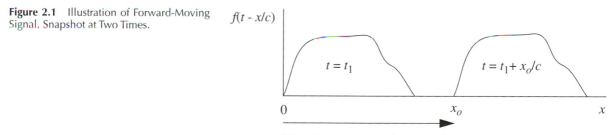

Time between snapshots $= x_o/c$

2. This definition of the characteristic impedance differs from the classical one used in Fourier transform analysis of transmission lines. The classical definition is $z_o = \sqrt{\dfrac{R + j\omega L}{G + j\omega C}}$. It is the ratio of voltage to current for an infinite length line at the frequency ω. The definition Equation (2-4) represents (+ or -) the ratio of (instantaneous) voltage to current for a propagating pulse applied independently to forward and backward propagating signals. It is numerically the same as the classical definition for a lossless line or a line at infinite frequency. See Appendix A.

Notation. Throughout this book the notation v_F and i_F will be used to represent forward-moving voltage and current signals (i.e., signals moving in the positive x direction; x is assumed to increase "to the right"). The notation v_B and i_B will be used to represent backward-moving signals; i.e., those moving to the left (in the negative x direction).

For a lossless line, forward and backward voltage and current signals can be defined such that

$$v_F(t - x/c) = z_o i_F(t - x/c) \tag{2-7}$$

and

$$v_B(t + x/c) = -z_o i_B(t + x/c) \tag{2-8}$$

EXERCISE 2-3. Show that Equations (2-7) and (2-8) are consistent with (i.e., satisfy) Equations (1-1) and (1-2). *Hint*: Set $R = 0$ and $G = 0$ and substitute.

The transmission line equations are linear; hence, if two solutions exist, then their sum is also a solution. The general solution for a lossless line can be written as the sum of the forward and backward signals. That is,

$$v(x, t) = v_F(t - x/c) + v_B(t + x/c) \tag{2-9}$$

and

$$i(x, t) = i_F(t - x/c) + i_B(t + x/c) \tag{2-10}$$

Equations (2-7) to (2-10) are fundamental to all of the analysis in this book. Only slight modifications are necessary for lossy lines; and for multi-conductor lines in layered media, these equations apply to individual propagation modes that will be defined later.

To obtain a meaningful solution of the transmission line equations, one must apply boundary conditions. That is, using information about the devices that are attached to the ends of the transmission line, one needs to determine the functions v_F, v_B, i_F, and i_B. In developing methods to do this, the term "departing signals" (voltage and/or current) will be used to refer to the forward signals at the left end of the line (v_F and i_F) and the backward signals at the right end of the line (v_B and i_B). They are the signals moving into the line at either of its ends. The term "arriving signals" will refer to the backward signals at the left end of the line (v_B and i_B) and the forward signals at the right end of the line (v_F and i_F). They are the signals arriving at either end of the line that have been propagating along the line.

2.3 TERMINATION IN THE CHARACTERISTIC IMPEDANCE Z_O

If a transmission line is terminated with a resistance equal to the characteristic impedance z_o, then there is no reflection from the end of the transmission line so terminated. That is, v_B and i_B are zero at the right end of a line if the line is terminated with z_o at its right end. Similarly, v_F and i_F are zero at the left end of a line if it is terminated at the left end with a

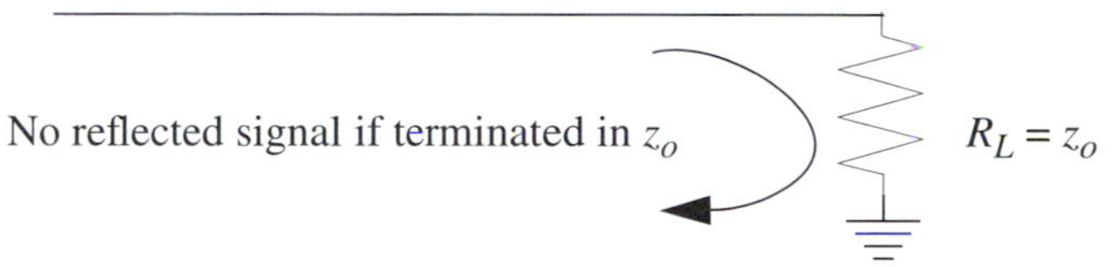

Figure 2.2 Termination in Characteristic
Impedance z_o.

No reflected signal if terminated in z_o $R_L = z_o$

resistor equal to z_o. Figure 2.2 illustrates a line terminated with its characteristic impedence at
its right end.

EXERCISE 2-4. For a transmission line terminated in its characteristic impedance at its right
end, the ratio of voltage to current is $\frac{v}{i} = z_o$. Show that in this case $v_B = 0$ and $i_B = 0$ at the
right end of the line. Use Equations (2-7) through (2-10). Also, for a transmission line terminated at its left end with a resistor equal to its characteristic impedance, the ratio of voltage to
current is $\frac{v}{i} = -z_o$ (note that positive currents flow into the left end, but out of the right end,
of a transmission line). Show for this case that $v_F = 0$ and $i_F = 0$.

2.4 TERMINATION WITH A RESISTIVE LOAD

For a line terminated with a resistive load, a reflection coefficient can be defined. That is, for
termination with resistance R_L, we have

$$\frac{\text{Departing voltage}}{\text{Arriving voltage}} = \left(\frac{v_B}{v_F}\right)_{\text{Right end}} = \left(\frac{v_F}{v_B}\right)_{\text{Left end}} = \frac{\frac{R_L}{z_o} - 1}{\frac{R_L}{z_o} + 1} \qquad (2\text{-}11)$$

The corresponding ratios of the currents are minus the voltage ratios (see Equations (2-7) and
(2-8)).

EXERCISE 2-5. Noting that at the right end, $\frac{v}{i} = R_L$ and at the left end $\frac{v}{i} = -R_L$, verify
that Equation (2-11) is correct. Use Equations (2-7) through (2-10).

2.5 TIME-STEPPING TRANSMISSION LINE SOLUTIONS

Typically in this book, time-stepping solutions will be discussed from which computer algorithms can be constructed to simulate the behavior of transmission lines. The general pattern
of a transmission line solution algorithm will be as follows

1. Set all functions to zero; that is, assume that all currents and voltages are zero for time previous to the start of the problem.
2. Find the transmission line terminal voltages or currents (both ends) using the equations of
   any circuit elements attached, combined with the value of the arriving signals. Initial con-

ditions (voltages and currents) must be available for capacitors and inductors. These initial conditions are updated by one time step at the end of this calculation for use in the next circuit update.

3. Calculate the departing signals by applying Equations (2-7) through (2-10) or equations derived from them.

4. Propagate the signals one time step along the transmission line, forward signals in the x direction, backward signals in the $-x$ direction. (In the lossless case they move along the line unchanged.) This will provide the arriving signals one time step later for use at the line ends for the next circuit solution.

5. Go to step 2 to begin the calculation over at the new time.

Each of these steps is implemented and clearly identified in the `main` function of Computer Code (2-1) at the end of this Section. Understanding these steps, as implemented in the code, will clarify the time-stepping solution method. The code simulates the response of the transmission line defined in Problem 2-1.

PROBLEM 2.1 Given a lossless transmission line 0.1 [m] long, $z_0 = 60$ [Ω] with propagation speed $c = 3 \times 10^8$ [m/s]. Let the delay time be $\tau = 0.1/c = 1/3$ [ns]. Let the left end of the line be driven by a unit step function voltage ($V = 0$, for $t < 0$, $V = 1$, $t > 0$).

  a. What is the inductance per unit length and the total inductance of the line?
  b. What is the capacitance per unit length and the total capacitance of the line?
  c. What is the current at the left end of the line and the voltage at the right end of the line for the first 2 ns (table or graph) if
   i. The line is terminated at the right end with $R = 60$ [Ω]?
   ii. The line is terminated at the right end with a short ($R = 0$)?
   iii. The line is terminated at the right end with $R = 180$ [Ω]?
  d. If the line were replaced by an inductor equal to the total inductance of the line, what is the current through the inductor for the first 2 ns? Compare this to the results of c.ii.

PROBLEM 2.2 Given a finite length transmission line with 1 volt applied at its left end for a very long time. Its right end is open circuit. Assume time enough has elapsed for all the transients to die out (the line is not quite lossless). The current through the line is zero and the voltage everywhere on it is 1 volt. What are the forward and backward currents and voltages? Does this result agree with Equation (2-11) (applied at the right end)?

2.6 NUMERICAL ALGORITHM — PROPAGATION

A signal propagating along a (lossless) line can be updated a time step by moving its value from memory location to memory location unchanged. However, this is unnecessary work; since the signal is unchanged, all that needs to be done is to increment an offset to indicate that the signal has moved along the line (at every point along the line). The method given here describes an implementation of this procedure using a circle (or ring) of storage locations to represent both forward and backward signals. (Even if the line is lossy and the signal changes as it propagates, this storage method can be used, but the signal must be modified at each location each time it propagates.) A ring of six storage locations is shown in Figure 2.3.

At $t = 0$, the departing signal at the left end of the line is stored at location 0 and the departing signal at the right end of the line is stored at location `Nhalf`. One time step later, the signal at location 0 is a forward signal located on the line at $x = $ `c*Tstep` and the signal at location `Nhalf` is a backward signal located at x = (`length` - `c*Tstep`). New departing signals (at t = `Tstep`) at the left and right ends are stored at locations 1 and (`Nhalf` + 1), respectively. At each succeeding time the departing signals are stored at locations incremented by one position each time step. At any given (computational) time, the locations counterclockwise around the storage ring from where the left-end departing signal is stored, contain forward signals that have propagated one or more time steps to the right. The locations counterclockwise from where the departing signal at the right end is stored, contain backward signals that have propagated one or more time steps to the left.

`StepOffset` points to one of the circular storage locations (ticks on the circle in Figure 2.3). It takes on values 0 through (2 * `Nhalf` - 1). Advancing time by `Tstep` is implemented by simply incrementing `StepOffset` to the next storage location. Departing signals at the left end are stored at location `StepOffset` and departing signals at the right end are stored at location (`StepOffset` + `Nhalf`).

The space between ticks represents the propagation time of `Tstep`. If only the transmission line must be considered, `Tstep` could be the delay time of the line. However, if circuit equations are to be solved using the same time step and the circuit has reactive elements, `Tstep` must be smaller than the smallest characteristic time of the circuit, or else the difference equations of the circuit solution can be unstable. Furthermore, `Tstep` must resolve the signal behavior.

The delay time represented by the circle below is in the range

$$(\text{Nhalf} - 1) \times \text{Tstep} \leq \text{delay time} < \text{Nhalf} \times \text{Tstep} \qquad (2\text{-}12)$$

This inequality defines `Nhalf`.

$$\text{Nhalf} = \text{int}\left(\frac{\text{delay time}}{\text{Tstep}}\right) + 1 \qquad (2\text{-}13)$$

where "int" is the integer function; it rounds to the next lowest integer.

Let the storage location to which `StepOffset` points represent time t; the location just counterclockwise from it then contains the signal that has propagated time `Tstep` along the line. The memory array represented by the circle contains locations for both forward and backward signals. The signal emerging from the "far" end of the line is in general an interpolation unless the delay time happens to be an exact multiple of `Tstep`.

Figure 2.3 Schematic of Circular Storage Array for Propagating Signal.

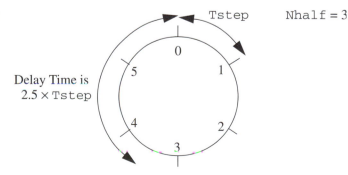

The following example program illustrates the principles. It performs a solution of *Problem* 2-1a, b, and c. Some boxes with explanations have been inserted to match parts of the program with parts of the problem and with the steps in the propagation solution listed in Section 2.5.

Computer Code 2-1. Propagation on Lossless Line.

```
/******       Main Program -- Problem 2-1       *****/
#include <math.h>
#include <stdio.h>
#include <stdlib.h>
// function prototypes
int InitPropTstep (  /*   Input variables      */
            float Length_Line,  // length of trans. line
            float Tstep,        // time step
            float Prop_Speed,   // Propagation speed

                    /*   Output variables    */
            int *Nhalf,         // ring half dimension
            float *wgt1,        // Interpolation weight
            float *wgt2  );     // Interpolation weight
int PropTstep (    /* Input variables  */
            float *Ring_ptr,    // ring pointer
            int Nhalf,          // number in half of ring
            float SigInForward, // forward from left
            float SigInBackward,// backward from right
            float Wgt1,         // interpolation wgt
            float Wgt2,         // interpolation wgt
               /* Input/Output variables  */
            int *TStepOffset,          // Time step offset
             /* Output variable  */
            float *SigOutBackward,    // backward at left
            float *SigOutForward);    // forward at right
/*------- Function Declaration -------------------------------- */
  void main ()
  {
/*   --- Local Variable Declarations -----------------------------
    Type    Name                Description & range
    ----    ----                ---------------------------- */
    float   length,         // length of line
            z0,             // characteristic impedance
            c,              // propagation speed
            Vleft,          // voltage at left end
            Lperm,          // inductance per meter
            Ltotal,         // inductance total
```

```
            Cperm,          // capacitance per meter
            Rload,          // load at right end of line
            Tlast,          // time limit for propagation
            Tstep,          // time step for propagation
            wgt1,           // interpolation weight
            wgt2,           // interpolation weight
            Tau,            // delay time of line
            Reflect,        // reflection coefficient
            Time;           // time in propagation
     float  vFleft,         // forward and backward voltages
            vBleft, vFright, vBright,
            Ileft,          // current left end
            Vright;         // terminal voltage right end
     int    Nhalf,          // half no. of locations in storage ring
            idx,            // for-loop index
            TStepOffset=0,  // offset to location in storage ring
            rcode;          // return code
     float  *Ring_ptr=NULL; // pointer to ring of memory
/*------------------------- Error Conditions Trapped ----------------
 Fatal/Nonfatal    Description
 --------------    -----------
 Fatal             error returned from InitPropTstep
                   error returned from PropTstep
----------------- Executable Code ------------------------------- */
     printf ("Problem 2-1  --  Propagation on Lossless Line\n\n");
//*** Input transmission line and problem parameters
length = 0.1F;      // [m]
z0 = 60.0F;         // [ohms]
c = 3e8F;           // [m/s]
```

Part A. *Inductance calculations.*

```
// a.
Lperm = z0/c;// [M/m]
Ltotal = Lperm * length;// [H]
printf ( "a. Inductance per meter = %e [H/m]\n", Lperm);
printf ( "   Total inductance = %e [H]\n", Ltotal);
```

Part B. *Capacitance calculations.*

```
// b.
Cperm = 1.0F/(z0 * c);
printf ( "b. Capacitance per meter = %e [C/m]\n", Cperm);
```

Part C. *Propagation calculation.*

```
// c.
// Set up propagation parameters
Vleft = 1.0F;              // [v] -- Unit step-function driver
Tau = length/c;           // [s] -- Delay time
Tlast = 2.0e-9F;          // [s]
Rload = 180.0F;           // [ohms]
//Eq. (2-11):
Reflect = (Rload/z0 - 1.0F)/(Rload/z0 + 1.0F);
Tstep = Tau/5.0F; // [s]
// calculate Nhalf, wgt1 and wgt2
rcode = InitPropTstep (/*   Input variables     */
            length,               // length of trans. line
            Tstep,                // time step
            c,                    // Propagation speed
                    /*   Output variables   */
            &Nhalf,               // ring half dimension
            &wgt1,                // Interpolation weight
            &wgt2  );             // Interpolation weight
if (rcode)
{
  printf ("Error returned from InitPropTstep, rcode = %i\n",rcode);
  goto Alldone;
}
```

Step 1. *Initialize the voltage on the line to zero.*

```
// Allocate memory for ring of memory & initialize to zero
Ring_ptr = (float*)malloc(sizeof(float) * 2 * Nhalf);
for (idx = 0; idx < 2 * Nhalf; idx++)
    *(Ring_ptr + idx) = 0.0F;
// Propagation loop
// Initializations
```

```
vBleft = 0.0F;
vFright = 0.0F;
Time = 0.0F;
while ( Time <= Tlast)
{
```

Step 2. *Calculate the voltages and currents at the terminals of line. That is, perform network solutions for the networks attached to the line terminations.*
Step 3. *Calculate departing signals (*vFleft *and* vBright*).*

```
// Left end of line -- calc. departing signal
vFleft = Vleft - vBleft; // Eq. (2-9) -- departing signal
// Right end of line -- Eq. (2-11)
vBright = Reflect * vFright;
// Current at left end
Ileft = (vFleft - vBleft)/z0;
// Votage at right end
Vright = vFright + vBright;
// Output to screen
printf ("Time = %e, I left end = %f, V right end = %f\n",
                Time, Ileft, Vright);
```

Step 4. *Propagate the signals one time step along the line.*

```
rcode = PropTstep (    /* Input variables  */
            Ring_ptr,              // ring pointer
            Nhalf,                 // number in half of ring
            vFleft,                // forward from left
            vBright,               // backward from right
            wgt1,                  // interpolation wgt
            wgt2,                  // interpolation wgt
                    /* Input/Output variables  */
            &TStepOffset,          // Time step offset
                /* Output variable  */
            &vBleft,               // backward at left
            &vFright);             // forward at right
if (rcode)
{
  printf("Error returned from PropTstep, rcode = %i\n", rcode);
  goto Alldone;
}
Time += Tstep;
```

> **Step 5.** *Go back to step 2.*

```
  }
  Alldone: ;
}
/* -----End of Procedure main -------------------------------- */
/*------- Function Declaration ---------------------------------
  *****      Initialize variables for PropTstep.      *****/
    int InitPropTstep (  /*   Input variables     */
              float Length_Line,  // length of trans. line
              float Tstep,        // time step
              float Prop_Speed,   // Propagation speed
                        /*   Output variables    */
              int *Nhalf,         // ring half dimension
              float *wgt1,        // Interpolation weight
              float *wgt2 )       // Interpolation weight
    {
/*   --- Local Variable Declarations ----------------------------
     Type      Name                 Description & range
     ----      ----                 -------------------------------
*/
     int       rcode;           // return code, =0, normal exit
     float     TDelay;          // time delay
     float     Texcess;         // extra time in half ring
/*---------------------- Error Conditions Trapped ----------------
   Fatal/Nonfatal Description
   -------------- -----------
   Fatal         InitPropTstep = -1, Tstep non-positive (input error)
                               = -2, Tstep too big
---------------- Executable Code ----------------------------- */
rcode = 0;    // initialize return code
if (Tstep <= 0)
{
  rcode = -1;
  goto InitAlldone;
}
TDelay = Length_Line/Prop_Speed;
if (Tstep > TDelay)
{
  rcode = -2;
  goto InitAlldone;
}
// Calculate Nhalf
*Nhalf = int(TDelay/Tstep) + 1;
```

```
// Calculate interpolation weights
Texcess = Tstep * *Nhalf - TDelay;
*wgt2 =  Texcess/Tstep;
*wgt1 =  1.0F - *wgt2;
InitAlldone:
return (rcode);
}
/* -----End of Procedure InitPropTstep -------------------------- */
/*------- Function Declaration --------------------------------- */
  /*****   Propagate one time step  *****/
  int PropTstep (    /* Input variables */
        float *Ring_ptr,         // ring pointer
        int Nhalf,               // number in half of ring
        float SigInForward,      // forward from left
        float SigInBackward,     // backward from right
        float Wgt1,              // interpolation wgt
        float Wgt2,              // interpolation wgt

                      /* Input/Output variables */
        int *TStepOffset,        // Time step offset
              /* Output variable  */
        float *SigOutBackward,   // backward at left
        float *SigOutForward)    // forward at right
{
/* --- Local Variable Declarations -------------------------------
    Type      Name                    Description & range
    ----      ----                    ------------------------- */
    int       Offset;               // offset to point in ring
    int       th;                   // 2*Nhalf[]
    int       rcode;                // return code, =0, normal exit
/*------------------------ Error Conditions Trapped --------------
 Fatal/Nonfatal    Description
 -------------     -----------
 Fatal             PropTstep          =-1, Nhalf < 2
---------------- Executable Code ------------------------------- */
rcode = 0;    // initialize return code
if (Nhalf < 2)
{
   rcode = -1;
   goto alldone;
}
th = 2 * Nhalf;
//*** Input signals at t = 0
*(Ring_ptr + *TStepOffset) = SigInForward;  // Forward wave input
Offset = (*TStepOffset + Nhalf) % th;       // Backward wave input
*(Ring_ptr + Offset) = SigInBackward;
```

```
//*** Increment timestep offset around circle
*TStepOffset = (*TStepOffset + 1) % th;
//*** Output signals at t = Tstep
// Forward signal
Offset = (*TStepOffset + Nhalf) % th;
*SigOutForward = Wgt1 * *(Ring_ptr + Offset);
Offset = (*TStepOffset + Nhalf + 1) % th;
*SigOutForward += Wgt2 * *(Ring_ptr + Offset);
// Backward signal
*SigOutBackward = Wgt1 * *(Ring_ptr + *TStepOffset);
Offset = (*TStepOffset + 1) % th;
*SigOutBackward += Wgt2 * *(Ring_ptr + Offset);
alldone:
  return(rcode);
  }
/* -----End of Procedure PropTstep ------------------------------- */
```

2.7 LOSSY LINES

If the parameters G or R are not zero, the transmission line has losses and signals will not propagate unchanged along the line. Forward and backward signals will have an additional dependence on x. For lines with loss, $v_F(t - x/c)$ becomes $v_F(x, t - x/c)$ and $v_B(t + x/c)$ becomes $v_B(x, t + x/c)$. As these signals propagate they change; the change is represented by an additional dependence on x. Equations (2-7) to (2-10) become

$$v_F(x,t - x/c) = z_o i_F(x,t - x/c) \tag{2-14}$$

$$v_B(x,t + x/c) = -z_o i_B(x,t + x/c) \tag{2-15}$$

$$v(x, t) = v_F(x,t - x/c) + v_B(x,t + x/c) \tag{2-16}$$

$$i(x, t) = i_F(x,t - x/c) + i_B(x,t + x/c) \tag{2-17}$$

By substituting Equations (2-14) and (2-15) into Equation (2-17), the current can be written

$$i(x, t) = z_o^{-1}[v_F(x,t - x/c) - v_B(x,t + x/c)] \tag{2-18}$$

> **Notation**. Subscripts 1 and 2 will be used to denote partial derivatives with respect to the first and second arguments, respectively, of the forward and backward voltage and current functions. That is, we define $v_{F1} \equiv \left(\dfrac{\partial v_F}{\partial x}\right)_{t - x/c}$ and $v_{F2} \equiv \left(\dfrac{\partial v_F}{\partial (t - x/c)}\right)_x$.

Using this convention, Equation (2-16) yields

$$v_x(x, t) = v_{F1}(x,t - x/c) + v_{B1}(x,t + x/c) - \frac{1}{c}[v_{F2}(x,t - x/c) - v_{B2}(x,t + x/c)] \tag{2-19}$$

$$v_t(x, t) = v_{F2}(x,t - x/c) + v_{B2}(x,t + x/c) \tag{2-20}$$

and Equation (2-18) yields

$$i_x(x, t) = z_o^{-1}[v_{F1}(x,t - x/c) - v_{B1}(x,t + x/c)] - \frac{z_o^{-1}}{c}[v_{F2}(x,t - x/c) \tag{2-21}$$
$$+ v_{B2}(x,t + x/c)]$$

$$i_t(x, t) = z_o^{-1}[v_{F2}(x,t - x/c) - v_{B2}(x,t + x/c)] \tag{2-22}$$

Substituting Equations (2-18), (2-19), and (2-22) into Equation (1-1), one obtains (arguments are omitted to save space)

$$v_{F1} + v_{B1} = -Rz_o^{-1}(v_F - v_B) + \left[\frac{1}{c} - Lz_o^{-1}\right](v_{F2} - v_{B2}) \tag{2-23}$$

Substituting Equations (2-16), (2-20), and (2-21) into (1-2) and multiplying by z_o, one obtains

$$v_{F1} - v_{B1} = -z_o G(v_F + v_B) + \left[\frac{1}{c} - z_o C\right](v_{F2} + v_{B2}) \tag{2-24}$$

From the definitions of c and z_o (Equations (2-3) and (2-4)), note that

$$\frac{1}{c} - Lz_o^{-1} \equiv 0 \tag{2-25}$$

$$\frac{1}{c} - z_o C \equiv 0 \tag{2-26}$$

Hence, Equations (2-23) and (2-24) simplify to

$$v_{F1} + v_{B1} = -Rz_o^{-1}(v_F - v_B) \tag{2-27}$$

and

$$v_{F1} - v_{B1} = -z_o G(v_F + v_B) \tag{2-28}$$

Adding and subtracting Equations (2-27) and (2-28), one obtains (Granzow and Jones, 1974, Equations 12 and 13)

$$v_{F1}(x,t - x/c) = -\alpha v_F(x,t - x/c) + \beta v_B(x,t + x/c) \tag{2-29}$$

and

$$v_{B1}(x,t + x/c) = \alpha v_B(x,t + x/c) - \beta v_F(x,t - x/c) \tag{2-30}$$

where

$$\alpha = \frac{1}{2}(Rz_o^{-1} + z_o G)$$

and

$$\beta = \frac{1}{2}(Rz_o^{-1} - z_o G)$$

A discussion of the physical meaning of Equations (2-29) and (2-30) is in order. First, consider Equation (2-29). Let a signal v_F propagate a short distance to the right at speed c (note that $(t - x/c)$ is constant). The derivative v_{F1} is defined as $v_{F1} = dv_F/dx$ along any path where $(t - x/c)$ is constant. Over a short propagation distance δx, this equation becomes $v_{F1} \cong \delta v_F/\delta x$. Therefore, as a forward signal propagates a small distance δx to the right, the change in v_F is $\delta v_F \cong \delta x v_{F1}$. According to equation (2-29), this change is given by two terms, the first is an attenuation term; it is given by the signal itself multiplied by $-\alpha \delta x$. If β were zero, the signal would attenuate exponentially as it propagates in the x direction. In fact, if $Rz_o^{-1} = Gz_o$, $\beta = 0$ and the transmission line is "distortionless," that is, a signal retains its original shape but attenuates as it propagates. The second term that changes v_F as it moves distance δx is $\beta \delta x$ multiplied by the backward signal at the given point on the line. That is, this term is a reflected signal (per unit length of propagation). Equation (2-30) has identical meaning to (2-29) except that propagation is in the $-x$ direction; the sign of the right-hand side is reversed so that the same physical processes occur as the signal propagates in the opposite direction. For v_B, the signal attenuates as it moves in the $-x$ direction.

Equations (2-29) and (2-30) are the same as those that describe the absorption and/or reflection of a stream of microscopic particles with motion at a constant speed in one dimension. They are called transport equations and this approach to particle transport is due to Boltzmann. Figure 2.4 illustrates graphically the attenuation and reflections processes.

The transport equations for the current that are equivalent to the voltage Equations (2-29) and (2-30) are

$$i_{F1}(x,t - x/c) = -\alpha i_F(x,t - x/c) - \beta i_B(x,t + x/c) \tag{2-31}$$

and

$$i_{B1}(x,t + x/c) = \alpha i_B(x,t + x/c) + \beta i_F(x,t - x/c) \tag{2-32}$$

If these equations are to represent a multi-wire line (Chapter 5), α and β are matrices; then α and β to be used in Equations (2-31) and (2-32) are given by

$$\alpha = \frac{1}{2}(z_o^{-1}R + Gz_o)$$

and

$$\beta = \frac{1}{2}(z_o^{-1}R - Gz_o)$$

Note the difference in the order of the factors in the parentheses as compared to those shown below Equation (2-30). If the matrix z_o^{-1} commutes with R and G commutes with z_o, then, of

Figure 2.4 Illustration of Signal Attenuating as It Propagates and Causes a Backward Propagating Signal by Distributed Backward Reflection.

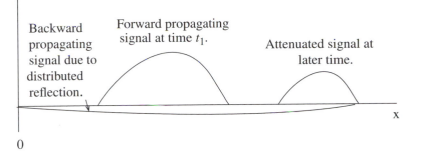

Backward propagating signal due to distributed reflection.

Forward propagating signal at time t_1.

Attenuated signal at later time.

0

x

course, the definitions of α and β are the same. However, in general these matrices do not commute. This will be discussed in Chapter 10.

The transport form of the transmission line equations is especially suitable for numerical time-stepping solutions. A simulated propagating signal can easily be modified step by step as it propagates using a small-increment approximation of Equations (2-29) and (2-30) or (2-31) and (2-32).

EXERCISE 2-6. Show that the potential $v(x,t)$ defined by Equation (2-16), where v_F and v_B satisfy Equations (2-29) and (2-30), satisfies Equation (2-1). *Hint*: Substitute v_{F1}, v_{B1} (2-29, 2-30) and Equation (2-22) back into Equation (2-19) to get Equation (1-1) and substitute v_{F1}, v_{B1} and Equation (2-20) back into Equation (2-21) to get Equation (1-2). Then differentiate (1-1) with respect to x and substitute (1-2) into (1-1) to get (2-1) (which is the way it was originally derived).

EXERCISE 2-7. Distortionless case ($\beta = 0$). Show that for a distortionless line the forward and backward signals can be written

$$v_F(x,t-x/c) \;=\; e^{-\alpha x} f_1(t-x/c) \tag{2-33}$$

and

$$v_B(x,t+x/c) \;=\; e^{+\alpha x} f_2(t+x/c) \tag{2-34}$$

where, f_1 and f_2 are arbitrary functions. That is, show that they satisfy Equations (2-29) and (2-30), respectively. (Though the distortionless case is a special case that is not usually encountered, one could approximately synthesize such a line using appropriate loading.)

EXERCISE 2-8. Show that if the current i and the voltage v are known at any point on a transmission line, then the forward and backward voltages and currents at that point are

$$v_F \;=\; \frac{1}{2}(v + z_o i) \tag{2-35}$$

$$v_B \;=\; \frac{1}{2}(v - z_o i) \tag{2-36}$$

$$i_F = \frac{1}{2}(i + z_o^{-1}v) \tag{2-37}$$

$$i_B = \frac{1}{2}(i - z_o^{-1}v) \tag{2-38}$$

Hint: Use Equations (2-14) to (2-17).

2.8 SMALL BACKWARD SIGNAL APPROXIMATION

A circuit board trace may be designed to carry a signal in the forward direction only. The trace may be impedance matched at the right end so that no backward (reflected) signal is launched from the right end. However, the backward signal that develops from distributed reflection of the forward signal may cause trouble to the driver. We would like to estimate this signal. Assume that the effect of the small backward signal on the forward signal is negligible; the forward signal is then affected only by the attenuation (α) term. Let $v_o(t)$ be the forward voltage departing from the left end. Then, the forward voltage anywhere on the trace is approximated by

$$v_F(x,t) \cong e^{-\alpha x}v_o(t - x/c) \tag{2-39}$$

The backward voltage is then determined by Equation (2-30) with Equation (2-39) substituted for v_F.

$$v_{B1}(x,t + x/c) = \alpha v_B(x,t + x/c) - \beta e^{-\alpha x}v_o(t - x/c) \tag{2-40}$$

Let $\tau = t + x/c$. For each value of τ (τ constant), Equation (2-40) is an ordinary differential equation in x that can be written

$$\frac{dv_B}{dx} - \alpha v_B = -\beta e^{-\alpha x}v_o(\tau - 2x/c) \tag{2-41}$$

At the right end $v_B = 0$; so the required boundary condition is $v_B(l,\tau) = 0$, where l is the length of the trace. The solution of Equation (2-41) with this boundary condition can be written

$$v_B(x,\tau) = \beta \int_x^l e^{\alpha(x - 2x')}v_o(\tau - 2x'/c)dx' \tag{2-42}$$

At the left end of the trace $x = 0$ and $\tau = t$. The arriving voltage at the left end is

$$v_B(0,t) = \beta \int_0^l e^{-2\alpha x'}v_o(t - 2x'/c)dx' \tag{2-43}$$

PROBLEM 2.3 Assume the small backward signal approximation:

a. If $v_o(t)$ is a unit step function, that is, $v_o(t) = u(t)$, where $u(t)$ is defined by

$$u(t) = \begin{cases} 0, & t < 0 \\ 1, & t \geq 0 \end{cases} \qquad (2\text{-}44)$$

show that the arriving voltage at the left end is

$$v_B(0,t) = \begin{cases} \dfrac{\beta}{2\alpha}(1 - e^{-\alpha ct}), & t < \dfrac{2l}{c} \\[3mm] \dfrac{\beta}{2\alpha}(1 - e^{-2\alpha l}), & t \geq \dfrac{2l}{c} \end{cases} \qquad (2\text{-}45)$$

Show that if α is small (as it usually is) and if the arguments of the exponentials are small compared to one, then Equation (2-45) can be written

$$v_B(0,t) \cong \begin{cases} \dfrac{\beta ct}{2}, & t < \dfrac{2l}{c} \\[3mm] \beta l, & t \geq \dfrac{2l}{c} \end{cases} \qquad (2\text{-}46)$$

b. If $v_o(t)$ is a unit amplitude square pulse with width t_w, show that the arriving voltage at the left end is

$$v_a(t) = v_B(0,t) - u(t - t_w)v_B(0, t - t_w) \qquad (2\text{-}47)$$

where $v_B(0,t)$ is the result for the unit step function driver $u(t)$.

c. Let $R = 10$ [Ω/m], $G = 0$, $z_o = 50$ [Ω], $l = 0.1$ [m]. For $v_o(t)$ equal to a five-volt step function, what is the maximum of the arriving voltage at the right end? At the left end?

2.9 NUMERICAL ALGORITHM — LOSSY PROPAGATION

A numerical method of approximating Equations (2-29), (2-30), (2-31) and (2-32) will be developed in this section. The same storage method will be used as was presented in Section 2.6. To help in visualization of the location of the signals vis-à-vis position in the circular storage array, refer to Figure 2.5. In Figure 2.5, Nhalf = 3 (as in Figure 2.3) for illustration only; Nhalf can be 2 or greater.

The numbers by the tick marks in Figure 2.5 are offsets from the pointer StepOff-set. Incrementing StepOffset (i.e., propagating one time step) is equivalent to decreasing the offset *from it* by one. Hence, the stored signal is effectively moved downward (in terms of its offset from StepOffset) as shown in Figure 2.5. In Figure 2.5, physical alignment of the forward and backward signals is illustrated. Memory positions of the forward and backward signals do not correspond to the same physical positions on the trace. When a signal is output, an interpolation is necessary (except for the special case when the delay of the line is an integer multiple of Tstep). Note that locations (0) and (3) (=Nhalf) appear twice on the diagram. In the propagation algorithm for lossless lines, an input signal is stored in (0) and (3) (=Nhalf), a propagation step is taken (StepOffset is incremented), then the interpolated output is calculated. When the interpolation is performed, (0) and (3) no longer contain the input because StepOffset has been incremented — they contain the propagated signal formerly at locations (1) and (4), respectively.

The shaded area in Figure 2.5 illustrates a region where forward and backward signals will be crossing during the next propagation step. That is, during the next propagation step while a signal is propagating backward from (3) to (2), a forward signal is propagating across the same physical segment of the trace from (interpolated) point (a) to (interpolated) point (b). The points along the top of the diagram (forward signal) have $(t - x/c)$ constant and the points along the bottom (backward signal) has $(t + x/c)$ constant. The transport equations (2-29) to (2-32) can be regarded as ordinary differential equations in x along these paths. Central differences in x can be used to numerically approximate their solutions.

The central difference equation for a forward voltage is an approximation to Equation (2-29); it can be written

$$\frac{v_F(x_2,\tau_F) - v_F(x_1,\tau_F)}{x_2 - x_1} = -\alpha\frac{v_F(x_2,\tau_F) + v_F(x_1,\tau_F)}{2} + \beta\frac{v_B(x_2,\tau_B) + v_B(x_1,\tau_B)}{2} \qquad (2\text{-}48)$$

where

$$\tau_F = t_1 - \frac{x_1}{c} = t_2 - \frac{x_2}{c}$$

$$\tau_B = t_1 + \frac{x_2}{c} = t_2 + \frac{x_1}{c}$$

There are two basic methods that can be used to solve the finite difference Equation (2-48), the "explicit" method and the "implicit" method. These two basic approaches are the subject of much literature and will not be discussed in depth here (see Richtmyer and Morton, 1967 or Smith, 1965). Using the explicit method, the derivative of the function is approximated, i.e., calculated explicitly, in terms of quantities known at the beginning of a given time step in the calculation. Applying the explicit method, the quantities on the right-hand side are set to currently known values stored in memory, i.e., values at time t_1.

Multiplying Equation (2-48) through by $(x_2 - x_1)$ and adding $v_F(x_1,\tau_F)$ to both sides, the new value of the propagated forward voltage at time t_2 can be written (where all the quantities on the right-hand side are those calculated at time t_1)

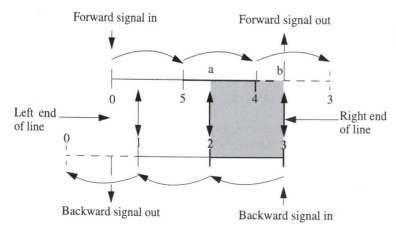

Figure 2.5 Physical Position of Signals in Circular Storage Array.

TABLE 2.1
Explicit and Implicit Difference Methods Compared.

Method	Advantages	Disadvantages
Explicit method	Computationally fast; no simultaneous equation solutions.	Gives divergent erratic results when $(x_2 - x_1)\alpha$ and $(x_2 - x_1)\beta$ are too large.
	Conceptually and algorithmically easy; easy to program.	
	Very accurate when coefficients $(x_2 - x_1)\alpha$ and $(x_2 - x_1)\beta$ are small.	
Implicit	Stable results even with $(x_2 - x_1)\alpha$ and $(x_2 - x_1)\beta$ large. (Can sometimes be used with much larger steps than explicit method.)	Computation time longer per step than explicit method
	Accurate results (even with $(x_2 - x_1)\alpha$ and $(x_2 - x_1)\beta$ large) if the functions are resolved by the time/space steps used.	Algorithmically complex.

$$v_F(x_2, \tau_F) = v_F(x_1, \tau_F) + (x_2 - x_1)\left[-\alpha \frac{v_F(x_2, \tau_F) + v_F(x_1, \tau_F)}{2} \right. \tag{2-49}$$
$$\left. + \beta \frac{v_B(x_2, \tau_B) + v_B(x_1, \tau_B)}{2}\right]$$

One might equally well have decided to set $v_F(x_2, \tau_F)$ and $v_B(x_1, \tau_B)$ on the right-hand side of Equation (2-48) to the new values at time t_2 (or to the average of their values at times t_1 and t_2). Then the (unknown) terms at time t_2 would have been moved to the left side of the equation and treated as unknowns. The difference equations built this way would contain two unknowns and are therefore dependent on each other; solution would then be effected by a simultaneous solution of the resulting set of linear equations. This is the implicit method.

The primary advantages and disadvantages of the implicit and explicit difference equation methods are summarized in Table 2.1. The explicit differencing method will be used exclusively in this book because it is the efficient method to solve low-loss transmission line equations.

EXERCISE 2-9. Show that the central difference approximation of the transport equation for backward voltages (2-30) yields

$$v_B(x_1, \tau_B) = v_B(x_2, \tau_B) + (x_2 - x_1)\left[-\alpha \frac{v_B(x_2, \tau_B) + v_B(x_1, \tau_B)}{2} \right. \tag{2-50}$$
$$\left. + \beta \frac{v_F(x_2, \tau_F) + v_F(x_1, \tau_F)}{2}\right]$$

Equations (2-49) and (2-50) are implemented in Computer Code 2-2, below. The terms that contain α and β on the right-hand side are calculated and put in temporary memory, then used to augment the values of v_F and v_B in the circular storage array. The current transport Equations (2-31) and (2-32) differ from those for voltage propagation only in that the sign of the β term is changed. The code below is written for voltage propagation; to use it for current propagation, input $-\beta$ for β.

Computer Code 2-2. Propagation on Lossy Line.

This computer code has essentially the same structure as function `PropTstep` in Computer Code 2-1. The calculation of attenuation and reflection have been added using the explicit method. Two additional input variables have been added; they are `alpha_dx` and `beta_dx`. They are simply α and β times the length of propagation over one time step; that is,

$$\text{alpha\_dx} = \alpha(x_2 - x_1) = \alpha c \text{Tstep} \tag{2-51}$$

and

$$\text{beta\_dx} = \beta(x_2 - x_1) = \beta c \text{Tstep} \tag{2-52}$$

The `main` program of Computer Code 2-1 can be used to call the new function with little revision. Simply replace `PropTstep` with `VoltPropTstep` and supply the two new input variables.

```
/*------- Function Declaration ---------------------------------- */
  /*****   Propagate voltage one time step lossy line   *****/
  // For propagation of current change sign of beta   at input
  int VoltPropTstep (    /* Input variables   */
              float *Ring_ptr,          // ring pointer
              int Nhalf,                // number in half of ring
              float SigInForward,       // forward from left
              float SigInBackward,      // backward from right
              float Wgt1,               // interpolation wgt
              float Wgt2,               // interpolation wgt
              float alpha_dx,           // alpha*dx
              float beta_dx,            // beta*dx
              /* Input/Output variables   */
              int *TStepOffset,         // Time step offset
              /* Output variables   */
              float *SigOutBackward,    // backward at left
              float *SigOutForward)     // forward at right
  {
/* --- Local Variable Declarations ----------------------------------
      Type     Name            Description & range
      ----     ----            ------------------------------ */
      int      Offset;         // offset to point in ring
```

```
      int       th;                  // 2*Nhalf[]
      int       rcode;               // return code, =0, normal exit
      int       idx;                 // for-loop index
      int       idx_beta;            // index for beta terms in rhtemp_ptr
      int       idxPlusNhalf;        // idx + Nhalf
      float     *rhtemp_ptr=NULL;    // temproary storage pointer
      float     MHalfAlpha_dx;       // minus 0.5 * alpha_dx
      float     HalfBeta_dx;         // 0.5 * beta_dx
/*------------------------- Error Conditions Trapped --------------
      Fatal/Nonfatal    Description
      -------------      -----------
      Fatal              PropTstep       =-1, Nhalf < 2
---------------- Executable Code ------------------------------- */
rcode = 0;    // initialize return code
   MHalfAlpha_dx = -0.5F * alpha_dx;
   HalfBeta_dx = 0.5F * beta_dx;
   if (Nhalf < 2)
   {
     rcode = -1;
     goto alldone;
   }
   th = 2 * Nhalf;
   // Allocate memory for temp storage of r.h. side of transport eqs.
   rhtemp_ptr = (float*)malloc(sizeof(float) * 2 * Nhalf);
   for (idx = 0; idx < 2 * Nhalf; idx++)
     *(rhtemp_ptr + idx) = 0.0F;
```

> *The difference equations require signals that will be lost when new data is input. These terms are calculated just before input to the circular storage ring overwrites the required data.*
> *Terms using data just input are calculated immediately after data is stored.*

```
// Calc. alpha term that depends on the last backward signal
// (before being overwritten by input)
*(rhtemp_ptr + 1) = *(Ring_ptr + *TStepOffset) * MHalfAlpha_dx;
// Input Forward signal at t = 0
*(Ring_ptr + *TStepOffset) = SigInForward;
// alpha and beta terms that use the left end input
*rhtemp_ptr = *(Ring_ptr + *TStepOffset) * MHalfAlpha_dx;
*(rhtemp_ptr + 1) += *(Ring_ptr + *TStepOffset) * HalfBeta_dx;
// Calc. alpha term that depends on the last forward signal
// (before being overwritten by input)
Offset = (*TStepOffset + Nhalf) % th;     // Backward wave input
*(rhtemp_ptr + Nhalf + 1) = *(Ring_ptr + Offset) * MHalfAlpha_dx;
// Input backward signal at t = 0
```

```
*(Ring_ptr + Offset) = SigInBackward;
// alpha and beta terms that use the right end input
*(rhtemp_ptr + Nhalf) = *(Ring_ptr + Offset) * MHalfAlpha_dx;
*(rhtemp_ptr + Nhalf + 1) += *(Ring_ptr + Offset) * HalfBeta_dx;
```

> *General loop to put* a *and* b *terms (attenuation and reflection terms) into tempo-rary memory.*

```
//*** Calculate r.h. side of transport eq. and put in rhtemp_ptr
// (idx = 0 are the special cases that were done above)
for (idx = 1; idx < Nhalf; idx++)
{
//***Backward signal sources
Offset = (*TStepOffset + idx) % th;
// Attenuation terms (alpha)
*(rhtemp_ptr + idx) += *(Ring_ptr + Offset) * MHalfAlpha_dx;
*(rhtemp_ptr + idx + 1) += *(Ring_ptr + Offset) * MHalfAlpha_dx;
// Reflection terms (beta)
idx_beta = (th + 1 - idx) % th;
*(rhtemp_ptr + idx_beta) +=   *(Ring_ptr + Offset) * HalfBeta_dx;
Offset = (Offset +1) % th;
*(rhtemp_ptr + idx_beta)+= *(Ring_ptr + Offset)*HalfBeta_dx*Wgt2;
idx_beta = (th - idx) % th;
if (idx < Nhalf - 1)
   *(rhtemp_ptr + idx_beta)+=
                             *(Ring_ptr + Offset)*HalfBeta_dx*Wgt1;
//***Forward signal sources
idxPlusNhalf = idx + Nhalf;
Offset = (*TStepOffset + idxPlusNhalf) % th;
// Attenuation terms (alpha)
*(rhtemp_ptr + idxPlusNhalf) +=
                             *(Ring_ptr + Offset) * MHalfAlpha_dx;
if (idx < Nhalf -1)
   *(rhtemp_ptr + idxPlusNhalf + 1) +=
                             *(Ring_ptr + Offset) * MHalfAlpha_dx;
else
*rhtemp_ptr += *(Ring_ptr + Offset) * MHalfAlpha_dx;
// Reflection terms (beta)
idx_beta = Nhalf + 1 - idx;
*(rhtemp_ptr + idx_beta) +=   *(Ring_ptr + Offset) * HalfBeta_dx;
Offset = (Offset +1) % th;
*(rhtemp_ptr + idx_beta)+= *(Ring_ptr + Offset)*HalfBeta_dx*Wgt2;
```

```
idx_beta = Nhalf - idx;
if (idx < Nhalf - 1)
   *(rhtemp_ptr + idx_beta)+=
                                *(Ring_ptr + Offset)*HalfBeta_dx*Wgt1;
}
//*** Apply Attenuation and Reflection terms to stored signals
for (idx = 0; idx < 2 * Nhalf; idx++)
{
   Offset = (*TStepOffset + idx) % th;
   *(Ring_ptr + Offset) += *(rhtemp_ptr + idx);
}
//*** Increment timestep offset around circle
*TStepOffset = (*TStepOffset + 1) % th;
//*** Output signals at t = Tstep
// Forward signal
Offset = (*TStepOffset + Nhalf) % th;
*SigOutForward = Wgt1 * *(Ring_ptr + Offset);
Offset = (*TStepOffset + Nhalf + 1) % th;
*SigOutForward += Wgt2 * *(Ring_ptr + Offset);
// Backward signal
*SigOutBackward = Wgt1 * *(Ring_ptr + *TStepOffset);
Offset = (*TStepOffset + 1) % th;
*SigOutBackward += Wgt2 * *(Ring_ptr + Offset);
alldone:
   free (rhtemp_ptr);
   rhtemp_ptr = NULL;
   return(rcode);
 }
/* -----End of Procedure VoltPropTstep ------------------------- */
```

Chapter 3
SOLUTIONS OF RESISTIVE NETWORKS

The solution of the interconnections of transmission lines and the solution of termination networks require algorithms to solve general networks of lumped elements. The required methods are built on the theory of the solution of the equations of resistive networks. In this chapter the classical theory of the solution of resistive networks is reviewed. Then a computer code is given that implements the solution algorithm. The routines in this computer code are incorporated into other codes, later in the book, to solve the equations at junctions of transmission lines and the equations at line terminations. In Part 2, the resistive network solution methods will serve as the launching point for time-stepping solutions of networks that contain reactive and nonlinear elements.

The topics discussed in this chapter are Kirchhoff's laws and the definitions of voltage and current sources. Thevenin and Norton equivalent circuits will be defined because they will be used later to represent the ends of transmission lines in termination circuits and junctions between lines. Finally, the node equation method to solve general resistive networks is reviewed and a general (resistive) network solving code is given and explained.

3.1 KIRCHHOFF'S LAWS

Kirchhoff's current law is a statement of the conservation of current (or, more fundamentally, charge). It simply states that the current flowing into a node equals the current flowing out of a node. More formally, the law may be stated as

> **Kirchhoff's Current Law:**
> *The sum of the currents flowing out of a circuit node equals zero.*

To employ this law, one must interpret "current flowing out of a circuit node" algebraically. That is, any current flowing *into* the node on a conductor is a negative current. Such an algebraic definition makes it easy to write the circuit equations as we will see below. For the simple example in Figure 3.1, Kirchhoff's current law can be written

$$i_1 + i_2 + i_3 + i_4 = 0 \qquad (3\text{-}1)$$

It's important to notice that this law implies that no current (or charge) accumulates on the node. That means that any *capacitance* of the node has been neglected. The node is treated as a point with no capacitance. If such a capacitance is to be included, it must be represented explicitly by a capacitor attached to the node.

The node of interest is (usually) not isolated as shown below, but is a connection point in a complex network. If a network has N such nodes, then the N node equations are the basis of the set of network equations from which the N node voltages can be determined by the simultaneous solution of N linear equations in N unknowns. This procedure will be carried out toward the end of the chapter.

Kirchhoff's voltage law states that the sum of the voltage drops around any closed path in a circuit equals the sum of the voltage increases. If a voltage "drop" is interpreted algebraically so that a rise is a negative drop, then Kirchhoff's voltage law can be written simply as

> **Kirchhoff's Voltage Law:**
> *The sum of the potential (or voltage) drops around any loop in a circuit equals zero.*

This is illustrated in Figure 3.2. The law states that the sum of the four voltages equals zero:

$$v_1 + v_2 + v_3 + v_4 = 0 \qquad (3\text{-}2)$$

This law needs some explanation that will show that it must be applied with care. It is based on static field theory, that is, on the idea that (scalar) potentials are single-valued functions of space. That if one carries a test charge from point to point and returns it to the place where it started, no net energy will be delivered to the test charge as it makes the trip. Such is not the case when time-changing magnetic fields are present. In fact, if a magnetic field threads through the loop shown in Figure 3.2, one can state that the sum of the potentials around the loop is given by the time derivative of the magnetic flux [in webers] that threads through the loop. Every such loop will have some magnetic flux threading through it due to its own loop current. If the loop current has a large time derivative, the voltage drop may be significant compared to other voltage drops. If this is true, it may be accounted for by including the self-inductance of the loop as a circuit element. If there is a significant time derivative of magnetic flux due to currents other than the loop current, it may be accounted for by including a mutual inductance element in the circuit.

Figure 3.1   Current Flow Out of a Node on Four Conductors.

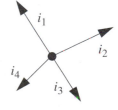

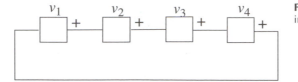

Figure 3.2 Four Circuit Elements Connected in Series.

Classical network loop equations are obtained by properly applying Kirchhoff's voltage law to a network. Consider the network shown in Figure 3.3. Three loops are shown representing closed paths in the network. Using the classical approach to the network loop equations, one defines a "loop current" for each of the paths. Let $i_1(t)$ be the current that follows loop p_1, let $i_2(t)$ be the current that follows loop p_2, and let $i_3(t)$ be the current that follows loop p_3. (This definition automatically conserves current and charge.) The currents through the branches of the circuit (where R_1, R_2, and R_3 are located) are given by the sum of the loop currents that pass through the branch. That is, the current through R_1 is $(i_1(t) + i_2(t))$, the current through R_2 is $(i_2(t) - i_3(t))$, and the current through R_3 is $(i_1(t) + i_3(t))$. The potential drops across the resistors are given by the value of resistors times the current through the resistors.

Applying Kirchhoff's voltage law to the loops, one obtains

Loop p_1:

$$R_1[i_1(t) + i_2(t)] + R_3[i_1(t) + i_3(t)] = v(t) \qquad (3\text{-}3)$$

collecting terms, one obtains

$$[R_1 + R_3]i_1(t) + R_1 i_2(t) + R_3 i_3(t) = v(t) \qquad (3\text{-}4)$$

Loop p_2:

$$R_1 i_1(t) + [R_1 + R_2]i_2(t) - R_2 i_3(t) = v(t) \qquad (3\text{-}5)$$

Loop p_3:

$$R_3 i_1(t) - R_2 i_2(t) + [R_2 + R_3]i_3(t) = 0 \qquad (3\text{-}6)$$

An attempt to find a unique solution of Equations (3-4) to (3-6) simultaneously would fail. The reason is that the equations are linearly dependent. (Subtract Equation (3-5) from (3-4) and Equation (3-6) is obtained.) There are only two linearly independent currents in the circuit of Figure 3.3. Any two of the loop equations will yield the network solution. Because of this redundancy in the choice of loop equations, it is more straightforward

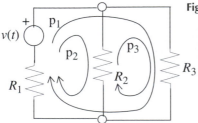

Figure 3.3 A Two-Loop Resistive Network.

to write computer algorithms to solve node equations rather than loop equations. Node equations can be defined without ambiguity by simply defining as a node any point where transmission line conductors connect to other transmission line conductors or circuit elements, or where circuit elements are connected to other circuit elements.

3.2 VOLTAGE AND CURRENT SOURCES

A *voltage source* is defined to be a (hypothetical) circuit element that maintains the defined voltage across its terminals regardless of the current that flows through it (or the impedance connected across its terminals). That means that any real physical voltage source must be represented with a more complex model such as a Thevenin equivalent circuit.

A *current source* is defined to be a (hypothetical) circuit element that maintains the defined current through it regardless of the voltage across its terminals (or the impedance connected across its terminals). Any physical current source must be represented by a more complex model such as a Norton equivalent circuit.

3.3 THEVENIN EQUIVALENT CIRCUITS

Thevenin concluded that any two-terminal network can be represented by a voltage source in series with an impedance. Such a Thevenin equivalent circuit is shown in Figure 3.4.
To represent general circuits, R could be the impedance looking into a network with reactive elements. We will restrict the analysis to resistive circuits in this chapter so that R is simply a real (nonnegative) number that represents a resistance in ohms. If the terminals (of a Thevenin circuit) are open as shown above, then the current $i(t) = 0$, the voltage drop across the resistor is zero, and the voltage across the terminals is V_{oc}. Hence, the voltage source is generally referred to as the open circuit voltage. The element R is the Thevenin equivalent impedance — in this chapter restricted to a resistance.

The Thevenin equivalent circuit in Figure 3.4 corresponds to the equation

$$v(t) = V_{oc}(t) - i(t)R \tag{3-7}$$

A network with many terminals can also be represented by a (generalized) Thevenin equivalent circuit. A Thevenin equivalent with 3 terminals plus ground is shown in Figure 3.5 for illustration.

Figure 3.4   Resistive Thevenin Equivalent Circuit.

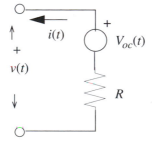

The equation that represents the circuit of Figure 3.5 is the matrix equation

$$\mathbf{v}(t) = \mathbf{V}_{oc}(t) - \mathbf{R}\mathbf{i}(t) \tag{3-8}$$

where $\mathbf{v}(t)$ is the column vector

$$\mathbf{v}(t) = \begin{bmatrix} v_1(t) \\ v_2(t) \\ v_3(t) \end{bmatrix}$$

$\mathbf{V}_{oc}(t)$ is the column vector

$$\mathbf{V}_{oc}(t) = \begin{bmatrix} V_{oc1}(t) \\ V_{oc2}(t) \\ V_{oc3}(t) \end{bmatrix}$$

$\mathbf{i}(t)$ is the column vector

$$\mathbf{i}(t) = \begin{bmatrix} i_1(t) \\ i_2(t) \\ i_3(t) \end{bmatrix}$$

and $\mathbf{R}$ is the symmetric matrix that represents the three-terminal (plus ground) resistive circuit.

3.4 NORTON EQUIVALENT CIRCUITS

Norton's equivalent circuit of a two-terminal network (resistive case) is shown in Figure 3.6. It consists of a current source in parallel with a conductance (G).

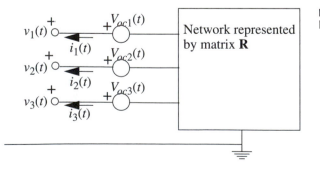

Figure 3.5   Three-Terminal Plus Ground Thevenin Equivalent Circuit.

Figure 3.6   Norton Equivalent Circuit for a Two-Terminal Network.

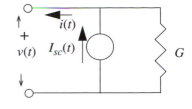

If the terminals are shorted, there is no voltage across the Norton conductance G, so that no current flows through it and $i(t) = I_{sc}(t)$. The current source $I_{sc}(t)$ is therefore the short-circuit current. The Norton equivalent circuit of Figure 3.6 is equivalent to the equation

$$i(t) = I_{sc}(t) - v(t)G \qquad (3\text{-}9)$$

The Norton equivalent circuit can be generalized to many terminals. A Norton equivalent circuit with three terminals is shown in Figure 3.7. The equation equivalent to this three-terminal network is

$$\mathbf{i}(t) = \mathbf{I}_{sc}(t) - \mathbf{G}\mathbf{v}(t) \qquad (3\text{-}10)$$

where $\mathbf{I}_{sc}(t)$ is the column vector

$$\mathbf{I}_{sc}(t) = \begin{bmatrix} I_{sc1}(t) \\ I_{sc2}(t) \\ I_{sc3}(t) \end{bmatrix}$$

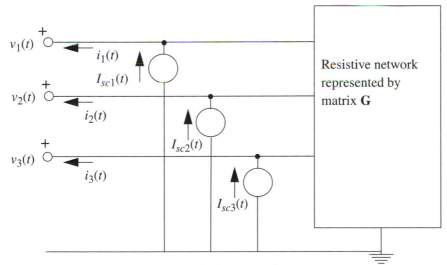

Figure 3.7   Norton Equivalent Circuit with Three Terminals.

The matrix $\mathbf{G}$ is the three-terminal resistive circuit matrix, and $\mathbf{i}(t)$ and $\mathbf{v}(t)$ are current and voltage column vectors as defined below Equation (3-8). The open circuit voltage of the Norton equivalent circuit is $\mathbf{G}^{-1}\mathbf{I}_{SC}(t)$ (obtain this by setting $\mathbf{i}(t) = \mathbf{0}$ in Equation (3-10)). A Norton equivalent circuit has a Thevenin equivalent with

$$\mathbf{V}_{oc}(t) = \mathbf{G}^{-1}\mathbf{I}_{sc}(t) \tag{3-11}$$

and

$$\mathbf{R} = \mathbf{G}^{-1} \tag{3-12}$$

EXERCISE 3-1.   Show that if the matrices $\mathbf{R}$ and $\mathbf{G}$ are not singular, Equations (3-11) and (3-12) imply that Equations (3-8) and (3-10) are equivalent (i.e., imply each other). That is, the Thevenin and Norton equivalent circuits of Equations (3-8) and (3-10) represent the same network if their variables are related by Equations (3-11) and (3-12).

When solving a network using loop equations as above (Figure 3.3 and Equations (3-3) through (3-6)), Thevenin sources are appropriate because voltage drops are being summed and a Thevenin voltage directly enters the equations. If a current source were present, the voltage across it would not be known *a priori* and loop equations could not be directly written. In this case, it would be appropriate to change the current source to a voltage source using Equations (3-11) and (3-12). Likewise, when solving a network using node equations, a Thevenin equivalent would not be appropriate because the current through the voltage source is unknown. A Norton equivalent circuit, on the other hand, directly yields the current to include in the sum of the currents at a node. Therefore, Thevenin equivalent circuits must be changed to Norton equivalent circuits using Equations (3-11) and (3-12) when the node equation approach is to be used.

3.5 GENERAL NETWORK SOLUTIONS USING NORTON EQUIVALENT CIRCUITS

Norton equivalent circuits and node equations will be used because network nodes can be uniquely defined as the junctions between circuit elements (or circuit elements and line ends) whereas current loops can be defined many ways for complex circuits. Algorithms to automatically generate node equations are therefore easily written. In this section, the setup of node equations for general networks of elements will be discussed. Each circuit element will be represented by a Norton equivalent circuit with a real (resistive) conductance. It will be shown in Part II of this book that for a time-stepping circuit solution, reactive elements can be approximated by this type of Norton equivalent circuit over a short time step. The source-current part of the Norton equivalent circuit changes from time step to time step for reactive elements reflecting the change in stored energy in the element, i.e., the initial current through an inductor or the initial voltage of a capacitor. Nonlinear passive elements can also be represented in this way, using the slope of its current-voltage curve as the conductance and the recently calculated value of current as the current source.

The network equations of a general network, where all the circuit elements are represented by Norton equivalent circuits, are determined by a straightforward application of Kirchhoff's current law. Kirchhoff's current law states that, for each node, the total current into the node is zero. One can also state this as: the sum of the currents into any node equals the sum of the currents out of the node. In Figure 3.8 two nodes (node i and node j) are shown with Norton circuits connected between them and from each of them to ground. For transmission line analysis, the Norton circuits may represent discreet circuit elements or may be transmission line ends.

At node i, the current flowing *out of* the node *through the conductors* (g_{ii} and g_{ij}) is

Current to node j

$$i_{gi} = v_i g_{ii} + (v_i - v_j) g_{ij} \qquad (3\text{-}13)$$

Current to ground

or, collecting terms multiplying each voltage, one has

$$i_{gi} = v_i(g_{ii} + g_{ij}) - v_j g_{ij} \qquad (3\text{-}14)$$

Suppose there are K nodes, Equation (3-14) generalizes to

$$i_{gi} = v_i \sum_{\substack{j=1}}^{K} g_{ij} - \sum_{\substack{j=1 \\ j \neq i}}^{K} v_j g_{ij} \qquad (3\text{-}15)$$

Equation (3-15) can be written in matrix notation as

$$\mathbf{i}_g = \mathbf{G}\mathbf{v} \qquad (3\text{-}16)$$

where $\mathbf{i}_g$ and $\mathbf{v}$ are column vectors given by

$$\mathbf{i}_g = [i_{g1}, i_{g2}, i_{g3}, \ldots, i_{gK}]^T$$

Figure 3.8   Schematic of a Portion of a General Network.

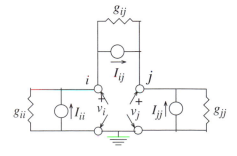

and

$$\mathbf{v} = [v_1, v_2, v_3, \ldots, v_K]^T$$

$\mathbf{G}$ is the matrix with diagonal elements

$$\mathbf{G}_{ii} = \sum_{j=1}^{K} g_{ij} \tag{3-17}$$

and off-diagonal elements

$$\mathbf{G}_{ij} = -g_{ij}, \quad i \neq j \tag{3-18}$$

The current flowing *into* node i *through the current sources* is

$$i_{si} = \sum_{j=1}^{K} \varepsilon_{ij} I_{ij} \tag{3-19}$$

where ε_{ij} is an indicator such that

$$\varepsilon_{ij} = \begin{cases} +1, & I_{ij} \text{ defined as positive into node } i, \\ -1, & I_{ij} \text{ defined as negative into node } i \end{cases}$$

The network node voltages are then determined by the solution of the matrix equation

$$\mathbf{G}\mathbf{v} = \mathbf{i}_s \tag{3-20}$$

where $\mathbf{i}_s$ is the column vector with elements i_{si}. In the applications of interest (unless nonlinear elements are in the network), the matrix $\mathbf{G}$ will be unchanged from time step to time step while the source currents (right-hand side of 3-20) change from time step to time step due to propagation on a transmission line or because they are time-dependent driver sources.

PROBLEM 3.1   Assume that the circuit of Figure 3-8 has only 2 nodes (as shown); let $i = 1$ and $j = 2$. Write out explicitly the matrix G and the column vector $\mathbf{i}_s$.

Computer Code 3-1. Solution of a Resistive Network.

The program listed below begins with the definition of two structures, one to be used in a linked list of current sources (I_{ij}s), the other to be used in a linked list of conductances (g_{ij}s). (These structures would ordinarily be defined in an include file, but they are defined in line, below, where they would normally be included.) The main program defines a network with two nodes. It allocates storage for three current sources and three conductances and assigns them values and node numbers. It then calls function `SolveNodeEqs` that solves for and returns the node voltages using the method described above. Both the input parameters and the resulting node voltages are printed to the screen. The allocated memory is not freed in this routine to save space; however, if this were a subroutine, the temporary memory (i.e., the linked lists) should be freed before exit to avoid a memory leak (i.e., retention of allocated memory no longer needed or used).

```c
#include <math.h>
#include <stdio.h>
#include <stdlib.h>
// These structures would normally be in an include file
// Note: Sources or conductances to ground will use node 0.
//        Circuit nodes are 1 to Nnodes.
struct ISource
{
    float Is;                  // Source current
    int NodeInto;              // Node no. pos. current enters
    int NodeOutof;             // Node no. pos. current exits
    struct ISource *next;      // Next link pointer
};
typedef struct ISource ISOURCE;        // type for current source
typedef struct ISource *ISOURCE_PTR;   // Link pointer
struct Gelement
{
    float g;                   // Conductance value
    int Node1;                 // Node no. its connected to
    int Node2;                 // Other node no. its connected to
    struct Gelement *next;     // Next link pointer
};
typedef struct Gelement GELEMENT;        // type for conductance g
typedef struct Gelement *GELEMENT_PTR;   // Link pointer
/* ---- Function prototypes ---------------------------------- */
    int IndexMatrix (int irow, int jcol, int nrows);
    int SimpleGauss (int nrows, int ncols, float *matrix);
    int SolveNodeEqs ( int Nnodes,          // number of nodes
                ISOURCE_PTR ISlist_ptr,  // pointer to current sources
                GELEMENT_PTR Glist_ptr,  // pointer to g element list
                    float *Vnodes );     // node voltages pointer
/* -----Function declaration main ---------------------------- */
    void main ()
    {
/* ----- Local Variables ------------------------------------- */
int          Nnodes;                  // number of nodes
ISOURCE_PTR  ISlist_ptr = NULL,  // pointer to current sources list
                LastIS_ptr = NULL; // this NULL used to terminate list
```

```
GELEMENT_PTR  Glist_ptr = NULL,   // pointer to G element list
              LastGptr = NULL;    // this NULL used to terminate list
float         *Vnodes = NULL;     // node voltages
int           rcode;              // return code from SolveNodeEqs
int           idx;                // for-loop index
/* ------------- Executable Code ------------------------------- */
// Test case for node equation solution
// Define case with two nodes, 3 g's and 3 source currents
// like Problem 3-2
Nnodes = 2;
// allocate memory for Vnodes
Vnodes = (float*)malloc(sizeof(float) * Nnodes);
// Make linked list of current sources
ISlist_ptr = (ISOURCE_PTR)malloc(sizeof(ISOURCE) * 3); //allocate 3
//Source 1
ISlist_ptr->Is = 0.1F;            //[A]
ISlist_ptr->NodeInto =   1;       //into node 1
ISlist_ptr->NodeOutof = 0;        //out of ground
ISlist_ptr->next = LastIS_ptr;    //first time = NULL -- IMPORTANT!
//Source 2
LastIS_ptr = ISlist_ptr;
ISlist_ptr++;                     // go to next link
ISlist_ptr->Is = 0.5F;            //[A]
ISlist_ptr->NodeInto =   2;       //into node 2
ISlist_ptr->NodeOutof = 0;        //out of ground
ISlist_ptr->next = LastIS_ptr;
//Source 3
LastIS_ptr = ISlist_ptr;
ISlist_ptr++;                     // go to next link
ISlist_ptr->Is = 1.0F;            //[A]
ISlist_ptr->NodeInto =   2;       //into node 2
ISlist_ptr->NodeOutof = 1;        //out of node 1
ISlist_ptr->next = LastIS_ptr;
// Make linked list of G's
Glist_ptr = (GELEMENT_PTR)malloc(sizeof(GELEMENT) * 3);// allocate 3
//G 1
Glist_ptr->g = 0.02F;             //[mhos]
Glist_ptr->Node1 = 1;             //node 1
Glist_ptr->Node2 = 0;             //ground
Glist_ptr->next = LastGptr;       // first time = NULL
//G 2
LastGptr = Glist_ptr;
Glist_ptr++;                      // go to next link
Glist_ptr->g = 0.05F;             //[mhos]
Glist_ptr->Node1 = 2;             //node 2
Glist_ptr->Node2 = 0;             //ground
```

```
Glist_ptr->next = LastGptr;
//G 3
LastGptr = Glist_ptr;
Glist_ptr++;                        // go to next link
Glist_ptr->g = 0.01F;              //[mhos]
Glist_ptr->Node1 = 2;             //node 2
Glist_ptr->Node2 = 1;             //node 1
Glist_ptr->next = LastGptr;
rcode = SolveNodeEqs ( Nnodes,         // number of nodes
                       ISlist_ptr,     // pointer to current sources
                       Glist_ptr,      // pointer to g element list
                       Vnodes );       // node voltages address
if (rcode)
{
   printf ("Error returned from SolveNodeEqs\n");
   goto DoneMain;
}
//Output data
// Input -- list G's
LastGptr = Glist_ptr;
while (LastGptr != NULL)
{
   printf ("Node1 is %i   Node2 is %i  g = %f\n",
             LastGptr->Node1, LastGptr->Node2, LastGptr->g);
LastGptr = LastGptr->next;
}
// Input -- list Is's
LastIS_ptr = ISlist_ptr;
while ( LastIS_ptr != NULL)
{
   printf ("NodeInto is %i  NodeOutof is %i  Is = %f\n",
   LastIS_ptr->NodeInto,LastIS_ptr->NodeOutof,LastIS_ptr->Is);
   LastIS_ptr = LastIS_ptr->next;
}
// Node voltages
for (idx = 0; idx < Nnodes; idx++)
{
   printf ("index is %i   Vnode = %f\n", idx, Vnodes[idx]);
}

DoneMain:   ;
// If this were a subroutine, memory should be freed here to avoid
// memory leak.
}
/* --------- End of main ------------------------------------ */
```

The function `SolveNodeEqs` (below) allocates memory for a matrix for solution of the circuit Equation (3-20). The solution code used expects the matrix **G**, Equations (3-17) and (3-18), to be in the left `Nnodes` columns of a matrix and the right-hand side of the linear equations (in this case vector $\mathbf{i}_s$) to be in the next column of the same matrix. Hence, a matrix is allocated with `Nnodes` rows and (`Nnodes` + 1) columns. A column matrix is allocated (`Check`) as a column of flags to make sure a conductance is connected to each node — if no conductance is attached to a node, its voltage will be indeterminate and the matrix *G* will be singular. The routine then loads the **G** matrix (Equations 3-17 and 3-18) and the source-current column vector $\mathbf{i}_s$ (with elements given in Equation 3-19) into the matrix. A check is made to make sure every node has at least one g_{ij} attached. Then the equations are solved by calling function `SimpleGauss`. The resulting node voltages are loaded in vector `Vnodes` for output.

```
/* -----Function declaration SolveNodeEqs ----------------------- */
   int SolveNodeEqs ( int Nnodes,        // number of nodes
                ISOURCE_PTR ISlist_ptr,// pointer to current sources
                GELEMENT_PTR Glist_ptr,// pointer to g element list
                float *Vnodes )        // node voltages pointer
{
/* ----- Local Variables ----------------------------------- */
   ISOURCE_PTR    IS_ptr = NULL;      // working ptr to source current list
   GELEMENT_PTR  G_ptr = NULL;        // working ptr to g list
   float         *matrix = NULL;      // matrix for solution
   int           *Check = NULL;       // check each node for g entry
   int           nrows, ncols;        // no. of rows & columns in matrix
   int           idx;                 // for-loop index
   int           rcode;               // return code from SimpleGauss
/* -------------- Executable Code ------------------------------- */
   if (Nnodes < 1)
   {
     rcode = -1;       // Bad value of Nnodes
     goto AllDone;
   }
   if ((ISlist_ptr == NULL) || (Glist_ptr == NULL))
   {
     rcode = -2;       // one of the lists is empty
     goto AllDone;
   }
   // Initializations
   nrows = Nnodes;
   ncols = Nnodes + 1;
   IS_ptr = ISlist_ptr;
   G_ptr = Glist_ptr;
   // Allocate memory for matrix and clear to zero
   matrix = (float*)malloc(sizeof(float) * nrows * ncols);
   for (idx = 0; idx < nrows * ncols; idx++)
```

```c
  *(matrix + idx) = 0.0F;
// Allocate memory for check column & clear to zero
Check = (int*)malloc(sizeof(int) * Nnodes);
for (idx = 0; idx < Nnodes; idx++)
  *(Check + idx) = 0;
// Fill G matrix
while (G_ptr != NULL)
{
// Accumulate in diagonal element for Node1
if ((G_ptr->Node1 > 0) && (G_ptr->Node1 <= nrows))
{
  *(matrix + IndexMatrix (G_ptr->Node1,G_ptr->Node1, nrows))
                      += G_ptr->g;
  *(Check + G_ptr->Node1 - 1) = -1; //something stored in this
                                    //diagonal element
}
// Accumulate in diagonal element for Node2
if ((G_ptr->Node2 > 0) && (G_ptr->Node2 <= nrows))
{
  *(matrix + IndexMatrix (G_ptr->Node2,G_ptr->Node2, nrows))
                      += G_ptr->g;
  *(Check + G_ptr->Node2 - 1) = -1; //something stored in this
                                    //diagonal element
}
// Accumulate in off-diagonal elements for Node1/Node2
if (  (G_ptr->Node1 > 0) && (G_ptr->Node1 <= nrows)
  && (G_ptr->Node2 > 0) && (G_ptr->Node2 <= nrows))
{
  *(matrix + IndexMatrix (G_ptr->Node1,G_ptr->Node2, nrows))
                      -= G_ptr->g;
  *(matrix + IndexMatrix (G_ptr->Node2,G_ptr->Node1, nrows))
                      -= G_ptr->g;
}
  G_ptr = G_ptr->next;
}
// Fill the source current vector
while (IS_ptr != NULL)
{
  if ((IS_ptr->NodeOutof > 0) && (IS_ptr->NodeOutof <= nrows))
    *(matrix + IndexMatrix (IS_ptr->NodeOutof,ncols, nrows))
                      -= IS_ptr->Is;
  if ((IS_ptr->NodeInto > 0) && (IS_ptr->NodeInto <= nrows))
    *(matrix + IndexMatrix (IS_ptr->NodeInto,ncols, nrows))
                      += IS_ptr->Is;
  IS_ptr = IS_ptr->next;
}
```

```
// Check to make sure something is connected to each
// node -- otherwise matrix is singular
for (idx = 0; idx < Nnodes; idx++)
{
  if (*(Check + idx) == 0)
  {
    rcode = -3;              // Nothing connected to a node
    goto AllDone;
  }
}
// Solve for voltages
rcode = SimpleGauss ( nrows, ncols, matrix);
if (rcode)
{
    rcode = -4;                    // error returned from SimpleGauss
    goto AllDone;
}
// Put results in output vector
for (idx = 1; idx <= Nnodes; idx++)
    *(Vnodes + idx - 1) = *(matrix + IndexMatrix (idx,ncols, nrows));
rcode = 0;
AllDone:
// Free memory
free (matrix);
free (Check);
return (rcode);
}
/* --------- End of SolveNodeEqs -------------------------------- */
```

The linear equation solver `SimpleGauss` is listed below. As the name implies it performs a Gauss elimination solution *without* pivoting. It is included because of simplicity and its usefulness for instruction. Production codes use more sophisticated linear algebra routines (see Anderson, et al., 1992, Dongara, et al., 1979 or Press, et al., 1986). The routine `SimpleGauss` uses (Gauss elimination) row operations that result in the inverse of the matrix that is stored in its first `nrows` columns. It will operate on any number of columns (input as `ncols`). On exit, the columns (`nrows` + 1) through `ncols` contain the solution vectors. In the illustration below, **G** is a 3-by-3 matrix, `nrows` = 3 and `ncols` = 5. Before calling `SimpleGauss`, `matrix` is loaded as shown below:

$$\texttt{matrix} = \begin{bmatrix} G_{11} & G_{12} & G_{13} & y_{11} & y_{12} \\ G_{21} & G_{22} & G_{23} & y_{21} & y_{22} \\ G_{31} & G_{32} & G_{33} & y_{31} & y_{32} \end{bmatrix} \tag{3-21}$$

On exit from `SimpleGauss`, the first three columns contain invalid data, the forth and fifth

columns contain $\mathbf{G}^{-1}\begin{bmatrix} y_{11} \\ y_{21} \\ y_{31} \end{bmatrix}$ and $\mathbf{G}^{-1}\begin{bmatrix} y_{12} \\ y_{22} \\ y_{32} \end{bmatrix}$, where, $\mathbf{G}^{-1}$ is the inverse of $\mathbf{G}$.

Note that the function `SimpleGauss` can be used to calculate the inverse of a matrix. To do this, one allocates `matrix` to hold two side-by-side `nrows`-by-`nrows` matrices. The left half is loaded with $\mathbf{G}$ and the right half is loaded with the identity matrix. The integer `ncols` is set to 2*`nrows`. On exit from `SimpleGauss`, the right half contains $\mathbf{G}^{-1}$.

```
/* ------- Function Declaration --------------------------------- */
  int SimpleGauss (int nrows, int ncols, float *matrix)
/* -------- Argument Variables ---------------------------------
      Type    Name    I/O    Description & range (normal use)
      ----    ----    ---    -------------------------------
      int     nrows   i      number of rows in matrix
      int     ncols   i      total number of columns in matrix
                             ncols = nrows + number of right-hand
                             columns.
                             Number of right-hand cols is 1 for a
                             simple linear equation solution or as
                             big as nrows for a matrix inversion.
      float   *matrix i/o    matrix plus right-hands side(s) on
                             entry.  Garbage (in place of matrix)
                             plus solutions on return.
                             result returned in icol = nrows +.. */
  {
/* ----- Local Variables --------------------------------------- */
  int      idx_diag,        // index of current diagonal element
           idx_col,         // index of current column
           idx_row;         // index of current row
  float    diag_element;    // temp. diagonal element
/* -------------- Executable Code ------------------------------ */
  // input error conditions
  if (ncols <= nrows)
   return (-1);
  if (nrows < 1)
   return (-2);
  // begin Gaussian elimination
  for (idx_diag = 1; idx_diag <= nrows; idx_diag++)
  {
    diag_element = *(matrix + IndexMatrix(idx_diag,idx_diag, nrows));
    if (diag_element == 0.0F)
      return (-3); //error -- divide by zero
```

```
      for (idx_col = idx_diag; idx_col < ncols; idx_col++)
        *(matrix + IndexMatrix(idx_diag,idx_col+1,nrows)) /= diag_element;
      for (idx_row = 1; idx_row <= nrows; idx_row++)
      {
        if (idx_row != idx_diag)
        {
        for (idx_col = idx_diag; idx_col < ncols; idx_col++)
              *(matrix + IndexMatrix(idx_row,idx_col + 1, nrows)) -=
              *(matrix + IndexMatrix(idx_row,idx_diag,nrows))   *
              *(matrix + IndexMatrix(idx_diag,idx_col + 1,nrows));
          }
        }
      }
      return (0);
  }
/* -----End of Procedure SimpleGauss ----------------------------- */
```

The function `IndexMatrix` calculates the index of matrix elements. A column-by-column ordering of elements is used.

```
/* ------- Function Declaration ------------------------------- */
  int IndexMatrix (int irow, int jcol, int nrows)
  {
/* -------- Argument Variables -------------------------------
      Type    Name      I/O      Description & range (normal use)
      ----    ----      ---      -------------------------------
      int     irow      I        Row index        1 to nm
      int     jcol      I        Column index     1 to array-size limit
      int     nrows     I        Number of rows
      int     IndexMatrix O      Linear index (offset) 0 to array size
/* ------------- Executable Code ----------------------------- */
    return (irow - 1 + nrows * (jcol - 1));
  }
/* -----End of Procedure IndexMatrix --------------------------- */
```

Chapter 4

BOUNDARY CONDITIONS — LINE END EQUIVALENT CIRCUITS

The solution of all but the simplest boundary equations at the ends of a transmission line will require that the end of the line be represented by a Thevenin or Norton equivalent circuit. Using a Thevenin equivalent circuit, the standard loop equations can be written, or, using a Norton equivalent circuit, standard node equations can be written and solved. Both approaches will be developed here. However, the node-equation approach is more easily adaptable to a general computer algorithm because the set of nodes is easily and uniquely defined as all the connecting points between electrical components (both transmission lines and discreet circuit elements). The current loops, for a loop-equation solution, can be defined in many different ways that are equally suitable. In subsequent sections of this book, the node-equation approach will be emphasized simply because of the ease in defining the circuit equations automatically. Both methods are, in principle, equally applicable to solving transmission line boundary problems.

4.1 THEVENIN EQUIVALENT CIRCUIT FOR A TRANSMISSION LINE

The Thevenin equivalent circuit for any two-terminal device is given by a voltage source in series with an impedance. For purposes of transmission line solutions, the impedance need only be a resistance because the (instantaneous) impedance looking into a transmission line is resistive.[1] Figure 4.1 shows the Thevenin equivalent circuit at the end of a single wire transmission line. The circuit is applicable to either end of the line. The values of the Thevenin voltage source and resistance will be derived below.

The circuit element values V_{oc} and Z are easily derived using Equations (2-14) to (2-17). For the open circuit case, $i(x,t) = i_F(x,t - x/c) + i_B(x,t + x/c) = 0$, hence,

---

1. The apparent disagreement of the resistive characteristic impedance defined here with the complex characteristic impedance defined in classical, Fourier-transform, transmission line theory is strictly one of definitions, especially as they relate to reflections. In the classical theory, an infinite line is *defined* as nonreflecting (even though on an instant-by-instant basis a reflected signal exists as defined in Equations (2-14) to (2-18)). The affect of the distributed reflections of Equations (2-29) to (2-32) are included in the classical characteristic impedance implicitly. In the present theory, all reflections must be accounted for as a function of time. It can be shown that both theories yield identical results for the same situations.

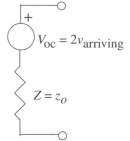

Figure 4.1   Thevenin Equivalent Circuit of Transmission Line End.

$$i_F(x,t - x/c) = -i_B(x,t + x/c) \tag{4-1}$$

Using Equations (2-14) and (2-15), then $v_F(x,t - x/c) = v_B(x,t + x/c)$ and Equation (2-16) implies that

$$V_{oc} = v(x,t) = \begin{cases} 2v_F(x,t - x/c), \text{ or} \\ 2v_B(x,t + x/c) \end{cases} \tag{4-2}$$

Because the source of the signal is the transmission line, it makes no physical sense to choose the departing signal. Hence, in both cases, the arriving signal is chosen. That is,

$$V_{oc} = 2v_B, \text{ left end} \tag{4-3}$$

and

$$V_{oc} = 2v_F, \text{ right end} \tag{4-4}$$

Note that the forward and backward signals correspond to the direction of energy flow (and that flow is at impedance z_o). The arriving signal always represents power flowing from the line to the terminals; the departing signal always represents power flowing from the terminals to the line. When the terminals of the line are shorted, Equation (2-16) implies that

$$v(x,t) = v_F(x,t - x/c) + v_B(x,t + x/c) = 0$$

hence,

$$v_F(x,t - x/c) = -v_B(x,t + x/c) \tag{4-5}$$

and Equations (2-14) and (2-15) imply that

$$i_F(x,t - x/c) = i_B(x,t + x/c) \tag{4-6}$$

If I_{sc} is defined as the short-circuit current positive to the right (out of the terminal) at the upper terminal in Figure 4.1, then

$$I_{sc} = \begin{cases} -2i_B(x,t + x/c), \text{ left end} \\ 2i_F(x,t - x/c), \text{ right end} \end{cases} \tag{4-7}$$

The impedance $Z = V_{oc}/I_{sc}$. Hence, using Equations (4-3), (4-4), (4-7), (2-14), and (2-15), the impedance $Z = z_o$.

Consider a line terminated with two Thevenin equivalent circuits as shown in Figure 4.2. Using the Thevenin equivalent circuits for the line ends, the configuration shown in Figure 4.2 becomes that shown in Figure 4.3. At the left end of the line, the terminal voltage $v(0,t)$ can be found. It is given by

$$v(0,t) = \frac{V_1(t)z_o + 2v_B(0,t)R_1}{R_1 + z_o} \tag{4-8}$$

From which the departing voltage is found to be

$$v_F(0,t) = v(0,t) - v_B(0,t) = \frac{V_1(t)z_o + v_B(0,t)(R_1 - z_o)}{R_1 + z_o} \tag{4-9}$$

At the right end of the line, the results are essentially identical with appropriate changes in subscripts and arguments

$$v(l,t) = \frac{V_2(t)z_o + 2v_F(l,t - l/c)R_2}{R_2 + z_o} \tag{4-10}$$

and the departing signal at the right end is

$$v_B(l,t) = \frac{V_2(t)z_o + v_F(l,t - l/c)(R_2 - z_o)}{R_2 + z_o} \tag{4-11}$$

EXERCISE 4-1. Solve the circuit equation at the left end of the transmission line (Figure 4.3) to verify that Equations (4-8) and (4-9) are correct.

Figure 4.2 Line Terminated with Thevenin Equivalent Circuits.

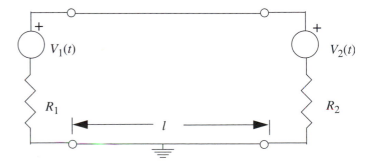

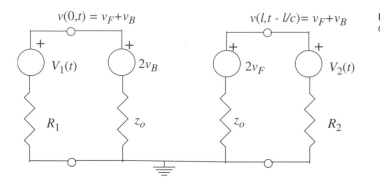

Figure 4.3 Use of Line-End Thevenin Equivalent Circuits.

PROBLEM 4.1 Refer to the transmission line shown in Figure 4.2. The transmission line has the following parameters:

$l = 0.1$ [m]

$c = 10^8$ [m/s]

(hence, $t_{delay} = l/c = 1$ [ns] for the line)

$z_o = 50$ [Ω]

$R = 10$ [Ω/m]

$G = 0$ [Ω^{-1}/m]

At the left end,

$V_1(t) = u(t) =$ unit step function

$R_1 = 5$ [Ω]

At the right end,

$V_2(t) = 0$

$R_2 = 50$ [Ω]

a. Calculate the voltage propagation every 0.5 [ns] from 0.0 to 2.5 [ns] using Equations (2-49) and (2-50) and at each time step use the line-end equivalent circuit to obtain the departing signals. Divide the line into two segments so that time steps are 0.5 [ns] and the x grid points are 0, 0.05, and 0.10 [m]. Fill in the missing voltages in Table 4-1.

Before starting, review the general pattern of transmission line solutions as given in Section 2.5. The equations mentioned in step 3 are replaced in this problem with Equations (4-9) and (4-11) (that were derived using the ones mentioned in step 3).

b. Calculate a row in the table at $t \rightarrow \infty$. That is, assume all the transients have decayed and the voltages and currents are d.c. *Hint:* Refer to *Exercise (2-8)*.

4.2 NORTON EQUIVALENT CIRCUIT FOR A TRANSMISSION LINE

The Norton equivalent circuit consists of a current source in parallel with a conductance. This circuit is shown in Figure 4.4. The parameters for the Norton equivalent circuit have already been determined in the preceding section except for the conductance G; it is simply the reciprocal of Z or $G = 1/z_o$.

We now attach a termination network at each end of a line (as in Figure 4.2) but use a Norton equivalent circuit instead of a Thevenin equivalent at each end. The Norton equivalent of the terminated line is shown in Figure 4.5. At the left end of the line, the line current is

TABLE 4.1
Problem 4-1.

Time [ns]	$x_1 = 0.00$ [m] Left termination			$x_2 = 0.05$ [m] Middle			$x_3 = l = 0.10$ [m] Right termination		
	$v_F(x_1)$	$v_B(x_1)$	$v(x_1)$	$v_F(x_2)$	$v_B(x_2)$	$v(x_2)$	$v_F(x_3)$	$v_B(x_3)$	$v(x_3)$
0.0	0.909	0	0.909	0	0	0	0	0	0
0.5									
1.0									
1.5									
2.0									
2.5									

$$i(0,t) = \frac{I_1(t)z_o^{-1} + 2i_B(0,t)G_1}{G_1 + z_o^{-1}} \tag{4-12}$$

The departing current at the left end is

$$i_F(0,t) = i(0,t) - i_B(0,t) = \frac{I_1(t)z_o^{-1} + i_B(0,t)(G_1 - z_o^{-1})}{G_1 + z_o^{-1}} \tag{4-13}$$

Similarly, at the right end

$$i(l,t) = \frac{-I_2(t)z_o^{-1} + 2i_F(l,t - l/c)G_2}{G_2 + z_o^{-1}} \tag{4-14}$$

The departing current at the right end is

$$i_B(l,t + l/c) = i(l,t) - i_F(l,t - l/c) = \frac{-I_2(t)z_o^{-1} + i_F(l,t - l/c)(G_2 - z_o^{-1})}{G_2 + z_o^{-1}} \tag{4-15}$$

EXERCISE 4-2.   Solve the circuit equation at the right end of the line shown in Figure 4.5 to verify that Equations (4-14) and (4-15) are correct.

EXERCISE 4-3.   Using a Norton equivalent circuit, show that if a line is terminated in its characteristic impedance that the departing current is zero.

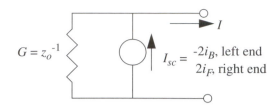

Figure 4.4 Norton Equivalent Circuit of Transmission Line End.

$G = z_o^{-1}$ $I_{sc} = \begin{array}{l} -2i_B, \text{ left end} \\ 2i_F, \text{ right end} \end{array}$

PROBLEM 4.2

For the transmission line defined for *Problem 4-1*, calculate the current propagation for $t = 0$ to $t = 2.5$ [ns]. Use the Norton equivalent circuits for the line and its terminations. Use Equations (2-49) and (2-50) with the sign of β reversed to propagate the forward and backward current. The Norton equivalent for the driver will have $I_1(t) = 0.2u(t)$ [A] and $G_1 = 0.2$ [℧]. Fill in Table 4-2 with the missing currents.

Do you know of any relationships between the various currents calculated in this problem and the voltages calculated (for the same model) in the preceding problem? If so, use them to see if your results for the current are consistent with the voltages.

4.3 JOINING TWO OR MORE TRANSMISSION LINES TOGETHER

Suppose two transmission lines (not alike) are connected in series as shown in Figure 4.6. The line ends can be represented by Thevenin or Norton equivalent circuits. It's obvious that, at the junction of the lines, the (instantaneous) impedance load seen by the left line is z_{o2} and the right line sees the impedance z_{o1}. When lines of different types are connected in this way, an impedance transformation occurs. In the illustration given here, an impedance increase occurs along with a voltage increase. Assume the following line parameters:

$l_1 = 0.1$ [m] $l_2 = 0.1$ [m]
$t_{d1} = l_1/c = 1.0$ [ns] $t_{d2} = l_2/c = 1.0$ [ns]
$z_{o1} = 30$ [ohms] $z_{o2} = 90$ [ohms]

Let a driver be connected at the left end of the configuration with driver impedance 5 [ohms] and (open circuit) voltage $V(t)$ (to be defined below). This is the usual way of specifying a driver; i.e., in terms of its Thevenin equivalent circuit. However, to use the node-equation approach, a Norton equivalent circuit is needed. Figure 4.7 shows the relationship between the Norton equivalent and the Thevenin equivalent for the same driver.

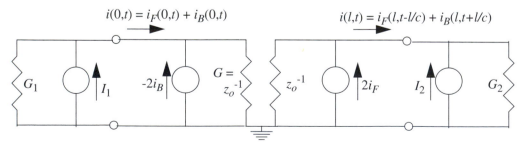

$i(0,t) = i_F(0,t) + i_B(0,t)$ $i(l,t) = i_F(l,t-l/c) + i_B(l,t+l/c)$

G_1 I_1 $-2i_B$ $G = z_o^{-1}$ z_o^{-1} $2i_F$ I_2 G_2

Figure 4.5 Use of Line-End Norton Equivalent Circuits.

TABLE 4.2
Problem 4-2.

Time [ns]	$x_1 = 0.00$ [m] Left Termination			$x_2 = 0.05$ [m] Middle			$x_3 = l = 0.10$ [m] Right Termination		
	$i_F(x_1)$	$i_B(x_1)$ (10^{-6})	$i(x_1)$	$i_F(x_2)$	$i_B(x_2)$ (10^{-6})	$i(x_2)$	$i_F(x_3)$	$i_B(x_3)$ (10^{-6})	$i(x_3)$
0.0	.0182	0	.0182	0	0	0	0	0	0
0.5									
1.0									
1.5									
2.0									
2.5									

The voltage $V(t)$ and impedance 5 [ohms] yield a Norton equivalent with parameters:

$G = 1/R = 1/5 = 0.2$ [℧]

$I(t) = V(t)/R = V(t)/5$ [A]

Therefore, at the left end of the pair of lines, the circuit shown below in Figure 4.8 must be solved.

The (instantaneous) solution at the left end is found (in the following order):
Node voltage at the left end:

$$V_{left} = \frac{\frac{V(t)}{5} - 2i_{B1}(t)}{0.2 + 0.03333} \tag{4-16}$$

($i_{B1}(t)$ is obtained as the result of propagation on line 1.)
Current at the left end of line 1:

$$i_1(t) = \frac{V(t)}{5} - V_{left}0.2 \tag{4-17}$$

Departing current at the left end of line 1:

$$i_{F1}(t) = i_1(t) - i_{B1}(t) \tag{4-18}$$

Figure 4.6   Two Transmission Lines in Series.

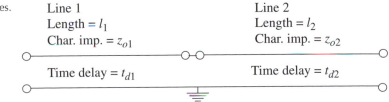

Line 1
Length = l_1
Char. imp. = z_{o1}
Time delay = t_{d1}

Line 2
Length = l_2
Char. imp. = z_{o2}
Time delay = t_{d2}

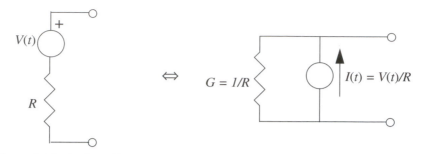

Thevenin Equivalent Circuit   Norton Equivalent Circuit

Figure 4.7   Thevenin Equivalent and Norton Equivalent of the Same Circuit.

The equivalent circuit at the middle junction is shown in Figure 4.9. The network solution at the middle junction is:
The middle node voltage is:

$$V_{mid} = \frac{2i_{F1}(t) + 2i_{B2}(t)}{0.03333 + 0.01111}$$ (4-19)

The current at the middle is:

$$i_1(t) = i_2(t) = 2i_{F1}(t) - V_{mid}0.03333$$ (4-20)

The departing signals are (for line 1 and line 2, respectively):

$$i_{B1}(t) = i_1(t) - i_{F1}(t)$$ (4-21)

and

$$i_{F2}(t) = i_2(t) - i_{B2}(t)$$ (4-22)

Assume the lines are impedance matched at the right end; that is, terminated in z_o of the right line, 90 [ohms]. Then, the right-end equivalent circuit is shown in Figure 4-10. The network solution at the right junction is:

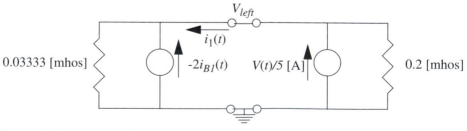

Figure 4.8   Left-End Equivalent Circuit.

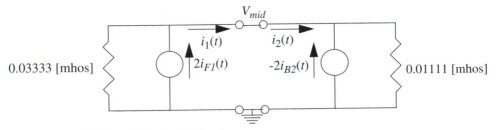

Figure 4.9   Middle-Junction Equivalent Circuit.

The node voltage:

$$V_{right} = \frac{2i_{F2}(t)}{0.01111 + 0.01111} \tag{4-23}$$

The current at the right end of line 2:

$$i_2(t) = 2i_{F2}(t) - V_{right}0.01111 = i_{F2}(t) \tag{4-24}$$

The departing current is:

$$i_{B2}(t) = i_2(t) - i_{F2}(t) = 0 \tag{4-25}$$

Each of the three node solutions requires, as a driving function, the arriving signal(s) from the line(s) connected to the node. These are supplied by propagating the signal on the lines using the methods developed in Chapter 2. Likewise, at each node, the departing signal(s) are calculated; these are supplied as input to the propagation algorithm and will appear as arriving signals at the other end of the line at a time t_d (= length/c) later.

To continue the illustration, assume that the function $V(t) = u(t)$ (the unit step function) and assume that the two lines are lossless. Table 4-3 can be generated by alternately propagating signals and applying the node-equation solutions given above.

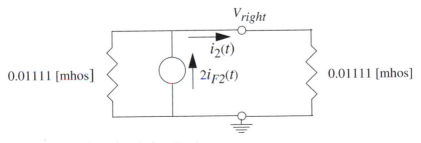

Figure 4.10   Right-End Equivalent Circuit.

PROBLEM 4.3 Perform the calculations to generate Table 4-3. Create an additional row in the table for $t \to \infty$, i.e., for the case when the d.c. solution is applicable.

TABLE 4.3
Illustration of Two Lossless Lines in Series.

Time [ns]	$x_1 = 0.00$ [m] Left Termination			$x_2 = l_1$ Middle				$x_3 = l_1 + l_2$ Right Termination	
t	v_{left}	i_{F1} dep.	i_{B1} arr.	$i_1 = i_2$	i_{F1} arr.	i_{B1} dep.	i_{F2} dep.	i_{F2} arr.	v_{right}
0.0	0.857	0.0286	0	0	0	0	0	0	0
1.0	0.857	0.0286	0	0.0143	0.0286	-0.0143	0.0143	0	0
2.0	0.980	0.004	-0.0143	0.0143	0.0286	-0.0143	0.0143	0.0143	1.287
3.0	0.980	0.004	-0.0143	0.002	0.004	-0.002	0.002	0.0143	1.287
4.0	0.874	0.027	-0.002	0.002	0.004	-0.002	0.002	0.002	0.180
5.0	0.874	0.027	-0.002	0.0135	0.027	-0.0135	0.0135	0.002	0.180
6.0	0.972	0.019	-0.0135	0.0135	0.027	-0.0135	0.0135	0.0135	1.215

Chapter 5
MULTI-WIRE LINES — SINGLE PROPAGATION SPEED

In this Section, multi-wire lines with a single propagation speed are treated. If the medium of the line (its dielectric) is homogeneous, then the capacitance and inductance matrices $\mathbf{C}$ and $\mathbf{L}$ and the propagation speed are related by

$$\mathbf{LC} = \mathbf{CL} = \frac{1}{c^2}\mathbf{I} \qquad (5\text{-}1)$$

where $\mathbf{I}$ is the identity matrix and c is the propagation speed. If the conductors are far from any dielectric boundary, Equation (5-1) may be approximately true and the line may be adequately simulated with the assumption that Equation (5-1) holds. In this section, it is assumed that Equation (5-1) holds for the lines discussed.

The transmission line Equations (1-1) and (1-2) can be written in vector form as

$$\mathbf{v}_x(x,t) = -\mathbf{L}\mathbf{i}_t(x,t) - \mathbf{R}\mathbf{i}(x,t) \qquad (5\text{-}2)$$

$$\mathbf{i}_x(x,t) = -\mathbf{C}\mathbf{v}_t(x,t) - \mathbf{G}\mathbf{v}(x,t) \qquad (5\text{-}3)$$

where
  $\mathbf{v}(x,t)$ is a column vector of the wire voltages [V]
  $\mathbf{i}(x,t)$ is a column vector of the currents [A]
  $\mathbf{R}$ is the matrix of wire resistances [ohms/m]
  $\mathbf{G}$ is the conductance matrix [$\mho$/m]
  $\mathbf{L}, \mathbf{C}, \mathbf{R},$ and $\mathbf{G}$ are all symmetric matrices.

5.1 PROPAGATION

With the speed of propagation c defined by Equation (5-1), the voltage and current are expressed as the sum of forward and backward signals. Let $\mathbf{v}(x,t)$ and $\mathbf{i}(x,t)$ be written in the form (vector form of Equations 2-16 and 2-17)

$$\mathbf{v}(x,t) \;=\; \mathbf{v}_F(x,t-x/c) + \mathbf{v}_B(x,t+x/c) \tag{5-4}$$

$$\mathbf{i}(x,t) \;=\; \mathbf{i}_F(x,t-x/c) + \mathbf{i}_B(x,t+x/c) \tag{5-5}$$

where the forward and backward currents and voltages are related by

$$\mathbf{v}_F(x,t-x/c) \;=\; \mathbf{Z}_o \mathbf{i}_F(x,t-x/c) \tag{5-6}$$

$$\mathbf{v}_B(x,t+x/c) \;=\; -\mathbf{Z}_o \mathbf{i}_B(x,t+x/c) \tag{5-7}$$

For multi-wire lines with a single propagation speed, the characteristic impedance is the symmetric matrix

$$\mathbf{Z}_O \;=\; c\mathbf{L} \tag{5-8}$$

Equations (5-4) through (5-7) can be substituted into Equations (5-2) and (5-3) in exactly the same way that was done in the single-wire case (see Equations 2-14 through 2-32) to obtain the transport equations for the forward and backward voltage vectors.

$$\mathbf{v}_{F1}(x,t-x/c) \;=\; -\mathbf{A}\mathbf{v}_F(x,t-x/c) + \mathbf{B}\mathbf{v}_B(x,t+x/c) \tag{5-9}$$

and

$$\mathbf{v}_{B1}(x,t+x/c) \;=\; \mathbf{A}\mathbf{v}_B(x,t+x/c) - \mathbf{B}\mathbf{v}_F(x,t-x/c) \tag{5-10}$$

where **A** and **B** are the matrices

$$\mathbf{A} \;=\; \tfrac{1}{2}\!\left(\mathbf{R}\mathbf{Z}_O^{-1} + \mathbf{Z}_O\mathbf{G}\right)$$

and

$$\mathbf{B} \;=\; \tfrac{1}{2}\!\left(\mathbf{R}\mathbf{Z}_O^{-1} - \mathbf{Z}_O\mathbf{G}\right)$$

The coefficients α and β, for the single-conductor lines, have been replaced with the matrices **A** and **B** for multi-conductor lines. The elements of **A** will be denoted α_{ij} and the elements of **B** will be denoted β_{ij}.

As it was pointed out earlier, below Equation (2-32), the coefficients α and β (now denoted **A** and **B**) appropriate to the current transport Equations (2-31) and (2-32) have $\mathbf{R}\mathbf{Z}_o^{-1}$ replaced by $\mathbf{Z}_o^{-1}\mathbf{R}$ and $\mathbf{Z}_o\mathbf{G}$ replaced by $\mathbf{G}\mathbf{Z}_o$. This commutation makes no difference for scalar quantities, but it does for matrices. A basic property of matrix multiplication is that for any matrices, say, **C** and **D**,

$$\mathbf{C}^T\mathbf{D}^T \;=\; (\mathbf{D}\mathbf{C})^T \tag{5-11}$$

where, the superscript T indicates the transpose of a matrix. Therefore, the coefficients **A** and **B** appropriate to the current transport equations are the transposes of those in the voltage transport equations, and we obtain

$$\mathbf{i}_{F1}(x,t - x/c) = -\mathbf{A}^T \mathbf{i}_F(x,t - x/c) - \mathbf{B}^T \mathbf{i}_B(x,t + x/c) \tag{5-12}$$

and

$$\mathbf{i}_{B1}(x,t + x/c) = \mathbf{A}^T \mathbf{i}_B(x,t + x/c) + \mathbf{B}^T \mathbf{i}_F(x,t - x/c) \tag{5-13}$$

The current transport equations are the same as the voltage transport equations with **A** replaced with $\mathbf{A}^T$ and **B** replaced with $-\mathbf{B}^T$. In general, **A** and **B** are not symmetric, i.e., $\mathbf{A} \neq \mathbf{A}^T$ and $\mathbf{B} \neq \mathbf{B}^T$.

EXERCISE 5-1.   Show that

$$\frac{1}{2}\left(\mathbf{Z}_o^{-1} \mathbf{R} + \mathbf{G}\mathbf{Z}_o \right) = \mathbf{A}^T \tag{5-14}$$

and

$$\frac{1}{2}\left(\mathbf{Z}_o^{-1} \mathbf{R} - \mathbf{G}\mathbf{Z}_o \right) = \mathbf{B}^T \tag{5-15}$$

Hint: Use Equation (5-11) and the symmetry of the commuted matrices.

5.1.1 Propagation Modes

The voltage and current vectors can contain any combination of forward and backward voltages and currents subject to the forward voltage-current relationship of Equations (5-6) and the backward voltage-current relationship of Equation (5-7). For wires (traces) that are near each other, the matrix $\mathbf{Z}_o$ has significant off-diagonal elements. Thus, if a signal exists such that all the wire currents are zero but one, then Equations (5-6) and (5-7) imply that (in general) there are voltages present on all the wires. Likewise, if all the voltages are zero but one, then there are currents on all the wires. A typical $\mathbf{Z}_o$ matrix for three traces on a circuit board is

$$\mathbf{Z}_o = \begin{bmatrix} 57.4 & 20.3 & 10.0 \\ 20.3 & 56.9 & 20.3 \\ 10.0 & 20.3 & 57.4 \end{bmatrix} \text{ [ohms]} \tag{5-16}$$

If all forward wire currents are zero except on wire 1 where $i_{F1} = 100$ [mA], the forward voltages are

$$\mathbf{v}_F = \begin{bmatrix} 57.4 & 20.3 & 10.0 \\ 20.3 & 56.9 & 20.3 \\ 10.0 & 20.3 & 57.4 \end{bmatrix} \begin{bmatrix} 0.1 \\ 0 \\ 0 \end{bmatrix} = \begin{bmatrix} 5.74 \\ 2.03 \\ 1.00 \end{bmatrix} \text{ [V]} \tag{5-17}$$

Similarly, if all forward wire voltages are zero except on wire 1 where $v_{F1} = 5$ [V], the forward current is

$$\mathbf{i}_F = \mathbf{Z}_o^{-1}\mathbf{v}_F = \begin{bmatrix} 0.0200 & -0.00674 & -0.00110 \\ -0.00674 & 0.0224 & -0.00674 \\ -0.00110 & -0.00674 & 0.0200 \end{bmatrix}\begin{bmatrix} 5.0 \\ 0 \\ 0 \end{bmatrix} = \begin{bmatrix} 0.1000 \\ -0.0337 \\ -0.0055 \end{bmatrix} \text{[A]} \quad (5\text{-}18)$$

The coupling between nearby wires illustrated above is due to the electric field and magnetic field fluxes that link the wires; these field couplings are mathematically input as the capacitance and inductance matrices. It will be shown that, if a multi-wire line is terminated in a network that represents its characteristic impedance (the matrix $\mathbf{Z}_O$), then the received signal at the end of the line is the forward signal; the received voltages and currents will not be intermixed. However, if the wires are terminated as individual single-wire transmission lines, then this coupling will intermix the signals causing crosstalk between the signal paths. This type of crosstalk will be referred to as termination crosstalk because it can be thought of as due to a less than ideal termination; it is not dependent on the length of the line. It occurs in lossless lines as well as those with loss.

Another type of crosstalk occurs on lossy lines; it is the wire-to-wire coupling due to the effect of the matrices $\mathbf{A}$ and $\mathbf{B}$ as a signal propagates in accordance with Equations (5-9), (5-10), (5-12), and (5-13). This crosstalk occurs per unit length as a signal propagates. It will be referred to as propagation crosstalk; it is dependent on the length of the line.

PROBLEM 5.1   Assume a 5 [V] signal departs the right end of a three-wire transmission line. The signal appears on wires 1 and 3; $v_{B2} = 0$. What are the three departing currents assuming the characteristic impedance of the line is given by Equation (5-16)?

5.1.2 Separation into Forward and Backward Signals

If the instantaneous voltage and current vectors are known at a given point on a multi-conductor line, the forward and backward voltage and current vectors are given by generalizations of Equations (2-35) to (2-38). The formulas for them are

$$\mathbf{v}_F = \frac{1}{2}(\mathbf{v} + \mathbf{Z}_o\mathbf{i}) \quad (5\text{-}19)$$

$$\mathbf{v}_B = \frac{1}{2}(\mathbf{v} - \mathbf{Z}_o\mathbf{i}) \quad (5\text{-}20)$$

$$\mathbf{i}_F = \frac{1}{2}(\mathbf{i} + \mathbf{Z}_o^{-1}\mathbf{v}) \quad (5\text{-}21)$$

$$\mathbf{i}_B = \frac{1}{2}(\mathbf{i} - \mathbf{Z}_o^{-1}\mathbf{v}) \quad (5\text{-}22)$$

EXERCISE 5-2.   Using Equations (5-4) through (5-7) derive Equations (5-19) through (5-22).

5.1.3 Lossless Line

The lossless line has $\mathbf{R}$, $\mathbf{G}$, $\mathbf{A}$, and $\mathbf{B}$ all equal to zero. The general solution of the vector transmission line equations is (the vector form of Equations 2-9 and 2-10)

$$\mathbf{v}(x, t) = \mathbf{v}_F(t - x/c) + \mathbf{v}_B(t + x/c) \tag{5-23}$$

and

$$\mathbf{i}(x, t) = \mathbf{i}_F(t - x/c) + \mathbf{i}_B(t + x/c) \tag{5-24}$$

where $\mathbf{v}_F$, $\mathbf{v}_B$, $\mathbf{i}_F$, and $\mathbf{i}_B$ are arbitrary functions subject only to the relationships between voltage and current given by Equations (5-6) and (5-7).

If these vectors represent an N-wire transmission line, each has N elements. Each of the vector Equations (5-23) and (5-24) then are equivalent to N *independent* scalar equations. Each of the scalar elements (each wire voltage or current) will propagate independently along the transmission line. (They may, or may not, be interdependent at the ends — depending on the terminating network.) Therefore, each wire voltage/current propagates identically to the voltage/current of a lossless, single-wire line that is described in Chapter 2. The methods and computer code given there can be used to simulate their propagation.

EXERCISE 5-3. For lossless transmission lines, verify that Equation (5-23) is consistent with Equations (5-4), (5-9), and (5-10) and that Equation (5-24) is consistent with Equations (5-5), (5-12), and (5-13).

5.1.4 Lossy Lines

For lossy lines, the signals change as they move along the line in accordance with Equations (5-9), (5-10), (5-12), and (5-13). Because $\mathbf{A}$ and $\mathbf{B}$ are not diagonal matrices, the signal on any of the wires affects the signal on the other wires to some amount. For illustration, consider a three-wire line with $\mathbf{Z}_O$ given by Equation (5-16) (its inverse is in the middle of Equation (5-18)). Let the resistance per unit length of wires 1, 2, and 3 be 8, 10, and 12 [Ω/m], respectively, and the ground return have negligible resistance. Then the $\mathbf{R}$ matrix is given by

$$\mathbf{R} = \begin{bmatrix} 8 & 0 & 0 \\ 0 & 10 & 0 \\ 0 & 0 & 12 \end{bmatrix} \tag{5-25}$$

Let the conductivity of the dielectric of the line be zero (i.e., $\mathbf{G}$ = matrix of all zeros). Then, $\mathbf{A} = \mathbf{B}$ and they are given by

$$\mathbf{A} = \mathbf{B} = \frac{1}{2}\mathbf{R}\mathbf{Z}_O^{-1} = \frac{1}{2}\begin{bmatrix} 8 & 0 & 0 \\ 0 & 10 & 0 \\ 0 & 0 & 12 \end{bmatrix}\begin{bmatrix} 0.0200 & -0.00674 & -0.00110 \\ -0.00674 & 0.0224 & -0.00674 \\ -0.00110 & -0.00674 & 0.0200 \end{bmatrix} \tag{5-26}$$

$$\mathbf{A} = \mathbf{B} = \begin{bmatrix} 0.08 & -0.02696 & -0.0044 \\ -0.0337 & 0.112 & -0.0337 \\ -0.0066 & -0.04044 & 0.12 \end{bmatrix} \quad [\text{m}^{-1}] \qquad (5\text{-}27)$$

The elements of these matrices have the dimension of reciprocal meters. The effect of a given element is equal to its value times the length of line over which it operates times the voltage or current on the wire corresponding to its column number (see Equations (5-9), (5-10), (5-12), and (5-13)). The signal on which the element has an effect is on the wire that corresponds to the voltage or current with the corresponding row index. Hence, the diagonal terms represent the effects of the signal of a given wire on the signal of the same wire; diagonal terms of $\mathbf{A}$ represent attenuation, while diagonal terms of $\mathbf{B}$ represent reflection along the same wire. The off-diagonal terms represent propagation crosstalk. The off-diagonal $\mathbf{A}$ terms represent an induced signal on another wire in the same propagation direction as the causing signal (forward-to-forward or backward-to-backward coupling); the off-diagonal $\mathbf{B}$ terms represent an induced signal in the reverse propagation direction from the causing signal (forward-to-backward or backward-to-forward coupling). One can think of the matrix $\mathbf{A}$ as representing attenuation in a generalized vector sense while the matrix $\mathbf{B}$ represents reflection in a generalized matrix sense.

PROBLEM 5.2   A two-wire line exists with:

Characteristic impedance, $\mathbf{Z}_o = \begin{bmatrix} 50 & 30 \\ 30 & 50 \end{bmatrix}$ $[\Omega]$

Wire resistances: $R_1 = 8$ $[\Omega/\text{m}]$, $R_2 = 10$ $[\Omega/\text{m}]$
Conductance $\mathbf{G}$ = zero matrix
Calculate $\mathbf{A}$ and $\mathbf{B}$ (since $\mathbf{G}$ is zero, they are equal).

PROBLEM 5.3   Assume a three-wire line with $\mathbf{A}$ and $\mathbf{B}$ given by Equation (5-27). A forward signal is launched from the left end with (forward voltages): $v_{F1} = 5$, $v_{F2} = 0$, $v_{F3} = 0$ [V]. The backward voltage signal is zero. Evaluate the right-hand sides of Equations (5-9) and (5-10). Explain the physical meaning of each element of the resulting two-column vectors.

5.1.5 Small Coupling Approximation of Propagation Crosstalk

The approximation discussed in this section is similar to the small backward signal approximation discussed in Section 2.2.8. Assume that a forward voltage signal is launched at the left end of a multi-wire transmission line; furthermore, assume that only one wire is driven and that all the other wires are held at ground (zero) potential at the left end. The right end is assumed to be terminated with a network that synthesizes the characteristic impedance matrix $\mathbf{Z}_0$ of the line so that at the right end there is no backward signal. The coupling (crosstalk) between the driven wire and the other wires and the reflected signal (due to the distributed reflection) on the driven wire are assumed small enough that their effect on the driven wire can be neglected. Then, the signal along the driven wire can be written explicitly as

$$v_{Fk}(x,t) \ = \ e^{-\alpha_{kk}x} v_o(t - x/c) \tag{5-28}$$

where wire k is the index of the driven wire and $v_o(t)$ is the voltage of its left end. The backward signals satisfy Equation (5-10) with the right-hand side of Equation (5-28) substituted for voltage $v_{Fk}(x,t)$. Writing vector element (row) i of Equation (5-10), but omitting the cross-coupling terms between the undriven wires, one obtains

$$v_{Bi1}(x,t + x/c) \ = \ \alpha_{ii} v_{Bi}(x,t + x/c) - \beta_{ik} e^{-\alpha_{kk}x} v_o(t - x/c) \tag{5-29}$$

Let $\tau = t + x/c$. For each value of τ (τ constant), Equation (5-29) is an ordinary differential equation in x that can be written

$$\frac{dv_{Bi}}{dx} - \alpha_{ii} v_{Bi} \ = \ -\beta_{ik} e^{-\alpha_{kk}x} v_o(\tau - 2x/c) \tag{5-30}$$

The solution of Equation (5-30) with the boundary condition $v_{Bi}(l,t + l/c) = 0$, where l is the length of the line, is

$$v_{Bi}(x,\tau) \ = \ \beta_{ik} e^{\alpha_{ii}x} \int_x^l e^{-(\alpha_{ii} + \alpha_{kk})x'} v_o(\tau - 2x'/c)\, dx' \tag{5-31}$$

where $\tau = t + x/c$.
At the left end of the line $x = 0$ and $\tau = t$. The arriving voltage at the left end is

$$v_{Bi}(0,t) \ = \ \beta_{ik} \int_0^l e^{-(\alpha_{ii} + \alpha_{kk})x'} v_o(t - 2x'/c)\, dx' \tag{5-32}$$

Recall the assumption that the left-end voltages are held to zero for $i \neq k$. Therefore, there is a forward signal launched at the left end given by

$$v_{Fi}(0,t) \ = \ -v_{Bi}(0,t) \ = \ -\beta_{ik} \int_0^l e^{-(\alpha_{ii} + \alpha_{kk})x'} v_o(t - 2x'/c)\, dx' \ , \ i \neq k \tag{5-33}$$

This equation supplies the boundary condition (at $x = 0$) for the forward crosstalk signal. The forward crosstalk signal is defined by Equation (5-9) with the right-hand side of Equation (5-28) substituted for voltage $v_{Fk}(x,t)$. Hence, for $i \neq k$, a row of Equation (5-9) is

$$v_{Fi1}(x,t - x/c) \ = \ -\alpha_{ii} v_{Fi}(x,t - x/c) - \alpha_{ik} e^{-\alpha_{kk}x} v_o(t - x/c) \tag{5-34}$$

Let $\tau = t - x/c$ and note that for τ constant, Equation (5-34) is an ordinary differential equation that can be written

$$\frac{dv_{Fi}}{dx} + \alpha_{ii} v_{Fi} \ = \ -\alpha_{ik} e^{-\alpha_{kk}x} v_o(\tau) \tag{5-35}$$

The solution of Equation (5-35) with the boundary condition at $x = 0$ given by Equation (5-33) can be written

$$v_{Fi}(x,\tau) = -\beta_{ik}e^{-\alpha_{ii}x}\int_0^l e^{-(\alpha_{ii}+\alpha_{kk})x'}v_0(\tau - 2x'/c)\,dx'$$

$$-\alpha_{ik}e^{-\alpha_{ii}x}v_0(\tau)\begin{cases} \dfrac{e^{(\alpha_{ii}-\alpha_{kk})x}-1}{\alpha_{ii}-\alpha_{kk}}, & \alpha_{ii} \neq \alpha_{kk} \\[2ex] x, & \alpha_{ii} = \alpha_{kk} \end{cases} \tag{5-36}$$

where $\tau = t - x/c$.

Equations (5-31) and (5-36) are first-order approximations of the crosstalk; that is, terms that involve the square and higher powers of the elements of the matrices **A** and **B** have been omitted.

A simple model of the crosstalk is obtained by letting $v_0(t) = u(t)$, the unit step function. (The resulting formula can be scaled to the actual signal voltage.) The integral in Equation (5-36) becomes

$$I_{ik}(\tau) = \int_0^l e^{-(\alpha_{ii}+\alpha_{kk})x'}u(\tau - 2x'/c)\,dx' \tag{5-37}$$

$$= \begin{cases} \dfrac{1}{\alpha_{ii}+\alpha_{kk}}\left[1 - e^{-(\alpha_{ii}+\alpha_{kk})\frac{c\tau}{2}}\right], & (\tau < 2l/c) \\[3ex] \dfrac{1}{\alpha_{ii}+\alpha_{kk}}\left[1 - e^{-(\alpha_{ii}+\alpha_{kk})l}\right], & (\tau \geq 2l/c) \end{cases} \tag{5-38}$$

and the forward voltage crosstalk, Equation (5-36), becomes

$$v_{Fi}(x,\tau) = -\beta_{ik}e^{-\alpha_{ii}x}I_{ik}(\tau)$$

$$-\alpha_{ik}e^{-\alpha_{ii}x}u(\tau)\begin{cases} \dfrac{e^{(\alpha_{ii}-\alpha_{kk})x}-1}{\alpha_{ii}-\alpha_{kk}}, & \alpha_{ii} \neq \alpha_{kk} \\[2ex] x, & \alpha_{ii} = \alpha_{kk} \end{cases} \tag{5-39}$$

PROBLEM 5.4   Consider a three-wire transmission line of length $l = 0.05$ [m], $c = 1.0\times10^8$ [m/s] and A and B given by Equation (5-27). At $t = 0$ a step function of amplitude 5 [V] is launched from the left end on wire 3; wires 1 and 2 are held at ground potential. Calculate the forward voltage on wire 2 at the right end at t = 0.5, 1.0 and 1.5 [ns] using the small coupling approximation, i.e., Equation (5-39).

5.1.6 Numerical Algorithm — Lossy Propagation with Crosstalk

In the single-wire case, the scalar voltage transport Equation (2-29) was approximated by the difference Equation (2-48). For the multi-wire case treated here, the vector voltage transport Equation (5-9) can be approximated by the vector difference equation:

$$\frac{\mathbf{v}_F(x_2,\tau_F) - \mathbf{v}_F(x_1,\tau_F)}{x_2 - x_1} = -\mathbf{A}\frac{\mathbf{v}_F(x_2,\tau_F) + \mathbf{v}_F(x_1,\tau_F)}{2} + \mathbf{B}\frac{\mathbf{v}_B(x_2,\tau_B) + \mathbf{v}_B(x_1,\tau_B)}{2} \quad (5\text{-}40)$$

where

$$\tau_F = t_1 - \frac{x_1}{c} = t_2 - \frac{x_2}{c}$$

$$\tau_B = t_1 + \frac{x_2}{c} = t_2 + \frac{x_1}{c}$$

Rearranging terms to form the explicit central difference equation, one obtains

$$\mathbf{v}_F(x_2,\tau_F) = \mathbf{v}_F(x_1,\tau_F)$$

$$+ (x_2 - x_1)\left[-\mathbf{A}\frac{\mathbf{v}_F(x_2,\tau_F) + \mathbf{v}_F(x_1,\tau_F)}{2} + \mathbf{B}\frac{\mathbf{v}_B(x_2,\tau_B) + \mathbf{v}_B(x_1,\tau_B)}{2} \right] \quad (5\text{-}41)$$

Similarly, for backward signals, the following difference approximation is obtained from the transport Equation (5-10)

$$\mathbf{v}_B(x_1,\tau_B) = \mathbf{v}_B(x_2,\tau_B)$$

$$+ (x_2 - x_1)\left[-\mathbf{A}\frac{\mathbf{v}_B(x_2,\tau_B) + \mathbf{v}_B(x_1,\tau_B)}{2} + \mathbf{B}\frac{\mathbf{v}_F(x_2,\tau_F) + \mathbf{v}_F(x_1,\tau_F)}{2} \right] \quad (5\text{-}42)$$

The analogous difference equations for the transport of the current vectors ($\mathbf{i}_F$ and $\mathbf{i}_B$) are the same as Equations (5-41) and (5-42) except $\mathbf{A}$ is replaced with $\mathbf{A}^T$ and $\mathbf{B}$ is replaced with $-\mathbf{B}^T$.

Equations (5-41) and (5-42) can be evaluated in a time-stepping method identical to that used to evaluate the scalar propagation difference equations. This is described in Section 2.9 and will not be repeated here. Computer codes 2-1 and 2-2 use the time-stepping method for single-wire lossless and lossy lines, respectively.

Computer Code 5-1.   Multi-Wire Lossy Propagation.

A propagation routine for a lossy multi-wire line based on the difference equations given above is listed below. It is a subroutine that can be called by the main routine of *Computer Code 2-1* or *2-2* with modifications to make the propagated current or voltage signal a vector instead of a scalar and `alpha_dx` and `beta_dx` matrices (rather than scalars). The new routine, named `VoltPropTstepMwires` is a version of `VoltPropTstep` modified to propagate a vector voltage (or current).

In the scalar routine `VoltPropTstep` there occurs multiplication of a signal by `alpha_dx` and `beta_dx` and accumulation (summation) of the result in several places. In the multi-wire routine `VoltPropTstepMwires` this becomes a matrix multiplication and accumulation of the vector result. To simplify the code, a subroutine was written to perform the operations of multiplication of a matrix by a column vector times a scalar multiplier and accumulation of the result in another (input/output) vector. This subroutine, included below, is named `AccumMatrixbyColumn`. It is called repeatedly in `VoltPropTstepM-wires`.

```
/*------- Function Declaration -------------------------------- */
  /***** Propagate voltage one time step lossy multi-wire line *****/
  // For propagation of current substitute alpha transpose for alpha
  // and minus beta transpose for beta.
  VoltPropTstepMwires (     /* Input variables  */
            int NumWires,             // number of wires
            float *Ring_ptr,          // ring pointer
            int Nhalf,                // number in half of ring
            float *SigInForwardVect,  // forward from left end
            float *SigInBackwardVect, // backward from right end
            float Wgt1,               // interpolation wgt
            float Wgt2,               // interpolation wgt
            float *alpha_dxMatrix,    // alpha*dx
            float *beta_dxMatrix,     // beta*dx
/* Input/Output variables  */
            int *TStepOffset,         // Time step offset
/* Output variable  */
            float *SigOutBackwardVect, // backward at left end
            float *SigOutForwardVect)  // forward at right end
  {
/* --- Local Variable Declarations --------------------------------
  Type        Name                 Description & range
  ----        ----                 -------------------------------
*/
  int       Offset;              // offset to point in ring
  int       th;                  // 2*Nhalf
  int       rcode;              // return code, =0, normal exit
  int       idx;                // for-loop index
  int       nwire;             // wire index
   int          idx_beta;                  // index for beta terms in
rhtemp_ptr
  int       idxPlusNhalf;       // idx + Nhalf
  float        *rhtemp_ptr=NULL;   // temporary storage pointer
```

```
   float         *MHalfAlpha_dxMat = NULL;// minus 0.5 * alpha_dxMat
   float         *HalfBeta_dxMat = NULL; // 0.5 * beta_dxMat
/*------------------------- Error Conditions Trapped --------------
 Fatal/Nonfatal    Description
 --------------    -----------
 Fatal             PropTstep           =-1, Nhalf < 2
---------------- Executable Code ------------------------------- */
rcode = 0;    // initialize return code
//*** Allocate matrices
MHalfAlpha_dxMat = (float*)malloc(sizeof(float)*NumWires*NumWires);
HalfBeta_dxMat = (float*)malloc(sizeof(float)*NumWires*NumWires);
for (idx = 0; idx < NumWires*NumWires; idx++)
{
  MHalfAlpha_dxMat[idx] = -0.5F * alpha_dxMatrix[idx];
  HalfBeta_dxMat[idx] = 0.5F * beta_dxMatrix[idx];
}
if (Nhalf < 2)
{
 rcode = -1;
 goto alldone;
}
th = 2 * Nhalf;
// Allocate memory for temp storage of r.h. side of transport eqs.
rhtemp_ptr = (float*)malloc(sizeof(float) * 2 * Nhalf*NumWires);
for (idx = 0; idx < 2 * Nhalf*NumWires; idx++)
 *(rhtemp_ptr + idx) = 0.0F;
// Calc. alpha term that depends on the last backward signal
// (before being overwritten by input)
AccumMatrixbyColumn ( NumWires,1.0F, MHalfAlpha_dxMat,
  (Ring_ptr + *TStepOffset*NumWires), (rhtemp_ptr + NumWires) );
// Input Forward signal at t = 0
for (idx = 0; idx < NumWires; idx++)
  *(Ring_ptr + *TStepOffset*NumWires + idx) = SigInForwardVect[idx];
// alpha and beta terms that use the left-end input
AccumMatrixbyColumn ( NumWires,1.0F, MHalfAlpha_dxMat,
  (Ring_ptr + *TStepOffset*NumWires), rhtemp_ptr );
AccumMatrixbyColumn ( NumWires,1.0F, HalfBeta_dxMat,
  (Ring_ptr + *TStepOffset*NumWires), (rhtemp_ptr + NumWires) );
// Calc. alpha term that depends on the last forward signal
// (before being overwritten by input)
Offset = (*TStepOffset + Nhalf) % th;       // Backward wave input
AccumMatrixbyColumn ( NumWires,1.0F, MHalfAlpha_dxMat,
  (Ring_ptr + Offset*NumWires),(rhtemp_ptr + (Nhalf + 1)*NumWires));
// Input backward signal at t = 0
for (idx = 0; idx < NumWires; idx++)
  *(Ring_ptr + Offset*NumWires + idx) = SigInBackwardVect[idx];
```

```
// alpha and beta terms that use the right-end input
AccumMatrixbyColumn ( NumWires,1.0F, MHalfAlpha_dxMat,
  (Ring_ptr + Offset*NumWires),(rhtemp_ptr + Nhalf*NumWires));
AccumMatrixbyColumn ( NumWires,1.0F, HalfBeta_dxMat,
  (Ring_ptr + Offset*NumWires),(rhtemp_ptr + (Nhalf + 1)*NumWires));
//*** Calculate r.h. side of transport eq. and put in rhtemp_ptr
// (idx = 0 are the special cases that were done above)
for (idx = 1; idx < Nhalf; idx++)
{
 //***Backward signal sources
 Offset = (*TStepOffset + idx) % th;
// Attenuation terms (alpha)
AccumMatrixbyColumn ( NumWires,1.0F, MHalfAlpha_dxMat,
  (Ring_ptr + Offset*NumWires),(rhtemp_ptr + idx*NumWires));
AccumMatrixbyColumn ( NumWires,1.0F, MHalfAlpha_dxMat,
  (Ring_ptr + Offset*NumWires),(rhtemp_ptr + (idx+1)*NumWires));
// Reflection terms (beta)
idx_beta = (th + 1 - idx) % th;
AccumMatrixbyColumn ( NumWires,1.0F, HalfBeta_dxMat,
  (Ring_ptr + Offset*NumWires),(rhtemp_ptr + idx_beta*NumWires));
Offset = (Offset +1) % th;
AccumMatrixbyColumn ( NumWires, Wgt2, HalfBeta_dxMat,
  (Ring_ptr + Offset*NumWires),(rhtemp_ptr + idx_beta*NumWires));
idx_beta = (th - idx) % th;
if (idx < Nhalf - 1)
  AccumMatrixbyColumn ( NumWires, Wgt1, HalfBeta_dxMat,
   (Ring_ptr + Offset*NumWires),(rhtemp_ptr + idx_beta*NumWires));
//***Forward signal sources
idxPlusNhalf = idx + Nhalf;
Offset = (*TStepOffset + idxPlusNhalf) % th;
// Attenuation terms (alpha)
AccumMatrixbyColumn ( NumWires, 1.0F, MHalfAlpha_dxMat,
  (Ring_ptr + Offset*NumWires),
  (rhtemp_ptr + idxPlusNhalf*NumWires));
if (idx < Nhalf -1)
  AccumMatrixbyColumn ( NumWires, 1.0F, MHalfAlpha_dxMat,
   (Ring_ptr + Offset*NumWires),
   (rhtemp_ptr + (idxPlusNhalf + 1)*NumWires));
else
  AccumMatrixbyColumn ( NumWires, 1.0F, MHalfAlpha_dxMat,
   (Ring_ptr + Offset*NumWires),rhtemp_ptr);
  // Reflection terms (beta)
idx_beta = Nhalf + 1 - idx;
AccumMatrixbyColumn ( NumWires, 1.0F, HalfBeta_dxMat,
   (Ring_ptr + Offset*NumWires),(rhtemp_ptr + idx_beta*NumWires));
Offset = (Offset +1) % th;
```

```
AccumMatrixbyColumn ( NumWires, Wgt2, HalfBeta_dxMat,
   (Ring_ptr + Offset*NumWires),(rhtemp_ptr + idx_beta*NumWires));
idx_beta = Nhalf - idx;
if (idx < Nhalf - 1)
AccumMatrixbyColumn ( NumWires, Wgt1, HalfBeta_dxMat,
   (Ring_ptr + Offset*NumWires),(rhtemp_ptr + idx_beta*NumWires));
}
//*** Apply Attenuation and Reflection terms to stored signals
for (idx = 0; idx < 2 * Nhalf; idx++)
{
  Offset = (*TStepOffset + idx) % th;
  for (nwire = 0; nwire < NumWires; nwire++)
   *(Ring_ptr + Offset*NumWires + nwire) +=
    *(rhtemp_ptr + idx*NumWires + nwire);
}
//*** Increment timestep offset around circle
*TStepOffset = (*TStepOffset + 1) % th;
//*** Output signals at t = Tstep
// Forward signal
Offset = (*TStepOffset + Nhalf) % th;
for (nwire = 0; nwire < NumWires; nwire++)
  SigOutForwardVect[nwire] = Wgt1 *
                            *(Ring_ptr + Offset*NumWires+nwire);
Offset = (*TStepOffset + Nhalf + 1) % th;
for (nwire = 0; nwire < NumWires; nwire++)
  SigOutForwardVect[nwire] +=Wgt2 *
                            *(Ring_ptr + Offset*NumWires+nwire);
// Backward signal
for (nwire = 0; nwire < NumWires; nwire++)
  SigOutBackwardVect[nwire] = Wgt1 *
                            *(Ring_ptr + *TStepOffset*NumWires+nwire);
Offset = (*TStepOffset + 1) % th;
for (nwire = 0; nwire < NumWires; nwire++)
  SigOutBackwardVect[nwire] += Wgt2 *
                            *(Ring_ptr + Offset*NumWires+nwire);
alldone:
  // Free temporary memory
  free (rhtemp_ptr);
  rhtemp_ptr = NULL;
  free (MHalfAlpha_dxMat);
  MHalfAlpha_dxMat = NULL;
  free (HalfBeta_dxMat);
  HalfBeta_dxMat = NULL;
  return(rcode);
}
/* -----End of Procedure VoltPropTstepMwires -------------------- */
```

```
/*------- Function Declaration --------------------------------- */
// Adds matrix times VectorIn times constant to VectInOut
  void AccumMatrixbyColumn ( /* Input variables  */
                    int dim,          // dimension of square matrix
                    float constant,   // constant multiplier
                    float *Matrix,    // input matrix
                    float *VectorIn,  // input vector
                    /* Input/Output  */
                    float *VectInOut )  // input and output vectors
{
/*    --- Local Variable Declarations -----------------------------
   Type        Name                  Description & range
   ----        ----                  ------------------------------ */
   int         idx,                  // for-loop index over columns/rows
               nrow;                 // row index
/*----------------------- Error Conditions Trapped ----------------
   Fatal/Nonfatal    Description
   --------------    -----------
---------------- Executable Code ------------------------------- */
for (nrow = 0; nrow < dim; nrow++)
{
  for (idx = 0; idx < dim; idx++)
    VectInOut[nrow] += Matrix[nrow + dim * idx] * VectorIn[idx]
                          *constant;
}
return;
}
/* ---------- End of function AccumMatrixbyColumn --------------- */
```

5.2 BOUNDARY CONDITIONS — LINE-END EQUIVALENT CIRCUITS

Equivalent circuits to represent the terminations of multi-wire transmission lines are generalizations of those presented for single-wire lines in Chapter 4. They are tools to facilitate
the numerical simulation of networks consisting of multi-wire transmission lines and circuit-
element networks attached to them. Methods to approximate the response of such networks
is the subject of Part II of this book. The generalized equivalent circuits are presented below
as groundwork for the network solutions and to increase understanding of transmission line
behavior.

5.2.1 Thevenin Equivalent Circuit

The Thevenin equivalent circuit for an end of a multi-wire transmission line is shown below
in Figure 5.1. The ambiguity of the sign of the currents shown at the line terminals is because

a transmission line current is defined as positive to the right; that is, it is positive if flowing (to the right) *out of* the terminals at the right end, and it's also positive if flowing (to the right) *into* the terminals at the left end. But in the equivalent circuit representation at either end, the current is assumed positive if flowing out. The voltage drop across an impedance has its plus sign at the tail of the current arrow for the current going through it. Therefore, the terminal voltage vector for the circuit shown below is

$$\mathbf{v} = 2\mathbf{v}_{arr.} - \mathbf{Z}_o\mathbf{i} \tag{5-43}$$

where $\mathbf{i}$ is positive out of the terminals (corresponding to the right end). If a termination (load) network is attached to the terminals and the network is a synthesis of $\mathbf{Z}_0$, the line is impedance matched. Then, the terminal voltage $\mathbf{v}$ can be written

$$\mathbf{v} = \mathbf{Z}_o\mathbf{i} \tag{5-44}$$

Setting $\mathbf{v}$ in Equation (5-43) to $\mathbf{v}$ in Equation (5-44), one finds that $\mathbf{v}_{arr.} = \mathbf{Z}_o\mathbf{i} = \mathbf{v}$ and therefore (using Equation 5-4), the departing signal is zero; i.e., the termination is nonreflecting.

EXERCISE 5-4. Show that if a resistive network whose matrix is given by $\mathbf{R}_L$ is connected to the terminals of the Thevenin equivalent circuit, the terminal voltage vector is

$$\mathbf{v} = 2\mathbf{R}_L(\mathbf{Z}_o + \mathbf{R}_L)^{-1}\mathbf{v}_{arr.} \tag{5-45}$$

PROBLEM 5.5 Consider a two-wire line with characteristic impedance matrix $\mathbf{Z}_o = \begin{bmatrix} 50 & 30 \\ 30 & 50 \end{bmatrix}$. Let the line be terminated with 50 [Ω] resistors to ground from both terminal 1 and terminal 2. Write out the matrix $(\mathbf{Z}_o + \mathbf{R}_L)$. Calculate the inverse of this matrix; i.e., write out the matrix $(\mathbf{Z}_o + \mathbf{R}_L)^{-1}$. If

Figure 5.1 Thevenin Equivalent for a Multi-Wire Transmission Line End.

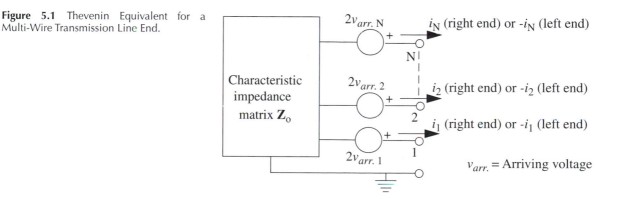

$$\mathbf{v}_{arr.} = \begin{bmatrix} 5 \\ 0 \end{bmatrix},$$ calculate the terminal voltage vector, $\mathbf{v}$. Calculate the departing voltage vector

$$\mathbf{v}_{dep.} = \mathbf{v} - \mathbf{v}_{arr.} \,.$$

5.2.2 Norton Equivalent Circuit

The Norton equivalent circuit for a multi-wire transmission line end is the generalization of the single-wire line Norton equivalent circuit discussed in Section 4.2. It is shown below in Figure 5.2. The short-circuit current is given by $\pm 2\mathbf{i}_{arr.}$, the sign depending on which end of the transmission line is to be represented. The figure indicates a positive short-circuit current flowing out of the terminals; however, a positive transmission line current flows into the terminals at the left end of the line, hence, the minus sign when the left end is to be represented. Referring to Figure 5.2, one can write the vector current flowing out of the terminals as

$$\mathbf{i} = 2\mathbf{i}_{arr.} - \mathbf{G}_o \mathbf{v} \tag{5-46}$$

where, the vector current $\mathbf{i}$ is taken as positive out of the terminals.

EXERCISE 5-5.   In parallel to the previous exercise, show that if a conductance load network that is represented by the conductance matrix $\mathbf{G}_L$ is connected to the line end, then the vector current flowing out of the transmission line into the network is given by

$$\mathbf{i} = 2\mathbf{G}_L (\mathbf{G}_o + \mathbf{G}_L)^{-1} \mathbf{i}_{arr.} \tag{5-47}$$

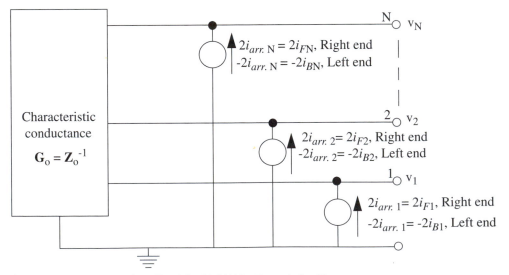

Figure 5.2   Norton Equivalent Circuit for Multi-Wire Transmission Line.

PROBLEM 5.6 For a two-wire line with characteristic impedance matrix $\mathbf{Z}_o = \begin{bmatrix} 50 & 30 \\ 30 & 50 \end{bmatrix}$ find the characteristic conductance matrix, i.e., the inverse of $\mathbf{Z}_o$. If the load is 50 [Ω] to ground on each of the two conductors, write the load conductance matrix $\mathbf{G}_L$. Find the matrix $(\mathbf{G}_o + \mathbf{G}_L)^{-1}$ and the matrix $\mathbf{G}_L(\mathbf{G}_o + \mathbf{G}_L)^{-1}$. For an arriving current vector $\mathbf{i}_{arr.} = \begin{bmatrix} 0 \\ 0.1 \end{bmatrix}$ [A], find the current out of the terminals into the load.

5.2.3 Reflection at a Resistive Termination

Let a line be terminated with a resistive network given by the matrix $\mathbf{R}_L$. Then the departing voltage vector is given as a matrix times the arriving voltage vector where the matrix is

$$\mathbf{T}_{Rv} = (\mathbf{R}_L - \mathbf{Z}_o)(\mathbf{R}_L + \mathbf{Z}_o)^{-1} \tag{5-48}$$

The matrix $\mathbf{T}_{Rv}$ is the voltage reflection matrix at the termination. The generalization of the scalar Equation (2-11) is

$$\mathbf{v}_{\text{Departing}} = \mathbf{T}_{Rv}\,\mathbf{v}_{\text{Arriving}} \tag{5-49}$$

or, depending on which end the termination is connected,

$$\mathbf{v}_F\big|_{\text{Left end}} = \mathbf{T}_{Rv}\,\mathbf{v}_B\big|_{\text{Left end}} \tag{5-50}$$

or

$$\mathbf{v}_B\big|_{\text{Right end}} = \mathbf{T}_{Rv}\,\mathbf{v}_F\big|_{\text{Right end}} \tag{5-51}$$

EXERCISE 5-6. Using Equations (5-19) and (5-20) derive Equations (5-50) and (5-51). *Hint*: Review *Exercise 2-5*.

EXERCISE 5-7. Using Equations (5-21) and (5-22) show that the current-reflection matrix is

$$\mathbf{T}_{Ri} = \left(\mathbf{R}_L^{-1} - \mathbf{Z}_o^{-1}\right)\left(\mathbf{R}_L^{-1} + \mathbf{Z}_o^{-1}\right)^{-1} \tag{5-52}$$

such that

$$\mathbf{i}_{\text{Departing}} = \mathbf{T}_{Ri}\,\mathbf{i}_{\text{Arriving}} \tag{5-53}$$

5.3 TERMINATION CROSSTALK BETWEEN TRACES

Wires or traces can be used as pairs and driven in differential mode or they can be driven with respect to a common reference, usually a ground plane, but sometimes a common reference trace or one wire in a bundle can be used as a common reference conductor. These signal paths are thought of as independent paths: after all, the electrical conductivity between paths is usually negligible. At very low frequencies (i.e., small time derivatives of the signals) they are independent paths for practical purposes. However, coupling exists between nearby conductors due to their mutual capacitance and inductance. The capacitance between conductors causes an induced current (per unit length) from one conductor to another proportional to the time derivative of their relative voltage. And the mutual inductance between them causes an induced relative voltage gradient along them proportional to the time derivative of the difference of their currents. Therefore, when time derivatives of signals are significant and the mutual capacitance/inductance is significant, the cross coupling will also be significant.

The capacitance/inductance between a pair of conductors is necessary for signal propagation along the pair of conductors. But if the conductors are intended to carry independent signals, the same coupling effect is uninvited and is often called crosstalk; in particular, this type of crosstalk will be called termination crosstalk in this book. Termination crosstalk is fundamentally different from crosstalk due to resistance or conductance along the line; that due to resistance or conductance along the line is simulated in the propagation algorithm. Termination crosstalk (even though it physically occurs due to the capacitance and inductance distributed along the line) is simulated at the termination and can be controlled by adjusting the termination network attached to the line ends.

Therefore, signal propagation along adjacent conductors and termination crosstalk between adjacent conductors is an identical physical effect to be distinguished only by its desirability. It is obvious, therefore, that the same mathematical/computational tools can be used to treat them both. And, of course, the termination network (and the geometry of the wires or traces themselves) can be designed to enhance desired coupling and reduced undesired coupling, i.e., the crosstalk.

In the models of terminations already developed, it has been emphasized that if a termination network is used that is equivalent to the characteristic impedance network, two things are true: (1) no signal is reflected back down the line, and, (2) the terminal voltage vector (or current vector) is equal to the arriving voltage vector (or current vector). On the other hand, Problems 5-5 and 5-6 illustrate what occurs if the termination consists of simple load resistors to ground from each of the conductors. In this case, the arriving signal on one wire is coupled to the terminal of the adjacent wire. This is an illustration of crosstalk.

It is obvious then, that termination crosstalk can be reduced by using a termination network that more nearly simulates the characteristic impedance matrix of nearby conductors. In Part II, methods will be developed to do this and to simulate the effects on crosstalk of various termination choices.

PART II

CIRCUIT SOLUTIONS AT LINE TERMINATIONS

Chapter 6

NETWORKS WITH REACTIVE AND NONLINEAR ELEMENTS

In Part I, Chapters 1 through 5, transmission line propagation algorithms were derived. These included Norton equivalent circuits for the line ends. A general method was given for solving networks that are represented by Norton equivalent circuits with nonreactive (resistive) conductances (Chapter 3). It was shown how to use these tools to solve general networks of interconnected lines with resistive networks at their terminations (Chapters 4 and 5).

In this Chapter it will be shown that reactive and nonlinear elements can be represented over short time steps by resistive Norton equivalent circuits. Hence, the methods given in Part I can be used in time-stepping algorithms for networks with reactive and nonlinear elements attached to transmission lines.

In Chapter 7, an algorithm for simultaneous solution of transmission line propagation, line junctions, and terminations is developed. Also, given in this Chapter is a method to determine the network of resistors required to match the characteristic impedance of a multi-wire line. That is, given a real, symmetric characteristic impedance matrix $\mathbf{Z}_O$, an algorithm is given to synthesize that matrix with a network of resistors. By terminating a multi-wire line with this network, the termination crosstalk can, in principle, be essentially eliminated. The method applies to the most general transmission lines analyzed in this book.

Finally, in Chapter 8, an implementing computer code is given that solves general networks of multi-wire single-speed transmission lines and circuit elements. In Part III of this book, this algorithm and code is generalized to networks with multi-speed lines (both lossless and lossy).

Chapter 9 contains examples of network solutions using the computer code of Chapter 8.

6.1 NETWORKS OF RESISTORS

Assume a general resistive network exists with K nodes. Let R_{ij} be the resistance [ohms] of the resistor connected between node i and node j if $i \neq j$. If $i = j$, let R_{ii} be the resistance of a resistor connected from node i to ground. Using the principles developed in Chapter 3, the elements of the conductance matrix $\mathbf{G}$ for the network can be written as:

Diagonal elements (Equation 3-17):

$$G_{ii} = \sum_{j=1}^{K} \frac{1}{R_{ij}} \tag{6-1}$$

and off-diagonal elements (Equation 3-18):

$$G_{ij} = \frac{-1}{R_{ij}}, \quad i \neq j \tag{6-2}$$

The matrix G is defined such that if a voltage source (vector) is attached to the nodes, the current that flows into the nodes is given by the vector

$$\mathbf{i} = \mathbf{Gv} \tag{6-3}$$

Such a network, with two terminals, is illustrated in Figure 6.1.
   Similarly, the resistance matrix for the network is defined such that

$$\mathbf{v} = \mathbf{Ri} \tag{6-4}$$

Therefore,

$$\mathbf{R} = \mathbf{G}^{-1} \tag{6-5}$$

EXERCISE 6-1.   If $R_{ij} = 0$, a short circuit exists between nodes i and j. How does one deal with the division by zero implied by Equations (6-1) and (6-2) in this case? *Hint*: How many nodes does the network with a short have?

PROBLEM 6.1   Given a two-terminal network with the following resistors: $R_{11} = 60\ [\Omega]$, $R_{22} = 40\ [\Omega]$, $R_{12} = 100\ [\Omega]$. Find the matrix G and the matrix $\mathbf{R} = \mathbf{G}^{-1}$.

6.2 SYNTHESIS OF A SYMMETRIC RESISTIVE CIRCUIT MATRIX

Suppose the characteristic impedance matrix of a transmission line is the symmetric matrix $\mathbf{Z_o}$. It's desired to find the network of resistors that synthesizes $\mathbf{Z_o}$ so that it can be used as an impedance matching load. One can calculate $\mathbf{G} = \mathbf{Z}_O^{-1}$ and then find the resistor network that represents $\mathbf{G}$. That is, the resistor values R_{ij} that yield $\mathbf{G}$ when substituted into Equations (6-1) and (6-2) can be found.

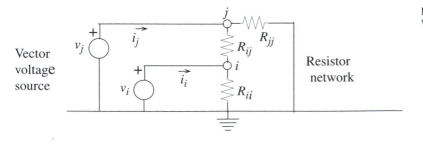

Figure 6.1   Resistor Network Driven by Voltage Source.

EXERCISE 6-2.   Show that a network with resistors given by Equation (6-6) yields the conductance matrix **G**. *Hint*: Substitute R_{ij} into Equations (6-1) and (6-2).

$$R_{ij} = \begin{cases} -\dfrac{1}{\mathbf{G}_{ij}}, & i \neq j \\[2ex] \dfrac{1}{\displaystyle\sum_k \mathbf{G}_{ik}}, & i = j \end{cases} \tag{6-6}$$

PROBLEM 6.2   The characteristic impedance matrix of a two-wire line is $\mathbf{Z}_o = \begin{bmatrix} 50 & 30 \\ 30 & 50 \end{bmatrix}$. What are the values of R_{11}, R_{22}, and R_{12}, the resistors that form a matching network for the line?

PROBLEM 6.3   A (nonsymmetric) two-wire line with conductors loosely coupled has a characteristic impedance matrix given by $\mathbf{Z}_o = \begin{bmatrix} 50 & 5 \\ 5 & 40 \end{bmatrix}$. What are the values of R_{11}, R_{22}, and R_{12}, the resistors that form a matching network for this line?

PROBLEM 6.4   Suppose that wire 2 of the two-wire line of the preceding problem is left open circuit at both ends (i.e., the current in wire 2 is very close to zero), what value of resistance should be used to terminate wire 1 to minimize reflection?

PROBLEM 6.5   Suppose that wire 2 of the two-wire line of the preceding problem is shorted to ground at both ends (i.e., the voltage of wire 2 is very close to zero), what value of resistance should be used to terminate wire 1 to minimize reflection? *Hint:* What is the solution of

$$\begin{bmatrix} v_1 \\ 0 \end{bmatrix} = \begin{bmatrix} 50 & 5 \\ 5 & 40 \end{bmatrix} \begin{bmatrix} i_1 \\ i_2 \end{bmatrix}$$

6.3 APPROXIMATE NORTON EQUIVALENT FOR A TWO-TERMINAL NETWORK

The Norton equivalent circuit shown below in Figure 6.2 represents the algebraic relationship

$$i(t + \delta t) = v(t + \delta t)g - I_{sc} \tag{6-7}$$

Equation (6-7) and the corresponding Norton equivalent circuit can be used to approximate the behavior of any smoothly behaving two-terminal network over the time interval $(t, t + \delta t)$.

Figure 6.2   Norton Equivalent Valid for Time Interval (t, t + dt).

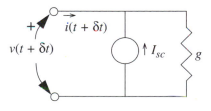

By smoothly behaving, we mean that the circuit does not display a discontinuous (or sudden) jump in the time interval. One method of defining the Norton equivalent is as a truncated Taylor series; then it's required that $i(t)$ and $v(t)$ be differentiable. The first two terms of such a series expanding i as a function of v would define g and I_{sc} as[1]

$$g = \frac{di}{dv}\bigg|_{\delta t = 0} = \frac{\dfrac{di}{dt}}{\dfrac{dv}{dt}}\bigg|_{\delta t = 0} \tag{6-8}$$

and

$$I_{sc} = v(t)g - i(t) \tag{6-9}$$

This two-term Taylor series defines the function as a tangent to the i-v curve. Often a better approximation is the chord approximation; see Figure 6-3. Using the chord approximation, g is defined by

$$g = \frac{i(t + \delta t) - i(t)}{v(t + \delta t) - v(t)} \tag{6-10}$$

and I_{sc} is given by Equation (6-9). The accuracy of either of the linear approximations is, of course, strongly dependent on the size of δt — making δt smaller increases the accuracy. (In the limit as $\delta t \to 0$, Equation (6-10) is essentially the definition of the derivative $\dfrac{di}{dv}$ and the chord and tangent "approximations" are equal.)

The usefulness of the Norton equivalent circuits defined in this Section is that they can easily be included in time-stepping solutions of large networks. Each time step, the current source I_{sc} normally must be recalculated. The conductance g normally does not change from time step to time step unless the circuit element being represented is nonlinear or the step size δt changes.

---

1. The current as a function of voltage has the Taylor series

$$i(v) = i(v_o) + \frac{di}{dv}\bigg|_{v = v_o}(v - v_o) + \dots$$

Let $v_o = v(t)$ and $v = v(t + \delta t)$, then

$$i(v(t + \delta t)) = i(v(t)) + \frac{di}{dv}\bigg|_{v = v(t)}[v(t + \delta t) - v(t)] + \dots$$

Expressing the current as a function of time, rather than as a function of voltage, we obtain

$$i(t + \delta t) = i(t) + \frac{di}{dv}\bigg|_t[v(t + \delta t) - v(t)] + \dots$$

Substituting g as given by Equation (6-8), we obtain

$$i(t + \delta t) \cong i(t) + v(t + \delta t)g - v(t)g$$

which is Equation (6-7) with I_{sc} given by Equation (6-9).

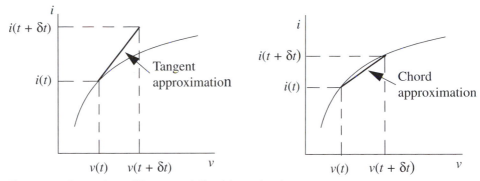

Figure 6.3   Comparison of Tangent and Chord Approximations.

PROBLEM 6.6   A capacitor connected directly to a resistor has voltage history

$$v(t) = v_o e^{-t/RC} \tag{6-11}$$

where, v_O is the voltage at $t = 0$. The current through the capacitor is

$$i(t) = C\frac{dv}{dt} = \frac{-v_o}{R}e^{-t/RC} \tag{6-12}$$

What is g and I_{sc} for the chord approximation (Equation (6-10))? Draw a sketch of the i-v curve. Does the negative value of g have a physical interpretation?

6.4 NORTON EQUIVALENT FOR A CAPACITOR

A capacitor behaves in accordance with the differential equation

$$i(t) = C\frac{dv(t)}{dt} \tag{6-13}$$

Over a short time step δt, this can be approximated by (the chord approximation of the derivative)

$$i(t + \delta t) \cong \frac{C}{\delta t}v(t + \delta t) - \frac{C}{\delta t}v(t) \tag{6-14}$$

This equation is identical to Equation (6-7) if g is set to g_C, where

$$g_C = C/\delta t \tag{6-15}$$

and I_{sc} is set to I_{scC}, where

$$I_{scC} = \frac{C}{\delta t}v(t) \tag{6-16}$$

Network solution algorithms are predicated on the assumption that all voltages and currents are known at time t and the network solution will update these quantities to time $t + \delta t$. Hence, $v(t)$ is assumed known. After solution of the network from t to $t + \delta t$, the node voltages are known at $t + \delta t$. By subtraction of the voltages of the two nodes where the capacitor is connected, the capacitor voltage $v(t + \delta t)$ is found; this should be saved and used as the initial value for solving the network over the next time step.

PROBLEM 6.7 Let a capacitor be discharging through a shunted resistor. The problem parameters are $C = 10{,}000$ [pF] $= 10^{-8}$ [F], $R = 100$ [Ω] and the initial voltage $v(0)$ is 5 [V]. For the time step $t = 0$ to $\delta t = 0.25 \times 10^{-6}$ [s], calculate the following: the RC decay time constant, g_C, I_{scC}, the value of $v(\delta t)$ using a. the Norton approximation for the capacitor, and, b. the exact exponential decay for the R-C circuit. What can be done to improve the agreement between the approximate model (a.) and the exact model (b.)?

6.5 NORTON EQUIVALENT FOR AN INDUCTOR

An inductor behaves according to the following differential equation

$$v = L\frac{di}{dt} \tag{6-17}$$

Over a short time step δt, this can be approximated by (the chord approximation)

$$v(t + \delta t) \cong L\frac{i(t + \delta t) - i(t)}{\delta t} \tag{6-18}$$

Solving for $i(t + \delta t)$, one obtains

$$i(t + \delta t) = \frac{\delta t}{L}v(t + \delta t) + i(t) \tag{6-19}$$

This equation is identical to Equation (6-7) if g is set to g_L, where

$$g_L = \delta t/L \tag{6-20}$$

and I_{sc} is set to I_{scL}, where

$$I_{scL} = -i(t) \tag{6-21}$$

It is assumed that $i(t)$ is known from the network solution over the previous time step (or as an initial condition). After the solution of the network from time t to $t + \delta t$, the new current through the inductor can be found at time $t + \delta t$ using Equation (6-19). That is, after the network solution, the node voltages are known at the two nodes where the inductor is connected so that the voltage across the inductor $v(t + \delta t)$ can be obtained as the difference of those two node voltages. Then, $i(t + \delta t)$ is obtained by substitution into Equation (6-19). It can then be saved and used as the initial current $i(t)$ for solution over the next time step.

PROBLEM 6.8 An inductor its driven by a driver as shown below in Figure 6.4. The driver is defined by $V_{oc}(t) = 5u(t)$ [V], where $u(t)$ is the unit step function and the driver impedance is $R = 5$ [Ω]. The inductor $L = 5\times10^{-6}$ [H].

    a. What is the initial voltage $v(0^+)$ and current $i(0^+)$?

    b. What is the exact solution for $v(t)$?

    c. Convert the driver to a Norton equivalent circuit. What is g and I_{sc} for the driver?

    d. Assume a time step of $\delta t = 0.2\times10^{-6}$ [s]. Find the Norton equivalent circuit of the load L for a time step from $t = 0$ to $t = \delta t$.

    e. Using the Norton equivalent for the driver (c.) and the Norton equivalent for the inductor (d.), calculate the voltage $v(\delta t)$.

    f. Calculate the exact solution at $t = \delta t$ using (b.).

6.6 NORTON EQUIVALENT FOR AN AC TERMINATION

A capacitor is often used in series with a resistor as a line termination. This reduces (often dramatically) the energy consumed by the circuit (especially in its quiescent state[2]). Hence, its cooling requirements are also reduced. Such an R-C series circuit used as a line termination has come to be known as an AC termination (see Johnson and Graham, 1993). A Norton equivalent circuit can easily be defined for such a circuit, thereby avoiding the use of separate elements for the resistor and capacitor (in a circuit model), simplifying the simulation of the network.

    The Thevenin equivalent circuit for the capacitor is given by

$$R_C = 1/g_C = \delta t/C \tag{6-22}$$

and

$$V_{ocC} = \frac{I_{scC}}{g_C} = v_C(t) \tag{6-23}$$

where g_C and I_{scC} are given by Equations (6-15) and (6-16) and v_C is the capacitor voltage. Let the resistor in the AC termination have resistance R. Then, the Thevenin resistance of the AC termination is

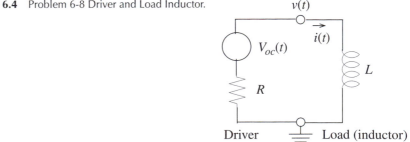

Figure 6.4   Problem 6-8 Driver and Load Inductor.

_____

2. The quiescent state is powered, but in the low state. It is usually the "zero" logical state of the circuit.

$$\frac{1}{g_{AC}} = R + R_C = R + \frac{\delta t}{C} \tag{6-24}$$

where g_{AC} is the Norton equivalent conductance. The parameters of the Norton equivalent of the AC termination are

$$g_{AC} = \frac{1}{R + \dfrac{\delta t}{C}} \tag{6-25}$$

and

$$I_{scAC} = g_{AC} V_{ocC} = \frac{v_C(t)}{R + \dfrac{\delta t}{C}} \tag{6-26}$$

The Norton equivalent Equation (6-7) for the AC termination is

$$i(t + \delta t) = v(t + \delta t) g_{AC} - I_{scAC} \tag{6-27}$$

It is assumed that the capacitor voltage at the beginning of the time step $v_C(t)$ is known. Thus, after network solution to time $t + \delta t$, $v_C(t + \delta t)$ must be calculated to be used as the initial condition for the next time step. The voltage $v_C(t + \delta t)$ is easily found directly from the circuit for the AC termination at time $t + \delta t$. The circuit of Figure 6-5 yields

$$v_C(t + \delta t) = v(t + \delta t) - Ri(t + \delta t) \tag{6-28}$$

Substituting $i(t + \delta t)$ using Equation (6-27), Equation (6-28) becomes

$$v_C(t + \delta t) = v(t + \delta t)(1 - Rg_{AC}) + RI_{scAC} \tag{6-29}$$

The voltage $v(t + \delta t)$ is the difference of the voltages of the nodes where the AC termination is attached after solution to time $t + \delta t$.

PROBLEM 6.9   Refer to the AC termination shown in Figure 6.6.
    a. What is the time constant (RC) of the termination?
    b. What is the Norton conductance g_{AC} for the AC termination?
    c. Assume the capacitor voltage is initially zero; what is the Norton current source I_{scAC} at
        $t = 0$ for the AC termination?

Figure 6.5   AC Termination Circuit.

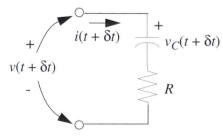

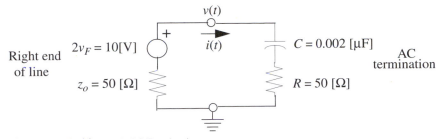

Figure 6.6   Problem 6-9, AC Termination.

  d. Let $\delta t = 1$ [ns], $v_C(0) = 0$ and the forward voltage $v_F = 5$ [V] (constant). Calculate $v(t)$ and $v_C(t)$
  for the first four time steps. *Hint*: Convert the circuit of Figure 6.6 to Norton equivalents using
  the Norton equivalent for the AC termination (Figure 6-2 and Equations (6-25) and (6-26), for
  the AC termination).

6.7 PERFORMANCE OF AN AC TERMINATION

  The AC termination is designed to load the line with its characteristic impedance for an initial
  instant of time at the arrival of a pulse, thus preventing an initial reflection. The impedance of
  the termination will subsequently increase from z_o to infinity (exponentially) as the R-C circuit relaxes. For lines with a small driver impedance, the RC time constant of the termination
  should be much larger than the delay time ($t_{delay} = length/c$) of the line for minimal
  overshoot due to the relaxation of the termination circuit.

  Consider a lossless line of length l driven by a step function of voltage (amplitude v_o)
  initiated at time $-l/c$ so that the pulse arrives at the right end at time zero. The driver impedance is assumed equal to zero to approximate the behavior of a line with a small driver
  impedance. See Figure 6-7. The arriving voltage at the right end is $v_F = v_o$; it will be constant for a period of time $2l/c$ from $t = 0$, the moment it arrives, until a reflected signal can
  propagate from the right end to the left end, reflect at the left end, and return to the right end.
  The equivalent circuit at the right end is shown in Figure 6.8.

Figure 6.7   Line with AC Termination.

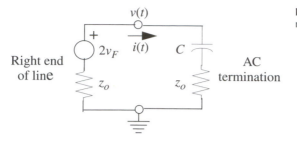

Figure 6.8 Right End of Line with AC Termination.

The loop equation for the circuit in Figure 6-8 is

$$2z_o i(t) + \frac{1}{C}\int_0^t i(t')dt' = 2v_F \tag{6-30}$$

with solution

$$i(t) = \frac{v_F}{z_o} e^{-t/2z_o C} \tag{6-31}$$

Then,

$$v(t) = z_o i(t) + \frac{1}{C}\int_0^t i(t')dt' \tag{6-32}$$

$$v(t) = v_F\left[2 - e^{-t/2z_o C}\right] \tag{6-33}$$

The reflected voltage is $v_B(t) = v(t) - v_F$, or

$$v_B(t) = v_F\left[1 - e^{-t/2z_o C}\right] \tag{6-34}$$

The voltage at the right end $v(t)$ is initially $v_F = v_o$ but then increases, asymptotically approaching $2v_o$. However, at time $2l/c$, the reflected signal from the left end arrives. It is negative ($v_F(t)$ decreases) and changes the slope of $v(t)$ so that $v(t)$ heads downward toward v_o. The largest value that $v(t)$ attains is its value at time $2l/c$. That maximum value is

$$v\left(\frac{2l}{c}\right) = v_o\left[2 - e^{-l/cz_o C}\right] \tag{6-35}$$

If $l/cz_o C$ is small (i.e., the RC time constant ($z_o C$) large compared to the delay time), then $e^{-l/cz_o C} \cong 1 - \frac{l}{cz_o C}$ and the overshoot is approximately

$$v_{overshoot} \cong v_o \frac{l}{c z_o C} . \tag{6-36}$$

If the driver impedance is not small, different behavior results. Suppose that the driver has impedance equal to z_o. Then, the forward signal $v_F = v_o/2$ and there is no reflection from the left end. The voltage at the right end is initially $v_o/2$; it then rises to approach v_o asymptotically with time constant equal to $2z_oC$. If that time constant is long, the signal will take a long time to reach the value v_o (that may be its intended high value).

PROBLEM 6.10 Assume the line parameters and termination of Problem 6-9 (with the exception of v_F). The line is lossless and is assumed driven by a 5 [V] step-function driver with a 5 [Ω] source impedance. Calculate the departing voltage taking into account the driver impedance. (This is the forward voltage signal v_F.) When this pulse arrives at the right end of the line, what is the expected overshoot (assume delay time $l/c = 5$ [ns]).

6.8 NONLINEAR TWO-TERMINAL CIRCUIT ELEMENTS

The integrated circuit (IC) components used in digital devices typically present a nonlinear load to the traces. An accurate simulation must account for the nonlinear behavior. A useful approach is to represent the nonlinear impedance by a linear, resistive Norton (or Thevenin) equivalent circuit over a short range. In a time-stepping solution, both the conductance g and the short-circuit current I_{sc} will change from time step to time step.

Consider the voltage-current (V-I) curve shown in Figure 6-9. The curve can be approximated by straight lines between data points as shown by the dashed lines. Assume one is given a table of voltage-current values that describe such a curve. The table entries are $(v_k, i_k, i = 1, K)$. Then, for a voltage v between v_k and v_{k+1}, the slope is the Norton conductance. Let g_{nl} be the Norton conductance for a nonlinear circuit element, then,

$$g_{nl} = \frac{i_{k+1} - i_k}{v_{k+1} - v_k} \tag{6-37}$$

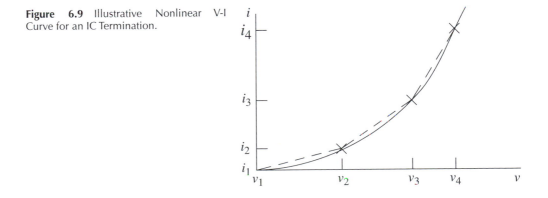

Figure 6.9 Illustrative Nonlinear V-I Curve for an IC Termination.

The (interpolated) current when $v_k \leq v \leq v_{k+1}$ is

$$i(t + \delta t) = i_k + g_{nl}[v(t + \delta t) - v_k] \tag{6-38}$$

The choice of k is made using $v(t)$; if $v(t + \delta t)$ is not in the range (v_k, v_{k+1}), an inaccuracy will result. This is a motivation to use small steps for the solution because one does not know the value of $v(t + \delta t)$ until after the choice of k is made and the network is solved from time t to time $t + \delta t$.

Comparing Equation (6-38) with Equation (6-7), one finds that the Norton equivalent short-circuit current is

$$I_{scnl} = g_{nl}v_k - i_k \tag{6-39}$$

The interfaces of commercial ICs often exhibit nonlinear i/v characteristics. An industry standard has been developed for modeling IC interfaces. It is the *I/O Buffer Information Specification* or IBIS. The standard is maintained by the industry group *IBIS Open Forum*. The specification document defines the format and contents of an IBIS file for the interface of a specific IC. The IBIS file with information, tables, etc., for modeling the interface of a specific device must be obtained from the manufacturer of the device or a supplier of such information licensed by the manufacturer. Up-to-date information on how to obtain the latest version of the IBIS specification can be obtained on the World Wide Web. Try the following Web sites[3].

> http://www.eia.org/ibis/ibis.htm
> http://www.eia.org Electronic Information Group
> http://www.vhdl.org/pub/ibis

or contact

> EIA/Electronic Information Group
> 2500 Wilson Blvd.
> Arlington, VA 22201
> (703) 907-7545

also through

> Global Engineering Documents
> 15 Inverness Way East
> Englewood, CO 80112
> (800) 854-7179

IBIS files are ASCII files; they can be generated or edited in a text editor. Because they must have a strictly defined format and contain exactly defined keywords, a special purpose

---

3. Courtesy of Bob Ross, Chairman, EIA/IBIS Committee, private communication.

editor (such as the one available at http://www.hyperlynx.com) is far easier to use. IBIS files are named with the extension `.ibs`. Provision is contained in an IBIS file for V-I curves (tables) for (keywords) Pullup, Pulldown, GND_clamp and POWER_clamp. The proper use of this data will not be described in this book; see the appropriate IBIS specification and other documents. The user's guide *BoardSim User's Guide*, HyperLynx, Inc., 1995, contains considerable information on IBIS files.

The nonlinear impedance methods given here can be used beyond the application of predefined voltage-current curves. The method has been successfully used to model arcs; the Norton parameters g and I_{sc} were determined step-by-step as functions of the current using a model that approximated the physics of the arc. If the conductance of a device depends on the density of charge carriers, the Norton parameters can be calculated by solving differential equations that determine the charge carrier density in a semiconductor or in a gas. This can be performed in a time-stepping algorithm as the propagation and termination calculations are performed; therefore, they can be interdependent with the transmission line currents or voltages. These methods are beyond the scope of this book and will not be discussed further.

Chapter 7
SIMULTANEOUS TRANSMISSION LINE NETWORK SOLUTIONS

In Chapter 3 it was shown how to solve general networks whose elements are resistive, two-terminal Norton equivalent circuits. This method was implemented in Computer Code 3-1 (Solution of Resistive Network). The method requires only slight modification to include the multi-terminal Norton equivalent circuits required for multi-wire transmission line ends (Section 5.2). This modification will be presented in this Chapter. Because essentially any circuit element can be represented by a resistive Norton equivalent over a short time interval, the method can be used for solution of an arbitrary network over a short time interval. It becomes the algorithm for each step of a general time-stepping solution.

It's been shown that the following elements have resistive Norton equivalent circuits valid over a short time interval:

Ends of multi-wire transmission lines (Section 5.2)
Capacitors (Section 6.4)
Inductors (Section 6.5)
AC terminations (Section 6.6)
Nonlinear elements (Section 6.8)

These devices can be interconnected in an arbitrary way to form networks that can be solved using a time-stepping method to be discussed.

7.1 MULTI-WIRE LINE TERMINATED IN A NETWORK

Assume that the multi-wire line to be terminated has N wires; its $\mathbf{G}_o$ matrix is $N \times N$ and, of course, it has an N-vector of arriving currents. The Norton equivalent current source is the N-vector given by plus or minus twice the arriving currents (see Figure 3.7). Suppose that the termination network has K "auxiliary" nodes, i.e., component connections that are not attached to any of the wires of the transmission line. The total number of nodes for the line termination is $N+K$. The solution of the network at each time step will require solution of $N+K$ linear equations in as many unknowns. These linear equations will be written as a single matrix equation. The vector of node voltages is the dependent variable to be obtained in the solution. The identifying node numbers, say n, correspond to those of the N-wire line for $1 \le n \le N$ and continue on with $N+1, N+2, \ldots, N+K$ for the auxiliary nodes. The situation is illustrated in Figure 7.1 where resistors are used to depict some of the circuit elements. The elements can be any type of two-terminal device and each resistor in the illustration represents a Norton equivalent circuit that approximates the behavior of the particular device. The devices can be attached between any two nodes or between any node and the reference conductor (ground). Let $\mathbf{G}$ be the $(N + K) \times (N + K)$ network conductance matrix for the termination. Then the network equation has the form

$$\mathbf{Gv} = \mathbf{i} \qquad (7\text{-}1)$$

where $\mathbf{v}$ is the $N+K$ column vector of node voltages and $\mathbf{i}$ is the vector of (Norton) source currents positive into the nodes. The solution can be written as

$$\mathbf{v} = \mathbf{G}^{-1}\mathbf{i} \qquad (7\text{-}2)$$

The matrix $\mathbf{G}$ has been defined in Section 3.5. It is given by Equations (3-17) and (3-18). In this section, $\mathbf{G}$ is discussed in terms of an algorithm for generating it. Let the matrix $\mathbf{G}$ be initially the zero matrix. Then values are added to it to represent the currents flowing from the nodes through the (Norton) conductances connected to other nodes and to ground (the Norton current sources do not contribute to $\mathbf{G}$, but contribute to the right-hand side vector $\mathbf{i}$, discussed below). Any circuit conductance matrix is defined to yield the column vector of currents into the nodes of the network it represents when it multiplies the column vector of voltages applied to the nodes. The node indices of the transmission line end have been made to correspond to a subset of the termination network's nodes. Thus, to include the effect of the line's characteristic conductance matrix, the matrix $\mathbf{G}_0$ is simply added to the appropriate submatrix of $\mathbf{G}$. With the indices defined as above, $\mathbf{G}_0$ is added to the upper-left submatrix of $\mathbf{G}$. In the next section, interconnections will be analyzed where the node indices of the line do not correspond to those of the termination matrix, but are mapped.

Figure 7.1   Termination of an N-Wire Line.

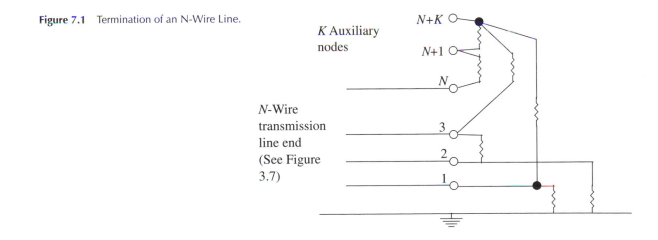

The conductances illustrated as resistors (in Figure 7.1 are included in the matrix $\mathbf{G}$ as described in Section 3.5. To the diagonal elements of $\mathbf{G}$, add all the conductance values that are connected to the corresponding node (see Equation 3-17). That is, to element $\mathbf{G}_{11}$ add all the conductances connected to node 1 (including both those to other nodes and those to ground) and similarly for all the other nodes. To the off-diagonal elements of $\mathbf{G}$, add minus the conductances that are attached between the nodes corresponding to its two indices (not those to ground). That is, to $\mathbf{G}_{ij}$, add minus all the conductances connected between node i and node j. See Equation (3-18). If two or more devices are connected to the same pair of nodes, the conductances are all added to the same off-diagonal element of $\mathbf{G}$.

The right-hand-side column vector $\mathbf{i}$ is the same as that described in Section 3.5. Each of the elements is the sum of the Norton current sources with a plus sign if the source is positive into the node and a minus sign if the source is positive out of the node (see Equation 3-19). Node-to-node current sources require two contributions, one at the node that the current arrow points into (positive contribution) and one at the node at the tail of the current arrow (negative contribution). Current sources between a node and ground require only one contribution, positive or negative depending on whether the current arrow points into or out of the node, respectively. The N-wire transmission line current sources are all defined as positive into their respective nodes and should be added to their corresponding elements in the $\mathbf{i}$ vector.

For the circuit shown in Figure 7.2, the characteristic conductance matrices (inverse of their characteristic impedance matrices) are:

$$\mathbf{G}_{o1} = \begin{bmatrix} 0.030 & -0.015 & -0.005 \\ -0.015 & 0.035 & -0.012 \\ -0.005 & -0.012 & 0.040 \end{bmatrix} \; [\mho], \text{Line 1}$$

$$(7\text{-}3)$$

and

$$\mathbf{G}_{o2} = \begin{bmatrix} 0.025 & -0.010 \\ -0.010 & 0.025 \end{bmatrix} \; [\mho], \text{Line 2} \qquad (7\text{-}4)$$

PROBLEM 7.1 For the network shown in Figure 7.2, let $r_1 = r_2 = r_3 = 50 \; [\Omega]$. What is the circuit conductance matrix $\mathbf{G}$ at the left end of the line?

PROBLEM 7.2 For the network shown in Figure 7.2, let $r_4 = 40 \; [\Omega]$, $r_5 = 100 \; [\Omega]$ and $r_6 = 50 \; [\Omega]$. What is the circuit conductance matrix $\mathbf{G}$ for the center part of the network (the junction of Lines 1 and 2 and resistors r_4 and r_5). *Hint:* The matrix $\mathbf{G}$ should be a 4×4 matrix. The auxiliary node is node 4.

PROBLEM 7.3 For the network shown in Figure 7.2, let the arriving voltages at the center part of the network be

$$\mathbf{v}_{arr1} = \begin{bmatrix} 5 \\ 0 \\ 0 \end{bmatrix} \; [\text{V}] \text{ and } \mathbf{v}_{arr2} = \begin{bmatrix} 0.5 \\ 0 \end{bmatrix} \; [\text{V}].$$ What are the arriving current vectors? What is the Norton current vector for the circuit that represents the junction (both lines combined, positive into the nodes)?

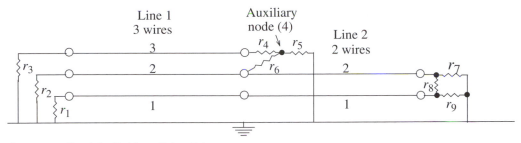

Figure 7.2   Circuit for Problems 7-1 to 7-4.

PROBLEM 7.4   For the network shown in Figure 7.2, let $r_7 = r_8 = 66.7$ [Ω] and $r_9 = 100$ [Ω]. What is the circuit conductance matrix for the right-end termination? What are the resistors required for a matching termination at the right end?

7.2 NETWORK OF MULTI-WIRE LINES AND OTHER NORTON CIRCUITS

Up to this point, the circuit analysis given has been centered on transmission lines and their termination equivalent circuits. Lumped circuit elements have been modeled and methods to write the network equations for lumped elements connected to the transmission line ends have been discussed. In this section, the point of view will change. Instead of considering the transmission line ends as the reference point for attaching lumped circuit elements, we will consider the transmission line ends simply as possible candidates for connection to the nodes of an electrical network. That is, suppose we begin with a collection of numbered connection points, or nodes, numbered 1 through N. These might be the numbered pins on a connector or numbered trace termination points on a circuit board; the order of the node numbering is arbitrary. Then, lumped elements and/or transmission line wire ends are connected to the nodes without any effort to make the node numbers correspond to the transmission line wire identification numbers. Node numbering will be independent of the numbering of the wires of any transmission line that may be connected to them.

The rows and columns of the circuit conductance matrix **G** will be based on the numbering of the nodes. The building of the **G** matrix and **i** column vector will proceed as described in the preceding section, except when adding the contribution of the $\mathbf{G_O}$ matrix of a transmission line end, the elements of $\mathbf{G_O}$ must be mapped to those of **G** according to what wire numbers of the transmission line end are connected to a given node. If transmission line wire 2 connects to node 10 and transmission line wire 3 connects to node 8, the value of $\mathbf{G}_{o2,2}$ must be added to $\mathbf{G}_{10,\ 10}$, $\mathbf{G}_{o3,3}$ must be added to $\mathbf{G}_{8,8}$, $\mathbf{G}_{o2,3}$ must be added to $\mathbf{G}_{10,8}$, and $\mathbf{G}_{o3,2}$ must be added to $\mathbf{G}_{8,10}$. Similarly, the column vector **i** is filled using the node indices; the Norton current sources due to the transmission line ends are mapped to the node numbers according to how they're connected.

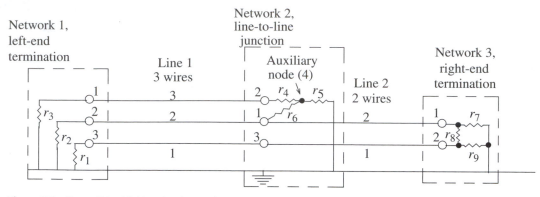

Figure 7.3   Figure 7.2 with New Node Numbers.

Each subnetwork to be solved is assumed to have nodes numbered starting with 1. The networks are also numbered. In the computer code to be discussed, each node is uniquely identified by a network number and a node number (within its network).

The node-number based system is illustrated in Figure 7.3. Figure 7.3 is the same as Figure 7.2 except that node numbers have been added that are different from the transmission line wire numbers connected to them. At the left-end termination, network 1, the circuit conductance matrix using the node numbers to label rows and columns of the matrix is

$$
\mathbf{G}_{net1} = \begin{bmatrix} \mathbf{G}_{o3,3}^{(1)} + \dfrac{1}{r_3} & \mathbf{G}_{o3,2}^{(1)} & \mathbf{G}_{o3,1}^{(1)} \\[2mm] \mathbf{G}_{o2,3}^{(1)} & \mathbf{G}_{o2,2}^{(1)} + \dfrac{1}{r_2} & \mathbf{G}_{o2,1}^{(1)} \\[2mm] \mathbf{G}_{o1,3}^{(1)} & \mathbf{G}_{o1,2}^{(1)} & \mathbf{G}_{o1,1}^{(1)} + \dfrac{1}{r_1} \end{bmatrix} \ [\mho] \tag{7-5}
$$

where $\mathbf{G}_{oi,j}^{(1)}$ are the elements of $\mathbf{G}_o$ for line 1 (let the superscript in parentheses be the line number). At the middle, network 2, the circuit conductance matrix is

$$
\mathbf{G}_{net2} = \begin{bmatrix} \mathbf{G}_{o2,2}^{(1)} + \mathbf{G}_{o2,2}^{(2)} + \dfrac{1}{r_6} & \mathbf{G}_{o2,3}^{(1)} & \mathbf{G}_{o2,1}^{(1)} + \mathbf{G}_{o2,1}^{(2)} & \dfrac{-1}{r_6} \\[3mm] \mathbf{G}_{o3,2}^{(1)} & \mathbf{G}_{o3,3}^{(1)} + \dfrac{1}{r_4} & \mathbf{G}_{o3,1}^{(1)} & \dfrac{-1}{r_4} \\[3mm] \mathbf{G}_{o1,2}^{(1)} + \mathbf{G}_{o1,2}^{(2)} & \mathbf{G}_{o1,3}^{(1)} & \mathbf{G}_{o1,1}^{(1)} + \mathbf{G}_{o1,1}^{(2)} & 0 \\[3mm] \dfrac{-1}{r_6} & \dfrac{-1}{r_4} & 0 & \dfrac{1}{r_4} + \dfrac{1}{r_5} + \dfrac{1}{r_6} \end{bmatrix} \tag{7-6}
$$

The right-hand-side column vector at the center is (assuming r_4, r_5, and r_6 have no source current associated with them).

$$\mathbf{i}_{net2} = 2 \begin{bmatrix} i^{(1)}_{arr2} - i^{(2)}_{arr2} \\ i^{(1)}_{arr3} \\ i^{(1)}_{arr1} - i^{(2)}_{arr1} \\ 0 \end{bmatrix} \tag{7-7}$$

At the right-end termination, network 3, the conductance matrix is

$$\mathbf{G}_{net3} = \begin{bmatrix} \mathbf{G}^{(2)}_{o2,\,2} + \dfrac{1}{r_7} + \dfrac{1}{r_8} & \mathbf{G}^{(2)}_{o2,\,1} - \dfrac{1}{r_8} \\ \mathbf{G}^{(2)}_{o1,\,2} - \dfrac{1}{r_8} & \mathbf{G}^{(2)}_{o1,\,1} + \dfrac{1}{r_8} + \dfrac{1}{r_9} \end{bmatrix} \tag{7-8}$$

Chapter **8**
COMPUTER ALGORITHM FOR GENERAL NETWORK SOLUTIONS

A general time-stepping network solving code has been written to illustrate the use of the methods derived to this point in the book. The code is included on the accompanying disk. It will not be listed here because of its length. However, its use and structure will be described in detail in the following sections. The code is in file `ch8_1.c`.

8.1 GENERAL STRUCTURE OF THE CODE

The input data and solution variables are essentially all stored in linked lists. The names and general descriptions of the linked lists are given in Table 8-1. Each time step, the code solves the equations for a list of "networks" that have attached to their nodes the various components defined in the linked lists: `ResistList`, `InductList`, `CapList`, `ACTermList`, `NonLinList`, `TranLineList`, `StepDriveList` (see Table 8-1).

The flow of data in the code is illustrated in Figure 8.1.

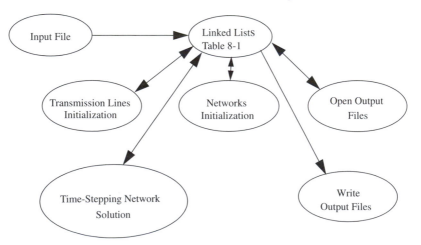

Figure 8.1 Data Flow in General Network Solution Code.

TABLE 8-1
Linked Lists in Code `ch8_1.c`.

Structure Type	Pointer Name	Description of Linked List
NETWORKEQS	`NetEqsList`	One link for each network. Contains all variables and data needed for solution of the given network.
RESISTORS	`ResistList`	One link for each resistor. Contains resistance and network and node numbers where connected.
INDUCTORS	`InductList`	One link for each inductor. Contains inductance, Norton I_{sc}, and connection points.
CAPACITORS	`CapList`	One link for each capacitor. Contains capacitance, Norton I_{sc}, and connection points.
ACTERMINATIONS	`ACTermList`	One link for each AC termination. Contains capacitance, resistance, Norton I_{sc}, and connection points.
NONLINEARVI	`NonLinList`	One link for each nonlinear device. Contains a voltage-current (V-I) table, Norton I_{sc}, and connection points.
TRANSMLINES	`TranLineList`	One link for each transmission line. Contains all parameters defining the transmission line, the signal propagating on it, and the Norton I_{sc} at both ends.
STEPDRIVES	`StepDriveList`	One link for each driver. Contains the turn-on time, rise time, and connection points.
OUTFILE	`OutFileList`	One link for each output file. Contains the time increment, identification of nodes, and filename.

8.2 DATA INPUT — NETWORK DEFINITION

The components, connections, drivers, and output files of the electrical network are completely described in a single input file. The input file is created using any editor that can write an ASCII file. (The reader may want to design software to write input files automatically.) The input file is organized into data blocks that begin with a header consisting of a keyword usually followed by the number items in the data block. There are 11 keywords. The use of each one will be discussed. Input files are created by keyboard input using a text editor (Figure 8.2.).

Any line that begins with two minus signs "--" in the input file is a comment line that is ignored by the program. These may be interspersed among the data lines in any way to make it easier to organize and read the input file. A comment line may be placed directly above a

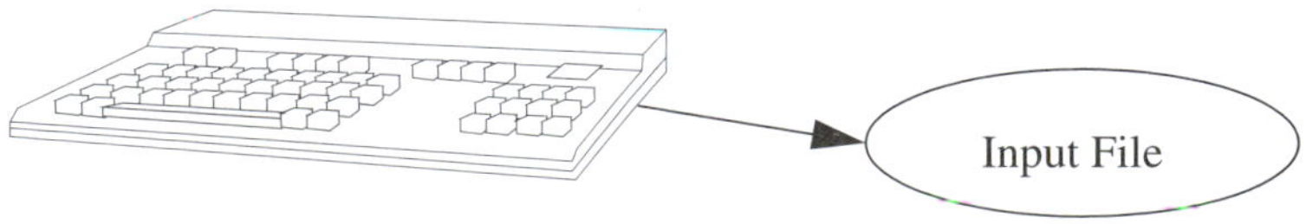

Figure 8.2 Preparing a Data File.

Input File

keyword header line without the two minus signs; such a comment line will be echoed to the screen, but will not be saved or used in any way by the program. Such lines should normally be used to tell the user how far the program has successfully read the input data in case an error causes premature exit from the program. The first data block in the input file must be the "Networks" block. The other data blocks may follow in any order; not all types of data block must be present, only those that contain data required to define the network need be included.

The input-file data blocks are defined in the Tables 8-2 through 8-11. The "Networks" data block must be the first data block in the input file. The last item input in each data block is stored in the link associated with the linked list pointer. The "->next" pointer points to succeeding items moving up the list of items input. The link associated with the first item input has ->next = NULL. This is used to identify the last item in the linked list. It is used to terminate loops through the linked lists. The numbers in the tables are arbitrary, sample numbers.

TABLE 8-2
Networks Data Block.

Line of Input File	Explanation	Where Stored
Reading networks data	Comment line echoed	Not Stored
[Networks] 2	Keyword, number of nets	NumNets
-- Number of nodes	Comment not echoed	Not Stored
3	Number of nodes, net 1	NetEqsList->next->Nnodes
3	Number of nodes, net 2	NetEqsList->Nnodes

TABLE 8-3
Time-Step Data Block.

Line of Input File	Explanation	Where Stored
Reading time step	Comment line echoed	Not Stored
[TimeStep] 0.2e-9	Keyword, time step [s]	Tstep
-- Tstop [s]	Comment not echoed	Not Stored
4.0e-9	Stop time of calculation [s]	Tstop

TABLE 8-4
Resistors Data Block.

Line of Input File	Explanation	Where Stored
Reading resistor list	Comment line echoed	Not stored
[Resistors] 2	Keyword, no. of resistors	Not stored, used for input
-- R[ohms], network no., node1, node2	Comment not echoed	Not stored
80, 2, 1, 0	Resistance [Ω], net no., node1, node 2	ResistList->next->R, ResistList->next->NetNum ResistList->next->node1 ResistList->next->node2
80, 2, 2, 0	Resistance [Ω], net no., node1, node 2	ResistList->R, ResistList->NetNum ResistList->node1 ResistList->node2

TABLE 8-5
Inductors Data Block.

Line of Input File	Explanation	Where Stored
Reading inductor list	Comment line echoed	Not stored
[Inductors] 2	Keyword, no. of inductors	Not stored, used for input
-- henries, network, node1, node2	Comment not echoed	Not stored
1.0e-7, 1, 3, 0	Inductance [H], net no., node1, node 2	InductList->next->L InductList->next->NetNum InductList->next->node1 InductList->next->node2
6.0e-6, 2, 1, 4	Inductance [H], net no., node1, node 2	InductList->L InductList->NetNum InductList->node1 InductList->node2

TABLE 8-6
Capacitors Data Block.

Line of Input File	Explanation	Where Stored
Reading capacitor list	Comment line echoed	Not stored
[Capacitors] 2	Keyword, no. of capacitors	Not stored, used for input
-- Farads, network, node1, node2	Comment not echoed	Not stored
40.0e-12, 1, 0, 1	Capacitance [F], net no., node1, node 2	CapList->next->C CapList->next->NetNum CapList->next->node1 CapList->next->node2
6.0e-9, 2, 1, 4	Capacitance [F], net no., node1, node 2	CapList->C CapList->NetNum CapList->node1 CapList->node2

TABLE 8-7
AC Terminations Data Block.

Line of Input File	Explanation	Where Stored
Reading AC terminations list	Comment line echoed	Not stored
[ACTerminators] 2	Keyword, no. of AC terminations	Not stored, used for input
-- Farads, ohms, network, node1, node2	Comment not echoed	Not stored
100.0e-11, 50.0 2, 1, 0	Capacitance [F], resistance [Ω], net no., node1, node 2	ACTermList->next->C ACTermList->next->R ACTermList->next->NetNum ACTermList->next->node1 ACTermList->next->node2
5.0e-12, 60.0, 2, 2, 0	Capacitance [F], resistance [Ω], net no., node1, node 2	ACTermList->C ACTermList->R ACTermList->NetNum ACTermList->node1 ACTermList->node2

TABLE 8-8
Nonlinear Element Data Block.

Line of Input File	Explanation	Where Stored
`Reading nonlinear VI tables`	Comment line echoed	Not stored
`[NonLinearVI] 1`	Keyword, no. of nonlinear elements	Not stored, used for input
`-- No. of pairs (1st table), network, node1, node2`	Comment not echoed	Not stored

`7, 2, 1, 0`	Number of voltage/current pairs, net no., node1, node 2	`NonLinList->LenTable`
		`NonLinList->NetNum`
		`NonLinList->node1`
		`NonLinList->node2`

`[TableVI]`	Keyword, header for each table of voltage/current pairs	Not stored
`0.0, 0.000`	Voltage [V], current [A]	`*NonLinList->V` `*NonLinList->I`
`0.5, 0.004`	Voltage [V], current [A]	`*(NonLinList->V + 1)` `*(NonLinList->I + 1)`
`1.0, 0.016`	Voltage [V], current [A]	`*(NonLinList->V + 2)` `*(NonLinList->I + 2)`
`2.0, 0.064`	Voltage [V], current [A]	`*(NonLinList->V + 3)` `*(NonLinList->I + 3)`
`4.0, 0.256`	Voltage [V], current [A]	`*(NonLinList->V + 4)` `*(NonLinList->I + 4)`
`8.0, 1.024`	Voltage [V], current [A]	`*(NonLinList->V + 5)` `*(NonLinList->I + 5)`
`16.0, 4.096`	Voltage [V], current [A]	`*(NonLinList->V + 6)` `*(NonLinList->I + 6)`

TABLE 8-9
Transmission Lines Data Block.

Line of Input File	Explanation	Where Stored
Reading transmission lines list	Comment line echoed	Not stored
[TLines] 1	Keyword, no. of transmission lines	Not stored, used for input
-- Speed, Length, NumberWires, LeftNetwork, RightNetwork	Comment not echoed	Not stored
1.0e8, 0.10, 3, 1, 2	Speed [m/s], length [m], no. of wires, net no. left end, net no. right end	TranLineList->Speed TranLineList->Length TranLineList->NumWires TranLineList->LeftEndNet TranLineList->RightEndNet
-- Left Nodes (for each trans. line wire)	Comment not echoed	Not stored
3	Node no., wire 1, left end	*TranLineList->LeftNodes
2	Node no., wire 2, left end	*(TranLineList->Left-Nodes+1)
1	Node no., wire 3, left end	*(TranLineList->Left-Nodes+2)
-- Right Nodes	Comment not echoed	Not stored
3	Node no., wire 1, right end	*TranLineList->RightNodes
2	Node no., wire 2, right end	*(TranLineList ->RightNodes+1)
1	Node no., wire 3, right end	*(TranLineList ->RightNodes+2)
-- Zzero 3 wires n*n = 9 elements	Comment not echoed	Not stored
-- Unpacked, column-by-column to the diagonal [ohms]	Comment not echoed	Not stored
57.4	Z_{o11} [Ω]	*TranLineList->Zo
20.3	Z_{o21} [Ω]	*(TranLineList->Zo + 1)
10	Z_{o31} [Ω]	*(TranLineList->Zo + 2)
20.3	Z_{o12} [Ω]	*(TranLineList->Zo + 3)
56.9	Z_{o22} [Ω]	*(TranLineList->Zo + 4)
20.3	Z_{o32} [Ω]	*(TranLineList->Zo + 5)
10	Z_{o13} [Ω]	*(TranLineList->Zo + 6)
20.3	Z_{o23} [Ω]	*(TranLineList->Zo + 7)
57.4	Z_{o33} [Ω]	*(TranLineList->Zo + 8)

TABLE 8-10
Driver Data Block.

Line of Input File	Explanation	Where Stored
Reading step-func-tion drive data	Comment line echoed	Not stored
[StepDrive] 1	Keyword, no. of drivers	Not stored, used for input
-- V, Z, Network, node1, node 2	Comment not echoed	Not stored
5, 5.0, 1, 1, 0	Voltage [V], Impedance [Ω], net no., node numbers	StepDriveList->V StepDriveList->Z StepDriveList->NetNum StepDriveList->node1 StepDriveList->node2
-- TurnOnTime, Rise-Time	Comment not echoed	Not stored
0, 1e-9	Driver time of turn-on [s], rise time [s]	StepDriveList->TimeOn StepDriveList->Trise

TABLE 8-11
Output Files Data Block.

Line of Input File	Explanation	Where Stored
Reading output file data	Comment line echoed	Not stored
[OutFiles] 3	Keyword, no. of output files	Not stored, used for input
-- Time Incr. net-work, node1, node 2	Comment not echoed	Not stored
0, 2, 0, 1	Output time increment [s], net no., node1 (pos.), node 2 (neg.)	OutFileList->next->next->TimeIncr OutFileList->next->next->NetNum OutFileList->next->next->node1 OutFileList->next->next->node2
-- output filename	Comment not echoed	Not stored
n1s2d1.out	Output filename	*OutFileList->next->next->OutFilename
0, 2, 2, 0	Output time increment [s], net no., node1 (pos.), node 2 (neg.)	OutFileList->next->TimeIncr OutFileList->next->NetNum OutFileList->next->node 1 OutFileList->next->node 2
-- output filename	Comment not echoed	Not stored
n1s2d2.out	Output filename	*OutFileList->next->OutFilename
0, 2, 3, 0	Output time increment [s], net no., node1 (pos.), node 2 (neg.)	OutFileList->TimeIncr OutFileList->NetNum OutFileList->node1 OutFileList->node2
-- output filename	Comment not echoed	Not stored
n1s2d3.out	Output filename	*OutFileList->OutFilename

8.3 INITIALIZING THE TRANSMISSION LINES

The following transmission line initializations are performed in `ReadInputFile`. The pointers `TranLineList->VFRightEnd`, `TranLineList->VFLeftEnd`, `TranLineList->IscTLleft` and `TranLineList->IscTLright` are allocated and their values set to zero.

The transmission line initializations below are performed in function `InitTransLines`. The characteristic conductance matrix $\mathbf{G_0}$ is created. First, memory is allocated for the $\mathbf{G_0}$ matrix at the pointer `TranLineList->Go`. The number of memory locations that are allocated is $(\text{NumWires})^2$ (`NumWires` is the number of wires in the line). The matrix $\mathbf{G_0}$ is calculated by inverting the matrix $\mathbf{Z_0}$. The routine `SimpleGauss` is used. The result is stored at the pointer `TranLineList->Go`. It is stored by columns, that is, down the first column, then the second, and so on to include the whole matrix.

The propagation parameters are created by calling function `InitPropTstep`. This routine calculates `Nhalf`, half the number of memory locations needed in the propagation ring. This integer is stored in `TranLineList->Nhalf`. The two propagation interpolation weights are also calculated and stored in `TranLineList->wgt1` and `TranLineList->wgt2`.

The propagation memory rings and the offsets in the memory ring (one for each wire of the transmission line) are allocated and initialized to zero. The memory rings are located at the pointer `TranLineList->PropCircle`. The number of locations allocated are (2 * Nhalf * NumWires). The propagation ring for wire *n* starts at pointer (`TranLineList->PropCircle + (n - 1) * 2 * Nhalf`). The offset from this memory location, at a given propagation step, is stored in location (`TranLineList->TStepOffset + (n - 1)`).

8.4 INITIALIZING THE NETWORKS

The pointer to the networks linked list is `NetEqsList`. This linked list is built in the routine `BuildNetEqsList`. The linked list is created in a for loop that loops over the networks (the number of networks is `NumNets`). After each link is allocated, the number of nodes for the given network is stored in `NetEqsList->Nnodes` and the net identification number is stored in `NetEqsList->NetNum`; it is assigned sequentially beginning with number one in the order that the networks appear in the input file. Memory is allocated for the node-voltage vector at the pointer `NetEqsList->Vnodes` and memory is allocated for the network equation-current vector in the pointer `NetEqsList->Icolumn`. These vectors are initialized to zero.

8.5 OPEN OUTPUT FILES

The output files are opened in function `OpenOutFiles`. In a loop over the linked list of output files, each output file is opened and the file pointer is stored in `OutFileList->outfptr`.

8.6 TIME-STEPPING LOOP

In the time-stepping loop, the first calculation is to increment the calculational time variable time by Tstep. Then the network conductance matrix **G** and the right-hand-side column vector **i** are loaded for each network.

8.6.1 Loading the Circuit Matrix **G** and Column Vector **i**

Loading the **G** matrix and the **i** column vector is done through a call to the routine Load-GandI. In LoadGandI, a loop through all the networks is entered. For each network, the following procedures take place:

1. The **G** matrix (pointer NetEqsList->Gmatrix) and the **i** column vector (pointer NetEqsList->Icolumn) for that network are initialized to zero.

2. For each resistor in the resistor linked list that is connected to the targeted network, the appropriate contribution(s) to the **G** matrix are made. For a given resistor, the contribution(s) are made by calling the subroutine GmatIcolAdd.

3. For each capacitor in the capacitor linked list that is connected to the targeted network, the appropriate contributions to the **G** matrix, and the **i** column vector are made. For a given capacitor, the contributions are made by calling the subroutine GmatIcolAdd. (The Norton short-circuit current I_{sc} for the capacitor has either been previously initialized or calculated in a call to UpdateIscLumped).

4. For each inductor in the inductor linked list that is connected to the targeted network, the appropriate contributions to the **G** matrix, and the **i** column vector are made. For a given inductor, the contributions are made by calling the subroutine GmatIcolAdd. (The Norton short-circuit current I_{sc} for the inductor has either been previously initialized or calculated in a call to UpdateIscLumped).

5. For each AC termination in the AC termination linked list that is connected to the targeted network, the appropriate contributions to the **G** matrix and the **i** column vector are made. For a given AC termination, the contributions are made by calling the subroutine GmatIcolAdd. (The Norton short-circuit current I_{sc} for the AC termination has either been previously initialized or calculated in a call to UpdateIscLumped).

6. For each nonlinear device in the nonlinear device linked list that is connected to the targeted network, the Norton conductance and short-circuit current is calculated using the previously calculated voltage across the device and the current-voltage (V-I) table for the device. These contributions are added to the **G** matrix and the **i** column vector. The contributions are made by calling the subroutine GmatIcolAdd. (In this case, the initial value of I_{sc} may not be zero, so it must be calculated before each network solution, even the first one.)

7. For each driver in the driver linked list that is connected to the targeted network, the Norton conductance and short-circuit current is calculated using the parameters stored for

the driver. These contributions are added to the **G** matrix and the **i** column vector. The contributions are made by calling the subroutine `GmatIcolAdd`.

8. For each transmission line end in the transmission line linked list that is connected to the targeted network, the transmission line characteristic conductance matrix (**G_O**) elements are mapped to and added to the appropriate network conductance matrix elements (**G**). The Norton I_{sc} for each transmission line wire is mapped to and added to the appropriate element of the column vector **i** of the network equation.

8.6.2 Solution of the Network Equations

Following the loading of the **G** matrix and **i** column vector, the network equations are solved in function `SolveCircuitEqs`. The linear equations are solved using `SimpleGauss`. The resulting node voltages are stored in the networks linked list at pointer `NetEqsList->Vnodes`.

8.6.3 Output Node Voltages

The node voltages requested for output in the outputs linked list are written to the appropriate files using the file pointers stored in `OutFileList->outfptr`.

8.6.4 Update I_{sc} for Capacitors, Inductors, and AC Termination

The Norton current sources for lumped elements (AC terminations, capacitors, and inductors) appropriate to the next time step are calculated in a call to `UpdateIscLumped`.

8.6.5 Update Transmission Lines

The final calculation in the time-stepping loop is to update the transmission line parameters. This is performed with a call to `UpdateIscTransLines`. For each wire, using the just-calculated network node voltages and the transmission line arriving voltages (calculated in the previous call to `UpdateIscTransLines`), the departing voltages are calculated ($v_{dep} = v - v_{arr}$). The departing voltages are input to the propagation routine `PropTstep`; it propagates the signals one time step. The output of `PropTstep` is the arriving voltages at the next time. A matrix multiplication of the vector of arriving voltages by **G_O** yields the arriving currents. The Norton equivalent currents (I_{sc}) for the transmission line ends are easily calculated from the arriving currents. These are used the next time step in the call to `LoadGandI`.

8.7 CLOSING THE OUTPUT FILES

Looping through the linked list of output files (pointer `OutFileList`), each output file is closed. The output files contain potential differences — the potential of node 1 minus that of node 2. Either node can be zero, i.e., ground potential.

Chapter 9

EXAMPLES OF SOLUTIONS USING COMPUTER CODE 8-1

The results of simulations are given in this Chapter. The input files are listed that were read by the computer code (`ch8_1.c`) described in Chapter 8. Three network configurations are simulated. Each network is simulated with somewhat different resistor values (or types) included to illustrate different effects. These are summarized as follows:

Single-wire line (Section 9.1)

- With matched load
- With nonlinear load (clamping effect)
- With AC termination (output similar to matched load)

Three-wire line (Section 9.2)

- Loads as shown
- Load removed from undriven traces at driven end
- Matrix load modified to single-wire loads

Branched Traces (Section 9.3)

- Loads as shown
- Load removed from undriven trace at driven end
- Matrix load modified to single-wire loads

9.1 SINGLE-WIRE LINE — VARIOUS TERMINATIONS

The single-wire line schematic is shown in Figure 9.1. The open circuit voltage $v_{oc}(t)$ is 5 [V] with a rise time of 1 [ns] (see input data file below). Three loads were used.

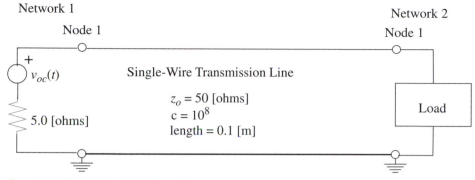

Figure 9.1   Single-Wire Line Schematic.

The loads were:

  Load 1, matching 50 [ohm] resistor

  Load 2, nonlinear resistor (see Figure 9.2)

  Load 3, AC termination (50 [ohms], 1 [nF])

  The input file for the matched, single-wire line (file `net1_1.in`) is listed below. For the other two cases, only the part of the input file that is different is listed.

```
-- Data file for ch8_1 Network Code
-- Matched load simple 1-wire line
--
Networks data.
[Networks] 2
-- Number of nodes each network
1
1
--
Timestep.
[TimeStep] 0.2e-9
-- Tstop [s]
   4.0e-9
--
Reading resistor list -- matched termination
[Resistors] 1
-- R[ohms], network no., node1, node2
    50,           2,         1,    0
--
Transmission lines list
[TLines] 1
```

```
-- Speed, Length, NumberWires, LeftNetwork, RightNetwork
   1.0e8,  0.10,      1,               1,         2
-- Left-end node number
 1
-- Right-end node number
 1
-- Single-element Zo, 50 [ohms]
50
--
Drive data
[StepDrive] 1
-- V, Z,  Network, node1, node2
   5, 5.0,   1,      1,      0
-- TurnOnTime,    RiseTime
     0,             1e-9
--
Output file data
[OutFiles] 1
-- Time Incr.  network, node1, node2
     0.2e-9,      2,       1,    0
-- output filename
net1_1.out
```

9.1.1 Output for the Line with Matching Load Resistor

In this case, the load is a (matching) 50 [ohm] resistor.

Output file `net1_1.out`

```
    Time            Voltage
2.000000e-010,   0.000000
4.000000e-010,   0.000000
6.000000e-010,   0.000000
8.000000e-010,   0.000000
1.000000e-009,   0.000000
1.200000e-009,   0.909091
1.400000e-009,   1.818182
1.600000e-009,   2.727273
1.800000e-009,   3.636364
2.000000e-009,   4.545455
2.200000e-009,   4.545455
2.400000e-009,   4.545455
2.600000e-009,   4.545455
2.800000e-009,   4.545455
3.000000e-009,   4.545455
```

```
3.200000e-009,   4.545455
3.400000e-009,   4.545455
3.600000e-009,   4.545455
3.800000e-009,   4.545455
4.000000e-009,   4.545455
4.200000e-009,   4.545455
```

9.1.2 Output for the Line with a Nonlinear Load Resistor

The input file for the same single-wire line with a nonlinear termination has the resistors data block replaced by the following nonlinear device data block (file net1_2.in):

```
Nonlinear VI tables
[NonLinearVI] 1
-- No. of pairs (1st table), network, node1, node2
             7,                   2,      1,      0
Table of V-I pairs -- approximates a square law
[TableVI]
--   Volts  Amps
      0.0,  0.000
      0.5,  0.004
      1.0,  0.016
      2.0,  0.064
      4.0,  0.256
      8.0,  1.024
     16.0,  4.096
```

The nonlinear voltage-current curve (piecewise linear approximation) of the input data table is shown in Figure 9.2. A better approximation to a given curve would contain more points.

This nonlinear device has a clamping effect on the voltage at the right end of the line. The output file (net1_2.out) is

```
        Time          Voltage
2.000000e-010,   0.000000
4.000000e-010,   0.000000
6.000000e-010,   0.000000
8.000000e-010,   0.000000
1.000000e-009,   0.000000
1.200000e-009,   1.298702
1.400000e-009,   1.540107
1.600000e-009,   2.074866
1.800000e-009,   2.357367
2.000000e-009,   2.670846
2.200000e-009,   2.670846
```

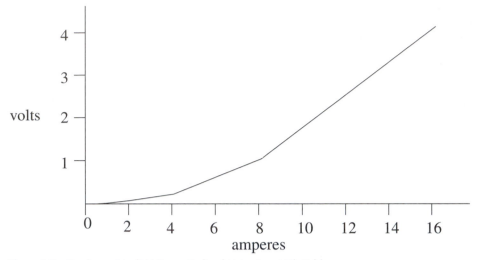

Figure 9.2 Nonlinear Load V-I Curve Defined Using Input File Table.

```
2.400000e-009,   2.670846
2.600000e-009,   2.670846
2.800000e-009,   2.670846
3.000000e-009,   2.670846
3.200000e-009,   2.560925
3.400000e-009,   2.749300
3.600000e-009,   2.854911
3.800000e-009,   3.031692
4.000000e-009,   3.199733
4.200000e-009,   3.199733
```

9.1.3 Line with AC Termination

The input file for the same single-wire line with an AC termination (file net1_3.in) has the termination device replaced with the following data block for an AC termination:

```
AC termination -- 50 [ns] RC time constant
[ACTerminators] 1
--   Farads,    ohms,   network,   node1, node2
     1.0e-9,    50.0,      2,         1,     0
```

The resulting output file (net1_3.out) is:

```
      Time            Voltage
2.000000e-010,    0.000000
4.000000e-010,    0.000000
6.000000e-010,    0.000000
```

```
8.000000e-010,  0.000000
1.000000e-009,  0.000000
1.200000e-009,  0.910090
1.400000e-009,  1.821994
1.600000e-009,  2.735709
1.800000e-009,  3.651231
2.000000e-009,  4.568556
2.200000e-009,  4.577591
2.400000e-009,  4.586608
2.600000e-009,  4.595607
2.800000e-009,  4.604588
3.000000e-009,  4.613550
3.200000e-009,  4.621677
3.400000e-009,  4.628298
3.600000e-009,  4.633414
3.800000e-009,  4.637024
4.000000e-009,  4.639130
```

The three single-line outputs listed above are shown below in graphical form; see Figure 9.3.

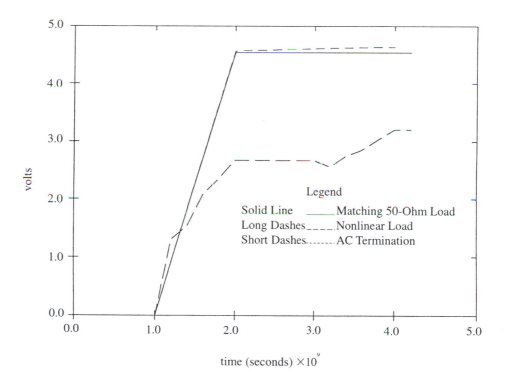

Figure 9.3 Single-Wire Line Output.

9.2 THREE-WIRE LINE — CONTROL OF CROSSTALK

The three-wire line shown below in Figure 9.4 is simulated as shown. Then the resistors R_1 and R_2 were removed for a second simulation to observe the effect on the crosstalk. A third simulation was performed with R_1 and R_2 included, but resistors R_3 and R_4 removed. The crosstalk increases for both the second and third cases.

   The data file for the circuit as shown in Figure 9.4 is given below (file net2_1.in).

```
-- Data file for ch8_1 Network Code
-- Three-wire line -- left clamped -- almost matched at right end
--
Networks data.
[Networks] 2
-- Number of nodes each network
3
3
--
Timestep and stoptime.
[TimeStep] 0.2e-9
-- Tstop [s]
4.0e-9
--
Resistor list
[Resistors] 7
-- R[ohms],    network no.,   node1, node2
          5,            1,        2,    0
          5,            1,        3,    0
         80,            2,        1,    0
        110,            2,        2,    0
         80,            2,        3,    0
        155,            2,        1,    2
        155,            2,        2,    3
```

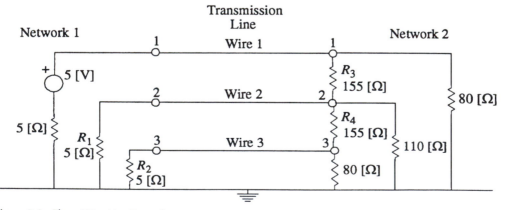

Figure 9.4   Three-Wire Line Example.

```
--
Reading transmission lines list
[TLines] 1
-- First transmission line
-- Speed, Length, NumberWires, LeftNetwork, RightNetwork
   1.0e8,  0.10,        3,              1,           2
-- Left Nodes (for each trans. line wire)
 1
 2
 3
 -- Right Nodes
 1
 2
 3
-- Zzero  3 wires -- n*n = 9 elements
57.4
20.3
10
20.3
56.9
20.3
10
20.3
57.4
--
Drive data
[StepDrive] 1
-- V, Z,  Network, node1, node2
   5, 5.0,    1,      1,     0
-- TurnOnTime,    RiseTime
       0,            1e-9
--
Reading output file data
[OutFiles] 3
-- Time Incr.  network, node1, node2
      0,          2,       1,    0
-- output filename
n2w1_1.out
-- Time Incr.  network, node1, node2
      0,          2,       2,    0
-- output filename
n2w2_1.out
-- Time Incr.  network, node1, node2
      0,          2,       3,    0
-- output filename
n2w3_1.out
```

The output files for the three cases are given below. For the circuit shown in Figure 9.4, the output is (at the right end, network 2):

Time [s]	Wire 1 Voltage [V]	Wire 2 Voltage [V]	Wire 3 Voltage [V]
2.000000e-010,	0.000000	0.000000	0.000000
4.000000e-010,	0.000000	0.000000	0.000000
6.000000e-010,	0.000000	0.000000	0.000000
8.000000e-010,	0.000000	0.000000	0.000000
1.000000e-009,	0.000000	0.000000	0.000000
1.200000e-009,	0.930595	0.019855	-0.022444
1.400000e-009,	1.861191	0.039710	-0.044888
1.600000e-009,	2.791786	0.059564	0.067332
1.800000e-009,	3.722381	0.079419	-0.089776
2.000000e-009,	4.652976	0.099274	-0.112220
2.200000e-009,	4.652976	0.099274	-0.112220
2.400000e-009,	4.652976	0.099274	-0.112220
2.600000e-009,	4.652976	0.099274	-0.112220
2.800000e-009,	4.652976	0.099274	-0.112220
3.000000e-009,	4.652976	0.099274	-0.112220
3.200000e-009,	4.635695	0.105985	-0.088189
3.400000e-009,	4.618413	0.112695	-0.064159
3.600000e-009,	4.601132	0.119406	-0.040129
3.800000e-009,	4.583851	0.126117	-0.016098
4.000000e-009,	4.566570	0.132827	0.007932
4.200000e-009,	4.566570	0.132827	0.007932

These three outputs are shown in graphical form in Figure 9.5.
With resistors R_1 and R_2 removed, the output at the right end is:

Time [s]	Wire 1 Voltage [V]	Wire 2 Voltage [V]	Wire 3 Voltage [V]
2.000000e-010,	0.000000	0.000000	0.000000
4.000000e-010,	0.000000	0.000000	0.000000
6.000000e-010,	0.000000	0.000000	0.000000
8.000000e-010,	0.000000	0.000000	0.000000
1.000000e-009,	0.000000	0.000000	0.000000
1.200000e-009,	0.934257	0.317572	0.133948
1.400000e-009,	1.868513	0.635143	0.267896
1.600000e-009,	2.802769	0.952715	0.401843
1.800000e-009,	3.737025	1.270286	0.535791
2.000000e-009,	4.671282	1.587858	0.669739
2.200000e-009,	4.671282	1.587858	0.669739
2.400000e-009,	4.671282	1.587858	0.669739
2.600000e-009,	4.671282	1.587858	0.669739
2.800000e-009,	4.671282	1.587858	0.669739

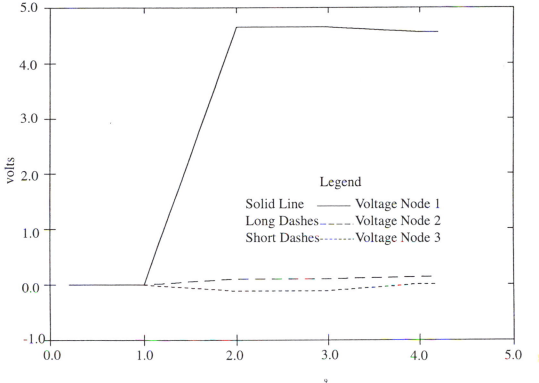

Figure 9.5   Three-Wire Line Output.

3.000000e-009,	4.671282	1.587858	0.669739
3.200000e-009,	4.659968	1.571036	0.638577
3.400000e-009,	4.648655	1.554214	0.607415
3.600000e-009,	4.637342	1.537392	0.576253
3.800000e-009,	4.626029	1.520570	0.545090
4.000000e-009,	4.614715	1.503747	0.513928
4.200000e-009,	4.614715	1.503747	0.513928

These three outputs are shown in graphical form in Figure 9.6.
With resistors R_3 and R_4 removed, the output at the right end is:

Time [s]	Wire 1 Voltage [V]	Wire 2 Voltage [V]	Wire 3 Voltage [V]
2.000000e-010	0.000000	0.000000	0.000000
4.000000e-010	0.000000	0.000000	0.000000
6.000000e-010	0.000000	0.000000	0.000000
8.000000e-010	0.000000	0.000000	0.000000
1.000000e-009	0.000000	-0.000000	-0.000000

1.200000e-009	1.078268	-0.134085	-0.057783
1.400000e-009	2.156535	-0.268170	-0.115566
1.600000e-009	3.234803	-0.402255	-0.173349
1.800000e-009	4.313070	-0.536341	-0.231132
2.000000e-009	5.391338	-0.670426	-0.288915
2.200000e-009	5.391338	-0.670426	-0.288915
2.400000e-009	5.391338	-0.670426	-0.288915
2.600000e-009	5.391338	-0.670426	0.288915
2.800000e-009	5.391338	-0.670426	0.288915
3.000000e-009	5.391338	-0.670426	-0.288915
3.200000e-009	5.217845	-0.488327	-0.227183
3.400000e-009	5.044354	-0.306229	-0.165451
3.600000e-009	4.870862	-0.124131	-0.103720
3.800000e-009	4.697371	0.057967	-0.041988
4.000000e-009	4.523880	0.240065	0.019744
4.200000e-009	4.523880	0.240065	0.019744

These three outputs are shown in Figure 9.7.

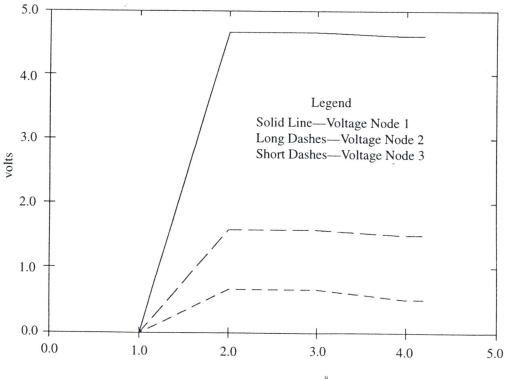

Figure 9.6   Three-Wire Line, R_1 and R_2 Removed.

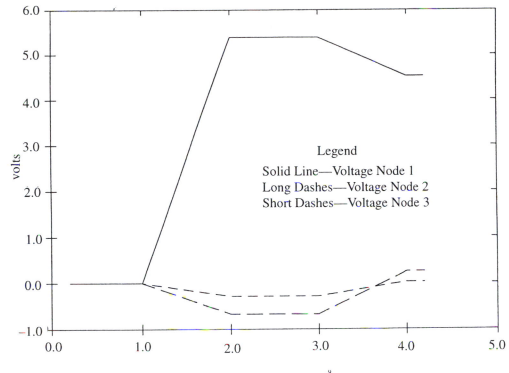

Figure 9.7   Three-Wire Line, R_3 and R_4 Removed.

9.3 BRANCHED TRACES

Example circuit with two one-wire transmission lines connected to the load end of a single two-wire line. The circuit is shown in Figure 9.8. The output voltage at networks 3 and 4 are given below. The first solution was performed with the circuit as shown in Figure 9.8. The second solution is with R_1 removed and a third solution was performed with R_2 removed. The input file for the three-transmission line circuit shown in Figure 9.8 is given below (file net3_1.in):

```
-- Data file for CPPII_1 Network Code
-- 3-line network -- left end clamped -- right ends matched
Networks data.
[Networks] 4
-- Number of nodes each network
2
2
1
1
```

```
--
Reading timestep.
[TimeStep] 0.2e-9
-- Tstop [s]
6.0e-9
--
Reading resistor list
[Resistors] 4
-- R[ohms], network no., node1, node2
         5,          1,        2,     0
       148,          2,        1,     2
      76.6,          3,        1,     0
      76.6,          4,        1,     0
--
Reading transmission lines list
[TLines] 3
-- First transmission line
-- Speed, Length, NumberWires, LeftNetwork,RightNetwork
     1.0e8,  0.10,       2,              1,           2
-- Left Nodes (for each trans. line wire)
 1
 2
 -- Right Nodes
 1
 2
-- Zzero  2 wires n*n = 4 elements
57.4
```

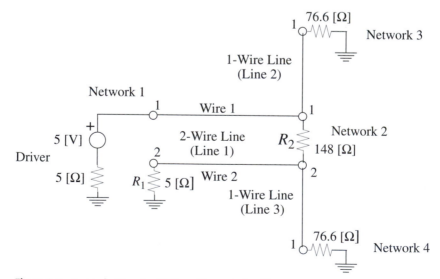

Figure 9.8   Example Circuit with Three Transmission Lines.

```
20.5
20.5
57.4
-- Second transmission line
-- Speed, Length, NumberWires, LeftNetwork,RightNetwork
    1.0e8,  0.10,      1,                 3,         2
-- Left Nodes (for each trans. line wire)
 1
-- Right Nodes
 1
-- Zzero  2 wires n*n = 4 elements
76.6
-- Third transmission line
-- Speed, Length, NumberWires, LeftNetwork,RightNetwork
    1.0e8,  0.10,      1,                 4,         2
-- Left Nodes (for each trans. line wire)
 1
-- Right Nodes
 2
-- Zzero  2 wires n*n = 4 elements
76.6
--
Reading step-function drive data
[StepDrive] 1
-- V, Z,  Network, node1, node2
    5, 5.0,   1,      1,      0
-- TurnOnTime,    RiseTime
      0,            0.5e-9
--
Reading output file data
[OutFiles] 2
-- Time Incr.  network, node1, node2
      0,           3,       1,     0
-- output filename
n3w1_1.out
-- Time Incr.  network, node1, node2
      0,           4,       1,     0
-- output filename
n3w2_1.out
```

The output voltages at networks 3 and 4 for the circuit as shown in Figure 9.8 are given below.

Time [s]	Voltage [V], Net 3	Voltage [V], Net 4
2.000000e-010	0.000000	0.000000
4.000000e-010	0.000000	0.000000

6.000000e-010	0.000000	0.000000
8.000000e-010	0.000000	0.000000
1.000000e-009	0.000000	0.000000
1.200000e-009	0.000000	0.000000
1.400000e-009	0.000000	0.000000
1.600000e-009	0.000000	0.000000
1.800000e-009	0.000000	0.000000
2.000000e-009	0.000001	0.000000
2.200000e-009	1.821177	0.042383
2.400000e-009	3.642354	0.084766
2.600000e-009	4.552942	0.105957
2.800000e-009	4.552942	0.105957
3.000000e-009	4.552942	0.105957
3.200000e-009	4.552942	0.105957
3.400000e-009	4.552942	0.105957
3.600000e-009	4.552942	0.105957
3.800000e-009	4.552942	0.105957
4.000000e-009	4.552942	0.105957
4.200000e-009	4.553124	0.119563
4.400000e-009	4.553308	0.133168
4.600000e-009	4.553399	0.139971
4.800000e-009	4.553399	0.139971
5.000000e-009	4.553399	0.139971
5.200000e-009	4.553399	0.139971
5.400000e-009	4.553399	0.139971
5.600000e-009	4.553399	0.139971
5.800000e-009	4.553399	0.139971
6.000000e-009	4.553399	0.139971
6.200000e-009	4.553501	0.139972

These two outputs are shown graphically in Figure 9.9. With R_1 removed, note the increase of the crosstalk to network 4:

Time [s]	Voltage [V], Net 3	Voltage [V], Net 4
2.000000e-010	0.000000	0.000000
4.000000e-010	0.000000	0.000000
6.000000e-010	0.000000	0.000000
8.000000e-010	0.000000	0.000000
1.000000e-009	0.000000	0.000000
1.200000e-009	0.000000	0.000000
1.400000e-009	0.000000	0.000000
1.600000e-009	0.000000	0.000000
1.800000e-009	0.000000	0.000000
2.000000e-009	0.000001	0.000000
2.200000e-009	1.835101	0.640686
2.400000e-009	3.670202	1.281373

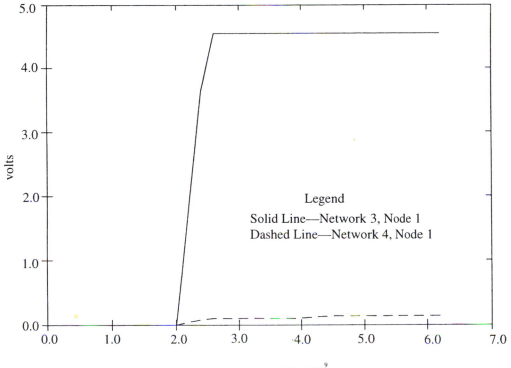

volts

time (seconds) $\times 10^{9}$

Legend

Solid Line—Network 3, Node 1
Dashed Line—Network 4, Node 1

Figure 9.9   Output for Branched Traces of Figure 9.8.

2.600000e-009	4.587751	1.601715
2.800000e-009	4.587751	1.601715
3.000000e-009	4.587751	1.601715
3.200000e-009	4.587751	1.601715
3.400000e-009	4.587751	1.601715
3.600000e-009	4.587751	1.601715
3.800000e-009	4.587751	1.601715
4.000000e-009	4.587751	1.601715
4.200000e-009	4.591775	1.588356
4.400000e-009	4.595799	1.574995
4.600000e-009	4.597811	1.568315
4.800000e-009	4.597811	1.568315
5.000000e-009	4.597811	1.568315
5.200000e-009	4.597811	1.568315
5.400000e-009	4.597811	1.568315
5.600000e-009	4.597811	1.568315
5.800000e-009	4.597811	1.568315
6.000000e-009	4.597811	1.568315
6.200000e-009	4.597707	1.568188

These two network voltages are shown graphically in Figure 9.10. With R_2 removed, note the increase of crosstalk to network 4.

Time [s]	Voltage [V], Net 3	Voltage [V], Net 4
2.000000e-010	0.000000	0.000000
4.000000e-010	0.000000	0.000000
6.000000e-010	0.000000	0.000000
8.000000e-010	0.000000	0.000000
1.000000e-009	0.000000	0.000000
1.200000e-009	0.000000	0.000000
1.400000e-009	0.000000	0.000000
1.600000e-009	0.000000	0.000000
1.800000e-009	0.000000	-0.000000
2.000000e-009	0.000001	-0.000000
2.200000e-009	2.120489	-0.256929
2.400000e-009	4.240976	-0.513857
2.600000e-009	5.301219	-0.642322
2.800000e-009	5.301219	-0.642322
3.000000e-009	5.301219	-0.642322

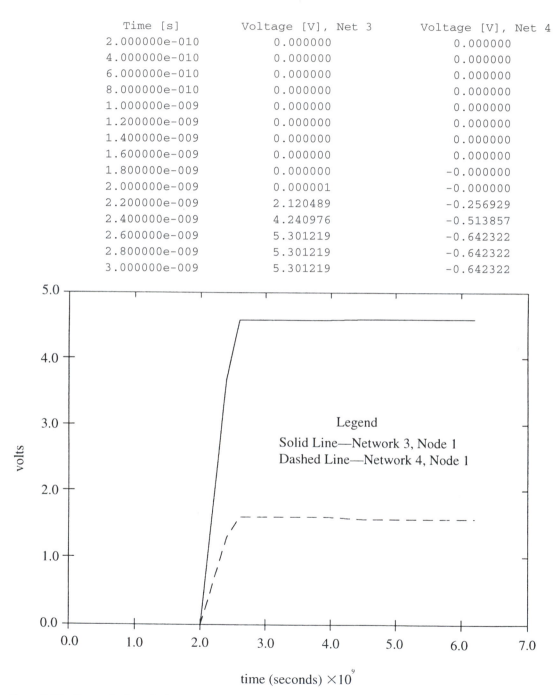

Legend
Solid Line—Network 3, Node 1
Dashed Line—Network 4, Node 1

Figure 9.10   Output for Branched Traces, R_1 Removed.

3.200000e-009	5.301219	-0.642322
3.400000e-009	5.301219	-0.642322
3.600000e-009	5.301219	-0.642322
3.800000e-009	5.301219	-0.642322
4.000000e-009	5.301219	-0.642321
4.200000e-009	4.991560	-0.318874
4.400000e-009	4.681902	0.004574
4.600000e-009	4.527073	0.166298
4.800000e-009	4.527073	0.166298
5.000000e-009	4.527073	0.166298
5.200000e-009	4.527073	0.166298
5.400000e-009	4.527073	0.166298
5.600000e-009	4.527073	0.166298
5.800000e-009	4.527073	0.166298
6.000000e-009	4.527074	0.166298
6.200000e-009	4.611422	0.082050

These two network voltages are shown graphically in Figure 9.11.

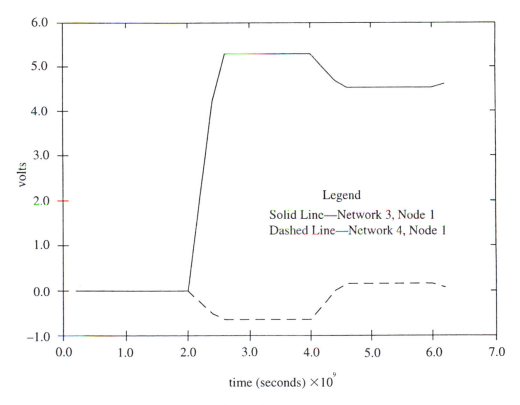

Figure 9.11   Output for Branched Traces, R_2 Removed.

PROPAGATION IN LAYERED MEDIA

Chapter 10
MODAL ANALYSIS IN LAYERED MEDIA

Traces embedded in a layered circuit board will support propagation at different speeds. If the trace has a high permittivity surrounding it, the propagation speed will be low; if the surrounding permittivity is low (e.g., that of free space), the propagation speed will be high, with an upper limit of the speed of light in free space. In a layered medium, it would be reasonable to think that propagation might occur at a continuum of speeds from that characteristic of the highest permittivity present to that of the lowest permittivity present. We shall find that that is not the case. There are a discreet number of propagation speeds present. The number of speeds is, in general, equal to the number of traces. Some of these speeds may be equal to each other. In this Chapter, a method for calculating the propagation speeds is derived. For each speed, a ratio of voltages or currents can be found that propagate at that speed. Separating signals into these propagation "modes" is another subject in this Chapter.

10.1 THE VECTOR WAVE EQUATIONS FOR LOSSLESS LINES

To analyze the multiple propagation speeds of multi-wire transmission lines in layered media, we begin with an examination of the lossless multi-conductor line. The wave equation for this type of transmission line can be derived by starting with the vector form of the transmission line Equations (5-2) and (5-3) with the matrices $\mathbf{G}$ and $\mathbf{R}$ set to zero. Differentiating the first of these with respect to x and the second with respect to t and substituting the second into the first, one obtains the vector wave equation for the voltage signal on a lossless multi-conductor line.

$$\mathbf{LCv}_{tt}(x,t) - \mathbf{v}_{xx}(x,t) = \mathbf{0} \tag{10-1}$$

The two equations can also be combined to obtain the vector wave equation for the current signal; it is

$$\mathbf{CLi}_{tt}(x,t) - \mathbf{i}_{xx}(x,t) = \mathbf{0} \tag{10-2}$$

For the lines that are analyzed in this section, the matrix **LC** is not (in general) of the form **I**/c^2 (see Equation (5-1)) nor is it diagonal. Furthermore, **LC** ≠ **CL** so that the wave equation satisfied by the voltage-vector signal is not identical to the wave equation satisfied by the current-vector signal. However, because **L** and **C** are both symmetric, it is easy to show that **LC** is the transpose of **CL**. That is, **L** symmetric implies that $\mathbf{L} = \mathbf{L}^T$ and **C** symmetric implies that $\mathbf{C} = \mathbf{C}^T$. Then, $(\mathbf{CL})^T = \mathbf{L}^T\mathbf{C}^T = \mathbf{LC}$. Obviously, the approach used to simulate the propagation on single speed lines must be modified somewhat for multi-speed lines.

10.2 EXAMPLE OF MULTI-SPEED LINE

To illustrate the situation, the two-wire line whose cross section is shown in Figure 10.1 is considered. Trace 2 is buried in a dielectric that has relative permittivity 5 where an electromagnetic (EM) wave travels at relative speed $(1/\sqrt{5}) \cong 0.447$ (relative to the speed in vacuum). Trace 1 is buried in a layer where the relative EM propagation speed is $1/\sqrt{2}$. One would expect a signal on trace 1, therefore, to travel at about relative speed $(1/\sqrt{2}) \cong 0.707$ and a signal on trace 2 to travel at relative speed of about 0.447.

However, the two traces influence each other because of the magnetic and electric field coupling (inductance and capacitance) between them; we shall find that the signal on each trace splits into two modes, each of which travels at a different speed (causing time dispersion of the signal). If the traces are weakly coupled, a given speed will dominate for a given trace and the effect of time dispersion will be small. But if the traces are strongly coupled, the effect of dispersion may be large. If more than two traces are near each other, the number of propagation modes is equal to the number traces. Because of geometric considerations, several of the propagating modes may have the same propagation speed. For the spacing shown below, Trace 1 carries a quite pure mode, only about 2% of its signal will be propagated in the slower mode. However, trace 2 splits into modes in about an 11-to-4 ratio, the dominant mode being the slower one. Methods to determine the splitting into modes and propagation of the modal signals are developed in the following sections.

For the geometry shown in Figure 10.1, the **C** and **L** matrices are

$$\mathbf{C} = \begin{bmatrix} 47.1{\times}10^{-12} & -21.7{\times}10^{-12} \\ -21.7{\times}10^{-12} & 133.7{\times}10^{-12} \end{bmatrix} \text{ [F/m]} \qquad (10\text{-}3)$$

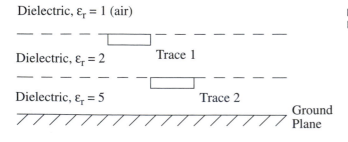

Dielectric, $\varepsilon_r = 1$ (air)

Dielectric, $\varepsilon_r = 2$ Trace 1

Dielectric, $\varepsilon_r = 5$ Trace 2

Ground Plane

Figure 10-1 Two Traces on a Two-Layered Circuit Board.

$$\mathbf{L} = \begin{bmatrix} 0.511\times10^{-6} & 0.160\times10^{-6} \\ 0.160\times10^{-6} & 0.370\times10^{-6} \end{bmatrix} \text{ [H/m]} \tag{10-4}$$

By direct matrix multiplication one obtains

$$\mathbf{LC} = \begin{bmatrix} 0.206\times10^{-16} & 0.1030\times10^{-16} \\ -0.493\times10^{-18} & 0.460\times10^{-16} \end{bmatrix} \text{ [s}^2\text{/m}^2\text{]} \tag{10-5}$$

$$\mathbf{CL} = \begin{bmatrix} 0.206\times10^{-16} & -0.493\times10^{-18} \\ 0.1030\times10^{-16} & 0.460\times10^{-16} \end{bmatrix} = (\mathbf{LC})^{\mathsf{T}} \text{ [s}^2\text{/m}^2\text{]} \tag{10-6}$$

To find a solution of Equation (10-1), an interesting bit of matrix theory will be used. Suppose that, for the 2-trace geometry, a constant voltage vector can be found, say $\mathbf{u}$, such that multiplying it by the matrix $\mathbf{LC}$ yields a vector proportional to $\mathbf{u}$ itself. This supposition can be written mathematically as the equation

$$\mathbf{LCu} = \gamma^2\mathbf{u} \tag{10-7}$$

It is assumed that the constant of proportionality γ^2 is real and positive; this assumption is valid if $\mathbf{L}$ and $\mathbf{C}$ are physically realizable inductance and capacitance matrices.

Note that, if such a voltage vector $\mathbf{u}$ is found, then $\mathbf{u}$ times any multiplicative constant also satisfies Equation (10-7). Now we define a scalar function $v'(x,t)$ and suppose that the vector of trace voltages is given by $\mathbf{v}(x,t) = v'(x,t)\mathbf{u}$. Substituting this into Equation (10-1) and using Equation (10-7), it is found that both components of the vector differential equation (all components, if there were more) yield the same scalar differential equation:

$$\gamma^2 v'_{tt}(x,t) - v'_{xx}(x,t) = 0 \tag{10-8}$$

This equation is, of course, the scalar wave equation and is identical to Equation (2-5) if the speed of propagation c is set to $1/\gamma$. The general solution is the same as that for a single-wire lossless line (see Equation (2-9)). For the two-trace geometry, there are two values of the constant γ^2 that satisfy Equation (10-7); their (positive) square roots will be denoted γ_1 and γ_2. They each have a corresponding vector $\mathbf{u}$, denoted $\mathbf{u}_1$ and $\mathbf{u}_2$. Trace voltages proportional to $\mathbf{u}_1$ propagate at speed $c_1 = 1/\gamma_1$ and voltages proportional to $\mathbf{u}_2$ propagate at speed $c_2 = 1/\gamma_2$. Any vector trace voltage $\begin{bmatrix} v_1(x,t) \\ v_2(x,t) \end{bmatrix}$ can be expressed as the sum of a voltage proportional to $\mathbf{u}_1$ plus a voltage proportional to $\mathbf{u}_2$. This will be demonstrated below. We now discuss methods for finding γ_1 and γ_2 and $\mathbf{u}_1$ and $\mathbf{u}_2$.

10.3 PROPAGATION MODES OF MULTI-SPEED LINES

For the two-trace line, Equation (10-7) can be written as

$$\begin{bmatrix} 0.206\times10^{-16} & 0.1030\times10^{-16} \\ -0.493\times10^{-18} & 0.460\times10^{-16} \end{bmatrix}\begin{bmatrix} u_1 \\ u_2 \end{bmatrix} - \begin{bmatrix} \gamma^2 & 0 \\ 0 & \gamma^2 \end{bmatrix}\begin{bmatrix} u_1 \\ u_2 \end{bmatrix} = \mathbf{0} \tag{10-9}$$

$$\begin{bmatrix} 0.206\times10^{-16}-\gamma^2 & 0.1030\times10^{-16} \\ -0.493\times10^{-18} & 0.460\times10^{-16}-\gamma^2 \end{bmatrix}\begin{bmatrix} u_1 \\ u_2 \end{bmatrix} = \mathbf{0} \tag{10-10}$$

where the vector **u** is written as

$$\mathbf{u} \equiv \begin{bmatrix} u_1 \\ u_2 \end{bmatrix}$$

Equation (10-10) has a nonzero solution if and only if the determinant of the matrix has value zero. That is, for a nonzero vector **u** to exist, the following equation must hold:

$$det\begin{bmatrix} 0.206\times10^{-16}-\gamma^2 & 0.1030\times10^{-16} \\ -0.493\times10^{-18} & 0.460\times10^{-16}-\gamma^2 \end{bmatrix} = 0 \tag{10-11}$$

$$(0.206\times10^{-16}-\gamma^2)(0.460\times10^{-16}-\gamma^2)+0.1030\times10^{-16}\times0.493\times10^{-18} = 0 \tag{10-12}$$

This quadratic equation (in γ^2) has the two solutions: $\gamma_1 = 4.56\times10^{-9}$ and $\gamma_2 = 6.77\times10^{-9}$. The propagation speeds for these two modes are $c_1 = 1/\gamma_1 = 2.19\times10^8$ [m/s] and $c_2 = 1/\gamma_2 = 1.48\times10^8$ [m/s]. Relative to the EM propagation speed in vacuum (speed of light, $\sim 3\times10^8$ [m/s]), c_1 is 0.731 and c_2 is 0.493.

The constants γ_1^2 and γ_2^2 are the *characteristic values* or *eigenvalues* of the matrix **LC**. The associated vectors $\mathbf{u}_1$ and $\mathbf{u}_2$ are the *characteristic vectors* or *eigenvectors* of the matrix **LC** (see Finkbeiner, 1978, Chapter 7). The propagation resulting from one of these eigenvalues and eigenvectors will be referred to as a *mode of propagation* or an *eigenmode*.

We proceed to find the characteristic vectors $\mathbf{u}_1$ and $\mathbf{u}_2$ of **LC** for the two-trace line. Because the determinant of the matrix of Equation (10-10) is zero, the two linear equations that it represents are linearly dependent. That is, a solution of one of them will be a solution of the other. Hence, the components of $\mathbf{u}_1$ (or $\mathbf{u}_2$) need only satisfy one of the equations; we have the freedom to set one of the components to an arbitrary (nonzero) value, then solve for the other one. Because multiplying both components by any factor still yields a vector that satisfies Equation (10-10), we can also normalize the vector in any way we choose; we can, for instance, normalize it so that the sum of the squares of the components equal one (i.e., make it a unit vector).

In Equation (10-10), let $\mathbf{u} = \mathbf{u}_1$, then its components are denoted $u_1 = u_{11}$, $u_2 = u_{21}$. Choosing the first of the two scalar equations represented by (vector) Equation (10-10) and arbitrarily setting $u_{11} = 1$, the following is obtained:

$$0.206 \times 10^{-16} - 2.08 \times 10^{-17} + 0.1030 \times 10^{-16} \times u_{21} = 0 \qquad (10\text{-}13)$$

$$u_{21} = 0.019 \qquad (10\text{-}14)$$

The same procedure for the second characteristic value or mode yields (with $u_{12} = 1$)

$$0.206 \times 10^{-16} - 4.58 \times 10^{-17} + 0.1030 \times 10^{-16} \times u_{22} = 0 \qquad (10\text{-}15)$$

$$u_{22} = 2.45 \qquad (10\text{-}16)$$

Normalizing $\mathbf{u}_1$ and $\mathbf{u}_2$ so they are unit vectors (i.e., divide both components by the square root of the sum of their squares), we obtain

$$\mathbf{u}_1 = \begin{bmatrix} 1.000 \\ 0.019 \end{bmatrix} \qquad (10\text{-}17)$$

$$\mathbf{u}_2 = \begin{bmatrix} 0.378 \\ 0.926 \end{bmatrix} \qquad (10\text{-}18)$$

It is convenient to define a matrix $\mathbf{U}$ with the vectors $\mathbf{u}_i$ as columns, so that, for the two-trace geometry, the $\mathbf{U}$ matrix is

$$\mathbf{U} = \begin{bmatrix} \mathbf{u}_1 & \mathbf{u}_2 \end{bmatrix} = \begin{bmatrix} 1.000 & 0.378 \\ 0.019 & 0.926 \end{bmatrix} \qquad (10\text{-}19)$$

Above Equation (10-8), a scalar function $v'(x,t)$ was defined that became the space and time dependence for propagation of a mode, i.e., a voltage vector given by $v'(x,t)\mathbf{u}$. A column vector of mode voltages can be defined as

$$\mathbf{v}'(x,t) = \begin{bmatrix} v'_1(x,t) \\ v'_2(x,t) \end{bmatrix} \qquad (10\text{-}20)$$

If the matrix $\mathbf{U}$ multiplies this vector of mode voltages, the vector of trace (or wire) voltages is obtained, that is,

$$\mathbf{v}(x,t) = \mathbf{U}\mathbf{v}'(x,t) = v'_1(x,t)\,\mathbf{u}_1 + v'_2(x,t)\,\mathbf{u}_2 = \begin{bmatrix} v'_1(x,t)1.000 + v'_2(x,t)0.378 \\ v'_1(x,t)0.019 + v'_2(x,t)0.926 \end{bmatrix} \qquad (10\text{-}21)$$

Hence, **U** is a transformation matrix from mode voltages to wire voltages. Multiplication by the inverse of the **U** matrix will then transform the wire voltages to mode voltages. That is,

$$\mathbf{v}'(x,t) = \mathbf{U}^{-1}\mathbf{v}(x,t) \tag{10-22}$$

where $\mathbf{v}(x,t)$ is the column vector of wire voltages. The inverse of **U** for the two-trace geometry is

$$\mathbf{U}^{-1} = \begin{bmatrix} 1.008 & -0.412 \\ -0.021 & 1.089 \end{bmatrix} \tag{10-23}$$

The vector of mode voltages is

$$\mathbf{v}'(x,t) = \begin{bmatrix} v_1(x,t)\,1.008 - v_2(x,t)\,0.412 \\ -v_1(x,t)\,0.021 + v_2(x,t)\,1.089 \end{bmatrix} \tag{10-24}$$

From Equation (10-24) we see that for a signal on trace 1, that is, the voltages are $v_1 = 1$ and $v_2 = 0$, the mode 1 voltage (v'_1) that propagates at speed c_1 is 1.008 and the mode 2 voltage that propagates at speed c_2 is -0.021. A voltage propagating on trace 1 will have very little time dispersion; 98% of the signal propagates at speed c_1. However, for a signal on trace 2, let the trace voltages be $v_1 = 0$ and $v_2 = 1$, then the mode 1 voltage is -0.412 [V] that propagates at speed c_1 and the mode 2 voltage is 1.089 [V] that propagates at speed c_2; a significant time dispersion may result. Part of the trace 2 signal (mode 1) will be delayed by time $t_{1delay} = l/c_1$ and part (mode 2) will be delayed by $t_{2delay} = l/c_2$, where l is the length of the pair of traces. Whether or not this time dispersion is troublesome will depend on the length of the trace and the required time resolution of the pulse.

PROBLEM 10.1 For the two-trace cross section discussed above, let the length of a pair of traces be $l = 0.1$ [m]. Assume a step function signal departs the left end at $t = 0$. What is the difference of the arrival times of the two modal step-function signals at the right end?

PROBLEM 10.2 Two closely coupled , (i.e., very close to each other) traces as shown below in Figure 10.2 have capacitance and inductance matrices:

$$\mathbf{C} = \begin{bmatrix} 72.5 & -35.1 \\ -35.1 & 72.5 \end{bmatrix} \text{ [pF/m]} \tag{10-25}$$

and

$$\mathbf{L} = \begin{bmatrix} 0.597 & 0.349 \\ 0.349 & 0.597 \end{bmatrix} \text{ [μH/m]} \tag{10-26}$$

1. Calculate the **LC** matrix.
2. Find the two propagation speeds (c_1 and c_2) and the associated eigenvectors ($\mathbf{u}_1$ and $\mathbf{u}_2$) normalized so that their absolute values are 1 (write them as the matrix **U**).
3. Find the inverse of **U**.
4. For some purposes, a balanced differential drive is used on two traces (i.e., the traces are driven with equal and opposite voltages). Would such a signal experience time dispersion if launched on the traces shown in Figure 10.2?
5. If a driver launched a signal on one of the traces, could time dispersion be a problem?

10.4 DIAGONALIZATION OF A MATRIX

The procedure developed in the preceding section is a special case of matrix *diagonalization*. Suppose we return again to the wave equation for the voltage vector, Equation (10-1), and simply multiply it by the transformation matrix $\mathbf{U}^{-1}$. We obtain

$$\mathbf{U}^{-1}\mathbf{LC}\mathbf{v}_{tt}(x,t) - \mathbf{U}^{-1}\mathbf{v}_{xx}(x,t) = \mathbf{0} \tag{10-27}$$

The matrix $\mathbf{U}\mathbf{U}^{-1}$ is the unit matrix and can be inserted into Equation (10-27) without changing it to obtain

$$[\mathbf{U}^{-1}\mathbf{LCU}]\left[\mathbf{U}^{-1}\mathbf{v}_{tt}(x,t)\right] - [\mathbf{U}^{-1}\mathbf{v}_{xx}(x,t)] = \mathbf{0} \tag{10-28}$$

Brackets have been added to separate the transformed matrix and the transformed voltage vector. We define the matrix Γ^2 as

$$\Gamma^2 \equiv \mathbf{U}^{-1}\mathbf{LCU} \tag{10-29}$$

and use $\mathbf{v}'(x,t) = \mathbf{U}^{-1}\mathbf{v}(x,t)$ as given by Equation (10-22) to write Equation (10-28) as

$$\Gamma^2\mathbf{v}'_{tt}(x,t) - \mathbf{v}'_{xx}(x,t) = \mathbf{0} \tag{10-30}$$

Figure 10.2   Two Closely Coupled Traces, Problem 10-2.

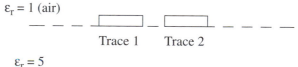

$\varepsilon_r = 1$ (air)

Trace 1   Trace 2

$\varepsilon_r = 5$
(dielectric)

Ground Plane

Performing the matrix multiplications on the right-hand side of Equation (10-29), and using Equation (10-7) for each of the eigenvalues, one obtains

$$
\Gamma^2 = \begin{bmatrix}
\gamma_1^2 & 0 & 0 & \ldots & 0 \\
0 & \gamma_2^2 & 0 & \ldots & 0 \\
0 & 0 & \gamma_3^2 & \ldots & 0 \\
. & . & . & \gamma_n^2 & . \\
0 & 0 & 0 & \ldots & \gamma_N^2
\end{bmatrix}
\tag{10-31}
$$

That is, Γ^2 is the diagonal matrix with the eigenvalues γ_n^2 on the diagonal. The matrix Γ is defined to be the diagonal matrix with the constants γ_n on the diagonal. Any matrix transformation of the form of Equation (10-29) (that transforms matrix **LC** into matrix Γ^2) is called a *similarity transformation*. If the new matrix Γ^2 is a diagonal matrix, the matrix is said to be *diagonalized* by the similarity transformation. A problem in matrix theory is that of determining whether a matrix can be diagonalized by a similarity transformation. We shall assume, on physical grounds, that the **LC** matrix is always *diagonable* (a contraction of *diagonalizable*). (See Finkbeiner, 1978, Chapter 6.) We will further assume that the eigenvalues of **LC** are all positive[1]. Because the matrix Γ^2 is a diagonal matrix, the vector wave Equation (10-30) is equivalent to N independent scalar wave equations or N modes, where N is the dimension of the vector equation. For the lossless multi-wire transmission line, propagation of each mode can be calculated as if each mode were an independent single-wire line. To obtain the physical wire voltages, the mode voltages must be recombined using the matrix **U** (Equation (10-21)).

There are several "canned" computer programs for performing matrix operations; they can greatly facilitate the development of simulation software. The programs can be grouped into two categories: (1) matrix inversion and linear equation solving routines, and (2) matrix eigenvector and eigenvalue calculating routines. For linear equation solving and matrix inversion, the LINPACK routines (in FORTRAN 66) are venerable workhorses (Dongara, et al., 1979). The author has translated some of these routines to C; they are included on the accompanying disk. For matrix eigenvalue calculations, EISPACK (in FORTRAN 66) is a traditional collection of routines (Smith et al., 1976). Both LINPACK and EISPACK have been updated with improved algorithms and programing methods (now in FORTRAN 77); they are the routines of LAPACK (for Linear Algebra Package; Anderson, et al., 1992). LINPACK, EISPACK, and LAPACK are public domain routines that can be obtained at low cost and used or distributed without copyright problems. There are other packages of routines such the IMSL library, now distributed with the Microsoft FORTRAN Power Station that will also

---

1. If an eigenvalue of a matrix **LC** were negative, the propagation speed would be imaginary and the corresponding mode (i.e., ratio of voltages or currents) would not propagate. The author conjectures that this cannot happen and that **LC** can be proven to be positive definite based on the properties of the **L** and **C** matrices. However, he's not aware of such a proof.

solve these matrix problems. The mathematics of performing such matrix solutions can be found in Golub and Van Loan, 1989.

For solving for the eigenvalues and eigenvectors of the matrix **LC**, the most suitable routine the author has found is in LAPACK. It is a routine designed specifically for finding the eigenvalues and eigenvectors of the product of two symmetric, real matrices (its name is SSPGV). For these matrices, the eigenvalues are always real, whereas general matrices have complex eigenvalues. The more specialized routine is more robust (and probably computationally faster) than general eigenvalue/eigenvector routines. This is exactly what's needed for the propagation problem. The LAPACK routines for doing this are included on the accompanying disk.

EXERCISE 10-1. Show that the matrix **LC** can be expressed as a similarity transformation of the matrix Γ^2; that is, show that

$$\mathbf{LC} = \mathbf{U}\Gamma^2\mathbf{U}^{-1} \tag{10-32}$$

Hint: Start with Equation (10-29).

EXERCISE 10-2. The vectors of current that travel at discreet speeds are eigenvectors of the matrix **CL**, show that the following similarity transformation diagonalizes the matrix **CL**:

$$\Gamma^2 = \mathbf{U}^T\mathbf{CL}(\mathbf{U}^{-1})^T \tag{10-33}$$

Hint: Take the transpose of Equation (10-29).

EXERCISE 10-3. Show that for any matrix that has an inverse, the transpose of the inverse is also the inverse of the transpose. That is, $(\mathbf{A}^{-1})^T = (\mathbf{A}^T)^{-1}$. *Hint*: Take the transpose of $\mathbf{I} = \mathbf{A}\mathbf{A}^{-1}$.

Computer Code 10-1.   Eigenvalues and Eigenvectors of the Matrix **LC**.

The trace cross sections shown in Figure 10.3 define a transmission line where the speeds of propagation vary by more than a factor of two. It was chosen to illustrate a computer program for calculating the eigenvalues and eigenvectors of the matrix **LC**.

The computer code consists of the main program that calls the subroutine eigenvectors. These two routines are contained in file ch10_1.c. The subroutine eigenvectors in turn calls a family of subroutines that are found in the file EISPACK.CPP. These routines are translations from FORTRAN to C of the corresponding EISPACK subroutines. The matrices **L** and **C** for the transmission line shown above are defined in the Local Variables Declarations of the main program; the main program is listed below. The output generated by this computer program is listed below the program listing.

```
/* ------ Procedure main    ----%proc%-------------------------------
   Description: Multiply L times C and calc. eigenvalues of LC
   Procedures Called
-------------------------------------------------------------------- */
```

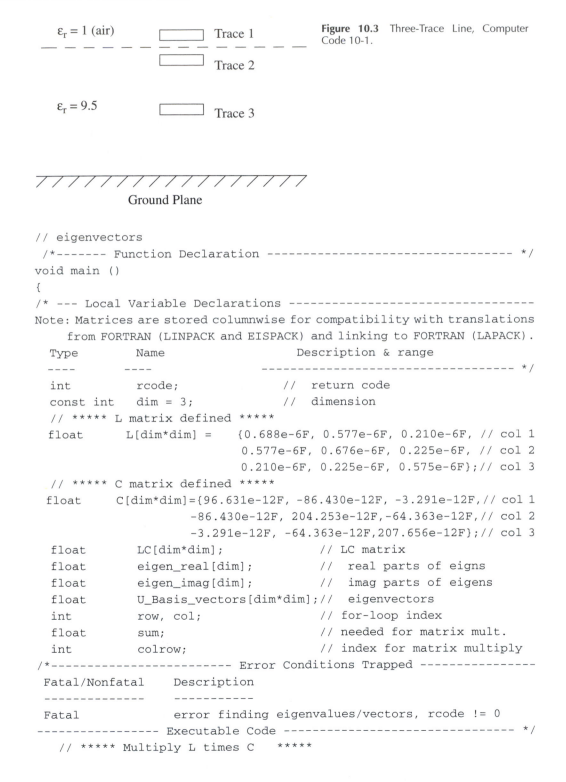

Figure 10.3 Three-Trace Line, Computer Code 10-1.

```
// eigenvectors
 /*------- Function Declaration -------------------------------- */
void main ()
{
/* --- Local Variable Declarations --------------------------------
Note: Matrices are stored columnwise for compatibility with translations
      from FORTRAN (LINPACK and EISPACK) and linking to FORTRAN (LAPACK).
   Type          Name               Description & range
   ----          ----               ------------------------------ */
   int           rcode;           //  return code
   const int     dim = 3;         //  dimension
   // ***** L matrix defined *****
   float        L[dim*dim] =    {0.688e-6F, 0.577e-6F, 0.210e-6F, // col 1
                                 0.577e-6F, 0.676e-6F, 0.225e-6F, // col 2
                                 0.210e-6F, 0.225e-6F, 0.575e-6F};// col 3
   // ***** C matrix defined *****
   float        C[dim*dim]={96.631e-12F, -86.430e-12F, -3.291e-12F,// col 1
                           -86.430e-12F, 204.253e-12F,-64.363e-12F,// col 2
                            -3.291e-12F, -64.363e-12F,207.656e-12F};// col 3
   float        LC[dim*dim];            // LC matrix
   float        eigen_real[dim];        //  real parts of eigns
   float        eigen_imag[dim];        //  imag parts of eigens
   float        U_Basis_vectors[dim*dim];//  eigenvectors
   int          row, col;               // for-loop index
   float        sum;                    // needed for matrix mult.
   int          colrow;                 // index for matrix multiply
 /*------------------------ Error Conditions Trapped ----------------
   Fatal/Nonfatal    Description
   -------------      ----------
   Fatal             error finding eigenvalues/vectors, rcode != 0
 ---------------- Executable Code ------------------------------- */
     // ***** Multiply L times C   *****
```

```
// Note: The function unp_idx uses input rows and col's staring at 1.
//        unp_idx (1, 1, dim) = 0
for (row = 1; row <= dim; row++)
{
   for (col = 1; col <= dim; col++)
   {
      sum = 0.0F;
      for (colrow = 1; colrow <= dim; colrow++)
         sum += L[unp_idx (row, colrow, dim)]
             * C[unp_idx (colrow, col, dim)];
      LC[unp_idx (row, col, dim)] = sum;
   }
}
// ***** Output LC matrix *****
printf (" matrix LC\n");
for (row = 1; row <= dim; row++)
{
   for (col = 1; col <= dim; col++)
      printf (" % e  ", LC[unp_idx (row, col, dim)]);
   printf ("\n");
}
// ***** Calculate eigenvalues/eigenvectors *****
rcode = eigenvectors (LC,
                      dim,
                      eigen_real,
                      eigen_imag,
                      U_Basis_vectors);
if (rcode)    // Check for error condition
   printf (" Return code = %i, Abnormal exit\n", rcode);
else
{
   // ***** Output eigenvalues *****
   printf("\n Eigenvalues of LC\n");
   printf("    real           imaginary      speeds[m/s]\n");
   for (row = 0; row < dim; row++)
   {
      printf ("   % e % e % e\n", eigen_real[row],
      eigen_imag[row], (float)(1.0/sqrt(eigen_real[row])));
   }
    // ***** Normalize the eigenvectors *****
   for (col = 1; col <= dim; col++)
   {
      sum = 0.0F;
      for (row = 1; row <= dim; row++)
         sum += U_Basis_vectors[unp_idx (row, col, dim)]
```

```
                    *U_Basis_vectors[unp_idx (row, col, dim)];
          sum = (float)sqrt(sum);
          for (row = 1; row <= dim; row++)
             U_Basis_vectors[unp_idx (row, col, dim)] /= sum;
       }
       // ***** Output the eigenvectors *****
       printf ("\n  U matrix -- columns are eigenvectors of LC\n");
       for (row = 1; row <= dim; row++)
       {
          for (col = 1; col <= dim; co
             printf ("% e  ",U_Basis_vectors[unp_idx (row, col, dim)]);
          printf ("\n");
       }
   }
}
/* -----End of Procedure main      ------------------------------- */
```

Output from Computer Code 10-1 is listed below. The program is written to write this output to the screen.

```
matrix LC
   1.592091e-017     4.487391e-017     4.206100e-018
  -3.411068e-018     7.372325e-017     1.314301e-018
  -1.046564e-018    -9.202100e-018     1.042294e-016

 Eigenvalues of LC
      real           imaginary        speeds[m/s]
    1.876524e-017   0.000000e+000    2.308463e+008
    7.128093e-017   0.000000e+000    1.184441e+008
    1.038274e-016   0.000000e+000    9.813954e+007

  U matrix -- columns are eigenvectors of LC
  9.979300e-001   -6.234413e-001   -6.611088e-002
  6.148775e-002   -7.476974e-001   -3.604357e-002
  1.884081e-002   -2.286253e-001   -9.971611e-001
```

The eigenvalue routines that are part of LAPACK that are very robust for finding the needed eigenvalues and eigenvectors are included as file EIGEN.FOR, and a program that uses them to perform the same calculation as the C program shown above is in file TESTEIG.FOR. These programs are, of course, in the FORTRAN language.

Chapter 11

CHARACTERISTIC IMPEDANCE OF MULTI-SPEED LINES

11.1 IMPEDANCE MATRIX FOR A SINGLE MODE

To find the characteristic impedance of a multi-speed transmission line, begin by considering a lossless line with a signal of the nth mode propagating in the forward direction. Let the vector of wire voltages for the nth mode be $\mathbf{v}_{Fn}(t - x/c_n)$

$$\mathbf{v}_{Fn}(t - x/c_n) \equiv v'_{Fn}(t - x/c_n)\, \mathbf{u}_n \tag{11-1}$$

And let the corresponding current vector be $\mathbf{i}_{Fn}(t - x/c_n)$. (One should note that the vector $\mathbf{v}_{nF}$ is proportional to the eigenvector $\mathbf{u}_n$, but the current vector $\mathbf{i}_{Fn}$ is not.) Substituting these into the first of the vector transmission line equations with $\mathbf{R} = \mathbf{0}$, Equation (5-2) becomes

$$-\frac{1}{c_n}\frac{d}{d\tau_F}\mathbf{v}_{Fn}(t - x/c_n) = -\mathbf{L}\frac{d}{d\tau_F}\mathbf{i}_{Fn}(t - x/c_n) \tag{11-2}$$

where $\tau_F = t - x/c_n$, the argument of both $\mathbf{v}_{Fn}$ and $\mathbf{i}_{Fn}$. Hence, for the nth mode, we have (assuming the constant of integration is zero)

$$\mathbf{v}_{Fn} = \mathbf{Z}_{on}\mathbf{i}_{Fn} \tag{11-3}$$

where

$$\mathbf{Z}_{on} = c_n \mathbf{L} = \frac{1}{\gamma_n}\mathbf{L} \tag{11-4}$$

(Compare this to $\mathbf{Z}_o$ for the single speed multi-wire transmission line, Equation (5-8).)

PROBLEM 11.1 Calculate $\mathbf{Z}_{o1}$, for the line defined by the matrices $\mathbf{C}$ and $\mathbf{L}$ of Equations (10-3) and (10-4).

11.2 IMPEDANCE MATRIX, COMBINED MODES

In order to sum the currents over the (voltage) modes, it's useful to write Equation (11-3) in the inverse form as

$$\mathbf{i}_{Fn} = \mathbf{G}_{on}\mathbf{v}_{Fn} \tag{11-5}$$

where

$$\mathbf{G}_{on} = \mathbf{Z}_{on}^{-1} = \gamma_n \mathbf{L}^{-1} \tag{11-6}$$

The total forward current is the sum of the forward mode currents

$$\mathbf{i}_F = \sum_{n=1}^{N} \mathbf{i}_{Fn} = \sum_{n=1}^{N} \mathbf{G}_{on}\mathbf{v}_{Fn} \tag{11-7}$$

Substituting $\mathbf{G}_{on}$ from Equation (11-6) and $\mathbf{v}_{Fn} = \mathbf{u}_n v'_{Fn}$ into Equation (11-7), we obtain

$$\mathbf{i}_F = \sum_{n=1}^{N} \gamma_n \mathbf{L}^{-1}\mathbf{u}_n v'_{Fn} \tag{11-8}$$

This is exactly the same as the matrix expression

$$\mathbf{i}_F = \mathbf{L}^{-1}\mathbf{U}\Gamma\mathbf{v}'_F \tag{11-9}$$

(Recall that Γ is the diagonal matrix of γ_ns and the columns of $\mathbf{U}$ are $\mathbf{u}_n$.)

Substituting $\mathbf{v}'_F = \mathbf{U}^{-1}\mathbf{v}_F$ (see Equation (10-22)), we obtain

$$\mathbf{i}_F = \mathbf{L}^{-1}\mathbf{U}\Gamma\mathbf{U}^{-1}\mathbf{v}_F \tag{11-10}$$

So that, including all the propagation modes, the characteristic conductance is

$$\mathbf{G}_O = \mathbf{L}^{-1}\mathbf{U}\Gamma\mathbf{U}^{-1} \tag{11-11}$$

with the inverse

$$\mathbf{Z}_O = \mathbf{U}\Gamma^{-1}\mathbf{U}^{-1}\mathbf{L} \tag{11-12}$$

EXERCISE 11-1.   Show that the matrix $\mathbf{M} = \mathbf{U}\Gamma^{-1}\mathbf{U}^{-1}$ is the square root of the inverse of **LC**.
Hint: Evaluate $\mathbf{M}^2\mathbf{LC}$, where **LC** is given by Equation 10-32.

EXERCISE 11-2.   Carry out the matrix multiplications in Equation (11-9) to show its equivalence to Equation (11-8).

If the characteristic impedance is derived using the dual approach, i.e., where the modes are defined in terms of the current vector as eigenvalues of **CL**, then the characteristic impedance is found to be the transpose of the matrix expression given in Equation (11-12)[1]. Since physically, these must be equal, the characteristic impedance matrix is necessarily a symmetric matrix. Such a symmetric impedance matrix can be synthesized using the methods of Section 6.2. Hence, even for multi-speed lines, a matching impedance network can be synthesized for termination.

It is a curious fact that, though the characteristic impedance matrix is different for each mode, that a single matrix, the one given in Equation (11-12) is correct for each mode and all linear combinations of modes. One can dissect the terms of Equation (11-12) and see that each mode is separated by the transformation matrix, the correct impedance matrix then multiplies each mode, and the result is then recombined and transformed back to the wire voltages. And all these operations combine into a single matrix, $\mathbf{Z}_o$.

Evaluating $\mathbf{Z}_o$ for the multi-speed line shown in Figure 10-1 (with matrix **L** given by Equation (10-4) and matrix **U** given by Equation (10-19)), one obtains

$$\mathbf{Z}_o = \begin{bmatrix} 1.000 & 0.378 \\ 0.019 & 0.926 \end{bmatrix} \tag{11-13}$$

$$\begin{bmatrix} 2.19 \times 10^8 & 0 \\ 0 & 1.48 \times 10^8 \end{bmatrix} \begin{bmatrix} 1.008 & -0.412 \\ -0.021 & 1.089 \end{bmatrix} \begin{bmatrix} 0.511 \times 10^{-6} & 0.160 \times 10^{-6} \\ 0.160 \times 10^{-6} & 0.370 \times 10^{-6} \end{bmatrix}$$

$$\mathbf{Z}_o = \begin{bmatrix} 107.5 & 24.3 \\ 24.3 & 54.8 \end{bmatrix} \, [\Omega] \tag{11-14}$$

PROBLEM 11.2 Calculate the characteristic impedance matrix for the line of Problem 10-2.

11.3 IMPEDANCE MATRIX IN THE MODAL BASIS

Inspection of the vector wave Equation (10-2) shows that to express it in a form such that it is equivalent to N independent scalar wave equations as was done for Equation (10-1)), the matrix **CL** must be diagonalized. The eigenvectors of **CL** will be the basis of current vectors

1. Parallels Equations (11-7) to (11-12):
$$\mathbf{v}_F = \sum_n \mathbf{v}_{Fn} = \sum_n \mathbf{Z}_{on} \mathbf{i}_{Fn}$$

Substitute $\mathbf{Z}_{on}$ (Equation (11-4)) and $\mathbf{i}_{Fn} = \tilde{\mathbf{u}}_n i'_{Fn}$, where $\tilde{\mathbf{u}}_n$ is the nth column of the matrix $(\mathbf{U}^{-1})^T$. One obtains
$$\mathbf{v}_F = \sum_n c_n \mathbf{L} \tilde{\mathbf{u}}_n i'_{Fn} = \mathbf{L}(\mathbf{U}^{-1})^T \Gamma^{-1} \mathbf{i}'_F = \mathbf{L}(\mathbf{U}^{-1})^T \Gamma^{-1} \mathbf{U}^T \mathbf{i}_F$$

Hence,
$$\mathbf{Z}_o = \mathbf{L}(\mathbf{U}^{-1})^T \Gamma^{-1} \mathbf{U}^T = \text{the transpose of the right-hand side of Equation (11-12).}$$

that propagate at discrete speeds. In *Exercise 10-2*, it was shown that the matrix whose columns are these eigenvalues is $(\mathbf{U}^{-1})^T$. Its inverse, multiplying a column vector of wire currents, will yield a column of modal currents. Hence, the vector of modal currents is

$$\mathbf{i'}_F(x,t) = \mathbf{U}^T \mathbf{i}_F(x,t) = \begin{bmatrix} i'_{F1}(x,t-x/c_1) \\ i'_{F2}(x,t-x/c_2) \\ \cdot \\ i'_{FN}(x,t-x/c_N) \end{bmatrix} \tag{11-15}$$

Each modal current element, $i'_{Fn}(x,t-x/c_n)$ must be proportional to the corresponding voltage mode $v'_{Fn}(x,t-x/c_n)$ that is propagating in the same direction at the same speed. Therefore, there is a *diagonal* (modal basis) impedance matrix $\mathbf{Z'}_o$ such that

$$\mathbf{v'}_F(x,t) = \mathbf{Z'}_o \mathbf{i'}_F(x,t) \tag{11-16}$$

Transforming Equation (11-16) back to the wire voltages and currents will yield it. Multiply Equation (11-16) by the matrix $\mathbf{U}$ and insert $\mathbf{U}^T(\mathbf{U}^{-1})^T (= \mathbf{I})$ on the right-hand side to obtain

$$\mathbf{v}_F(x,t) = \mathbf{U}\mathbf{v'}_F(x,t) = \left[\mathbf{U}\mathbf{Z'}_o\mathbf{U}^T\right]\left[(\mathbf{U}^{-1})^T\mathbf{i'}_F(x,t)\right] \tag{11-17}$$

Therefore,

$$\mathbf{Z}_o = \mathbf{U}\mathbf{Z'}_o\mathbf{U}^T \tag{11-18}$$

and, solving for $\mathbf{Z'}_o$,

$$\mathbf{Z'}_o = \mathbf{U}^{-1}\mathbf{Z}_o(\mathbf{U}^{-1})^T \tag{11-19}$$

And, substituting Equation (11-12) for $\mathbf{Z}_o$,

$$\mathbf{Z'}_o = \Gamma^{-1}\mathbf{U}^{-1}\mathbf{L}(\mathbf{U}^{-1})^T \tag{11-20}$$

PROBLEM 11.3 Calculate $\mathbf{Z'}_o$ for the line of *Problem 10-2*.

PROBLEM 11.4 For the transmission line of Figure 10.1 with $\mathbf{C}$ and $\mathbf{L}$ matrices given by Equations (10-3) and (10-4), the matrix $\mathbf{U}^{-1}$ is given by Equation (10-23). The transpose of $\mathbf{U}^{-1}$ has column vectors that are eigenfunctions of $\mathbf{CL}$ and, therefore, correspond to propagation modes of current. The $n = 1$ current propagation mode is proportional to the first column of $(\mathbf{U}^{-1})^T$ and is given by $\mathbf{i}_1 = \begin{bmatrix} 1.008 \\ -0.412 \end{bmatrix}$. Calculate (a) $\mathbf{v}_a = \mathbf{Z}_o\mathbf{i}_1$, and, (b) $\mathbf{v}_b = \mathbf{Z}_{o1}\mathbf{i}_1$. Note that $\mathbf{Z}_{o1}$ was calculated in Problem 11-1 and $\mathbf{Z}_o$ is given by Equation (11-14).

Chapter 12

TRANSPORT ON LOSSY MULTI-SPEED LINES

On lossless, multi-speed lines, the modes propagate unchanged at their respective speeds. Propagation on these lines is simulated by transforming the departing signals to their modal basis and applying the propagation (time delay) algorithm to each of the modal signals. Each element of the modal signal vector arrives unchanged at the far end after the delay time related to its propagation speed by $t_{delay,n} = l/c_n$, where l is the length of the line and c_n is the propagation speed of the nth mode.

If the multi-speed line is lossy, the modal signal will change as it propagates along the line. The signal change is described by a transport equation. The (lossy line) transport equations for single speed lines was developed in Part I. For single-wire lines the transport equations for voltage are given in Equations (2-29) and (2-30) and for multi-wire lines they are given in Equations (5-9) and (5-10). The transport equations for *mode* voltages of a multi-speed line will have the same form as the transport equations for *wire* voltages of a single-speed line. A signal will experience attenuation and reflection as it propagates. The attenuation and reflection coefficients will be matrices and therefore both will couple signals between *modes* just as signals were coupled between *wires* in the lossy multi-wire single-speed line.

12.1 TRANSMISSION LINE EQUATIONS IN THE MODAL BASIS

The derivation of the transport equation for the modes begins with the vector form of the transmission line Equations (5-2) and (5-3). Transforming the first of these to the modal basis is done by multiplying the equation through by the matrix $\mathbf{U}^{-1}$ and inserting the matrix $(\mathbf{U}^{-1})^T \mathbf{U}^T = \mathbf{I}$ in front of the current vectors. This yields

$$\frac{\partial \mathbf{v}'}{\partial x} = -\mathbf{L}'\frac{\partial \mathbf{i}'}{\partial t} - \mathbf{R}'\mathbf{i}' \tag{12-1}$$

where

$$\mathbf{v'} = \mathbf{U}^{-1}\mathbf{v}(x,t),$$

$$\mathbf{i'} = \mathbf{U}^{T}\mathbf{i}(x,t),$$

$$\mathbf{L'} = \mathbf{U}^{-1}\mathbf{L}(\mathbf{U}^{-1})^{T}, \text{ and}$$

$$\mathbf{R'} = \mathbf{U}^{-1}\mathbf{R}(\mathbf{U}^{-1})^{T}$$

Similarly, the second of the transmission line equations, Equation (5-3), is transformed to the modal basis by multiplying the equation through by $\mathbf{U}^{T}$ and inserting the matrix $\mathbf{UU}^{-1} = \mathbf{I}$ in front of the voltage vectors to obtain

$$\frac{\partial \mathbf{i'}}{\partial x} = -\mathbf{C'}\frac{\partial \mathbf{v'}}{\partial t} - \mathbf{G'v'} \qquad (12\text{-}2)$$

where $\mathbf{v'}$ and $\mathbf{i'}$ are given below Equation (12-1), and

$$\mathbf{C'} = \mathbf{U}^{T}\mathbf{CU}$$

and

$$\mathbf{G'} = \mathbf{U}^{T}\mathbf{GU}$$

12.2 TRANSPORT EQUATIONS IN THE MODAL BASIS

The mode voltage vector for forward signals has the form

$$\mathbf{v'}_F = \begin{bmatrix} v'_{F1}(x,t - x/c_1) \\ v'_{F2}(x,t - x/c_2) \\ \cdot \\ \cdot \\ v'_{FN}(x,t - x/c_N) \end{bmatrix} \qquad (12\text{-}3)$$

The notation $\mathbf{v'}_{F1}$ (subscript 1) will be used to denote the vector of derivatives of Equation (12-3) with respect to the first argument of each of its elements. The notation $\mathbf{v'}_{F2}$ (subscript 2) will be used to denote the vector of derivatives of Equation (12-3) with respect to the second argument of each of its elements. Using this notation, we can write

$$\frac{\partial \mathbf{v}'_F}{\partial x} = \mathbf{v}'_{F1} - \Gamma \mathbf{v}'_{F2} \qquad (12\text{-}4)$$

and

$$\frac{\partial \mathbf{v}'_F}{\partial t} = \mathbf{v}'_{F2} \qquad (12\text{-}5)$$

Similarly, for backward signals, we have

$$\frac{\partial \mathbf{v}'_B}{\partial x} = \mathbf{v}'_{B1} + \Gamma \mathbf{v}'_{B2} \qquad (12\text{-}6)$$

and

$$\frac{\partial \mathbf{v}'_B}{\partial t} = \mathbf{v}'_{B2} \qquad (12\text{-}7)$$

The similar expressions for the modal forward and backward current vectors are

$$\frac{\partial \mathbf{i}'_F}{\partial x} = \mathbf{i}'_{F1} - \Gamma \mathbf{i}'_{F2}, \qquad (12\text{-}8)$$

$$\frac{\partial \mathbf{i}'_F}{\partial t} = \mathbf{i}'_{F2}, \qquad (12\text{-}9)$$

$$\frac{\partial \mathbf{i}'_B}{\partial x} = \mathbf{i}'_{B1} + \Gamma \mathbf{i}'_{B2}, \qquad (12\text{-}10)$$

and

$$\frac{\partial \mathbf{i}'_B}{\partial t} = \mathbf{i}'_{B2} \qquad (12\text{-}11)$$

Adding the forward and backward modal signals, we obtain

$$\frac{\partial \mathbf{i}'}{\partial x} = \mathbf{i}'_{F1} + \mathbf{i}'_{B1} - \Gamma \mathbf{i}'_{F2} + \Gamma \mathbf{i}'_{B2}, \qquad (12\text{-}12)$$

$$\frac{\partial \mathbf{i}'}{\partial t} = \mathbf{i}'_{F2} + \mathbf{i}'_{B2}, \qquad (12\text{-}13)$$

and

$$\frac{\partial \mathbf{v}'}{\partial x} = \mathbf{v}'_{F1} + \mathbf{v}'_{B1} - \Gamma \mathbf{v}'_{F2} + \Gamma \mathbf{v}'_{B2}, \qquad (12\text{-}14)$$

$$\frac{\partial \mathbf{v}'}{\partial t} = \mathbf{v}'_{F2} + \mathbf{v}'_{B2}, \qquad (12\text{-}15)$$

Substituting Equations (12-13) and (12-14) into Equation (12-1), one obtains

$$\mathbf{v}'_{F1} + \mathbf{v}'_{B1} = \Gamma \mathbf{v}'_{F2} - \Gamma \mathbf{v}'_{B2} - \mathbf{L}'(\mathbf{i}'_{F2} + \mathbf{i}'_{B2}) - \mathbf{R}'(\mathbf{i}'_F + \mathbf{i}'_B) \qquad (12\text{-}16)$$

Substituting (see Equation (11-16))

$$\mathbf{i}'_F = \mathbf{Z}'^{-1}_o \mathbf{v}'_F \qquad (12\text{-}17)$$

and

$$\mathbf{i}'_B = -\mathbf{Z}'^{-1}_o \mathbf{v}'_B \qquad (12\text{-}18)$$

into Equation (12-16), one obtains

$$\mathbf{v}'_{F1} + \mathbf{v}'_{B1} = \Gamma \mathbf{v}'_{F2} - \Gamma \mathbf{v}'_{B2} - \mathbf{L}'\mathbf{Z}'^{-1}_o (\mathbf{v}'_{F2} - \mathbf{v}'_{B2}) - \mathbf{R}'\mathbf{Z}'^{-1}_o (\mathbf{v}'_F - \mathbf{v}'_B) \qquad (12\text{-}19)$$

By performing the matrix multiplications, one can show that $\mathbf{L}'\mathbf{Z}'^{-1}_o = \Gamma$; hence, the terms $\mathbf{v}'_{F2}$ and $\mathbf{v}'_{B2}$ vanish from Equation (12-19) and it reduces to

$$\mathbf{v}'_{F1} + \mathbf{v}'_{B1} = -\mathbf{R}'\mathbf{Z}'^{-1}_o (\mathbf{v}'_F - \mathbf{v}'_B) \qquad (12\text{-}20)$$

Substituting Equations (12-12) and (12-15) into Equation (12-2), one obtains

$$\mathbf{i}'_{F1} + \mathbf{i}'_{B1} = \Gamma \mathbf{i}'_{F2} - \Gamma \mathbf{i}'_{B2} - \mathbf{C}'(\mathbf{v}'_{F2} + \mathbf{v}'_{B2}) - \mathbf{G}'(\mathbf{v}'_F + \mathbf{v}'_B) \qquad (12\text{-}21)$$

And substituting using Equations (12-17) and (12-18) and multiplying through by $\mathbf{Z}'_o$, Equation (12-21) becomes

$$\mathbf{v}'_{F1} - \mathbf{v}'_{B1} = \mathbf{Z}'_o \Gamma \mathbf{Z}'^{-1}_o \mathbf{v}'_{F2} \qquad (12\text{-}22)$$
$$+ \mathbf{Z}'_o \Gamma \mathbf{Z}'^{-1}_o \mathbf{v}'_{B2} - \mathbf{Z}'_o \mathbf{C}'(\mathbf{v}'_{F2} + \mathbf{v}'_{B2}) - \mathbf{Z}'_o \mathbf{G}'(\mathbf{v}'_F + \mathbf{v}'_B)$$

One can show that $\mathbf{Z}'_o \mathbf{C}' = \mathbf{Z}'_o \Gamma \mathbf{Z}'^{-1}_o = \Gamma$; hence, the terms $\mathbf{v}'_{F2}$ and $\mathbf{v}'_{B2}$ vanish from Equation (12-22) and it reduces to[1]

$$\mathbf{v}'_{F1} - \mathbf{v}'_{B1} = -\mathbf{Z}'_o \mathbf{G}'(\mathbf{v}'_F + \mathbf{v}'_B) \qquad (12\text{-}23)$$

Adding and subtracting Equations (12-20) and (12-23), one obtains transport equations essentially identical in form to Equations (5-9) and (5-10) except for the primes that indicate the modal voltage basis. However, they must be interpreted quite differently, since each ele-

---

1. Because $\mathbf{Z}'_o$ is a diagonal matrix, it is a trivial exercise to show, by direct multiplication, that $\mathbf{Z}'_o \Gamma \mathbf{Z}'^{-1}_o = \Gamma$.

By substitution using $\mathbf{Z}'_o$, Equation (11-20), and $\mathbf{C}'$ below Equation (12-2),

$$\mathbf{Z}'_o \mathbf{C}' = \Gamma^{-1}\mathbf{U}^{-1}\mathbf{L}(\mathbf{U}^{-1})^{\mathrm{T}} \mathbf{U}^{\mathrm{T}}\mathbf{C}\mathbf{U} = \Gamma^{-1}\mathbf{U}^{-1}\mathbf{L}\mathbf{C}\mathbf{U} = \Gamma^{-1}\Gamma^2 = \Gamma.$$

ment of $\mathbf{v}'_F$ and $\mathbf{v}'_B$ propagates at its own speed; attenuation and reflection coefficients must be applied to modify signals that are moving at different speeds. The voltage transport equations in the modal basis are

$$\mathbf{v}'_{F1} = -\mathbf{A}'\mathbf{v}'_F + \mathbf{B}'\mathbf{v}'_B \tag{12-24}$$

and

$$\mathbf{v}'_{B1} = \mathbf{A}'\mathbf{v}'_B - \mathbf{B}'\mathbf{v}'_F \tag{12-25}$$

where $\mathbf{A}'$ and $\mathbf{B}'$ are the matrices

$$\mathbf{A}' = \frac{1}{2}\left(\mathbf{R}'\mathbf{Z}'^{-1}_O + \mathbf{Z}'_O\mathbf{G}'\right) \tag{12-26}$$

and

$$\mathbf{B}' = \frac{1}{2}\left(\mathbf{R}'\mathbf{Z}'^{-1}_O - \mathbf{Z}'_O\mathbf{G}'\right) \tag{12-27}$$

PROBLEM 12.1 Assume $\mathbf{Z}'_O = \begin{bmatrix} 158.8 & 0 \\ 0 & 47.9 \end{bmatrix}$, $\mathbf{U}^{-1} = \begin{bmatrix} 0.707 & 0.707 \\ 0.707 & -0.707 \end{bmatrix}$, each conductor has a resistance of 10 [Ω/m] and the matrix $\mathbf{G} = 0$. Calculate the matrix $\mathbf{A}' = \mathbf{B}'$.

12.3 TRANSPORT DIFFERENCE APPROXIMATION IN THE MODAL BASIS

As before, the difference equations are based on equal time steps, say δt. However, the propagation distance over a single time step will be different for each mode; for the nth mode, the propagation distance over one time step is

$$\delta x_n = c_n \delta t \tag{12-28}$$

This means that the spatial grid points for each mode will be at different locations along the line. Calculating the product of matrices $\mathbf{A}'$ and $\mathbf{B}'$ times the mode voltage vector (see Equation (12-24)) will require interpolations to obtain the resultant coupling at the correct spatial points.

The difference approximation of Equation (12-24) will be written on a row-by-row basis rather than in matrix form. That is, the nth row (i.e., the nth mode) of Equation (12-24) can be approximated by the difference equation

$$\frac{v'_{Fn}(x_{2n},\tau_{Fn}) - v'_{Fn}(x_{1n},\tau_{Fn})}{x_{2n} - x_{1n}}$$

$$= -a'_n \, v'_F \left((x_{2n} + x_{1n})/2,\tau_{Fn}\right) + b'_n \, v'_B \left((x_{2n} + x_{1n})/2,\tau_{Bn}\right) \tag{12-29}$$

where

$$\tau_{Fn} = t_1 - \frac{x_{1n}}{c_n} = t_2 - \frac{x_{2n}}{c_n}$$

$$\tau_{Bn} = t_1 + \frac{x_{2n}}{c_n} = t_2 + \frac{x_{1n}}{c_n}$$

$$x_{2n} - x_{1n} = c_n \delta t$$

The vector $\mathbf{a'}_n$ is the nth *row* of the matrix $\mathbf{A'}$; when it multiplies the column vector of forward mode voltages it has the form of an inner product. The kth term of the inner product represents the coupling of the kth mode forward signal to the forward signal of mode n (with the sign reversed). The coupling coefficient is a'_{nk}. When $k = n$, the coupling is to the nth mode itself; it represents an attenuation of the forward signal of the nth mode. When $k \neq n$, it couples the signal from each forward mode to every other forward mode signal, but with opposite sign. The amount of signal coupled is proportional to δx_n, the length of a computational segment for mode n.

The vector $\mathbf{b'}_n$ is the nth row of the matrix $\mathbf{B'}$; when it multiplies the column vector of backward mode voltages it, too, has the form of an inner product. The kth term of the inner product represents the coupling of a bit of the kth mode backward signal to the nth mode forward signal. It is a generalized reflection coefficient. When $k = n$, it reflects a bit of signal from the nth mode backward signal to the nth mode forward signal, a true reflection. When $k \neq n$, it couples a bit of signal from each mode's backward signal to every other mode's forward signal.

The x argument (first argument) of $\mathbf{v'}_F((x_{2n} + x_{1n})/2, \tau_F)$, simply means that each element of $\mathbf{v'}_F$ is to be evaluated by interpolation at the point $x = (x_{2n} + x_{1n})/2$ using the values stored at the bracketing locations for each mode. Suppose that for mode k, v'_{Fk} is stored at points x_i such that $x_i \leq x \leq x_{i+1}$, then the interpolated value of v'_{Fk} is

$$v'_{Fk}((x_{2n} + x_{1n})/2, \tau_F) = \frac{x_{i+1} - x}{x_{i+1} - x_i} v'_{Fk}(x_i, \tau_F) + \frac{x - x_i}{x_{i+1} - x_i} v'_{Fk}(x_{i+1}, \tau_F) \quad (12\text{-}30)$$

where

$$\tau_F = t_1 - \frac{x_i}{c_k} = t_2 - \frac{x_{i+1}}{c_k}$$

Chapter 13
SMALL COUPLING APPROXIMATION OF PROPAGATION CROSSTALK

The small coupling approximation is a more complex problem for multi-speed lines than it is for single-speed lines (see Section 5.1.5). A voltage signal is assumed to depart the left end of the line on a single wire (or trace). Driving a single wire of the line creates a signal on all of the modes each of which propagates at its own speed. Then, each mode couples a crosstalk signal to all of the modes. This crosstalk signal can then be transformed back to trace voltages. To create an approximate model for these processes, we define a "primary" voltage signal and a "secondary" voltage signal (the crosstalk signal). The primary signal is the signal due to a single trace or wire being driven by an arbitrary voltage function. It is transformed to modes whose propagation is approximated with attenuation only, no crosstalk or reflection. The secondary signal is that due to both the attenuation and crosstalk effects of the matrices $\mathbf{A}'$ and $\mathbf{B}'$ multiplying the primary signal in the transport Equations (12-24) and (12-25). The secondary signal is small if the terms of $\mathbf{A}'$ and $\mathbf{B}'$ times the line length are small. (They are zero for a lossless line.) After determination of the secondary signal in the modal basis, it is transformed back to the wire basis.

13.1 DEFINITION OF THE PRIMARY SIGNAL

Assume a signal departs the left end of the line on wire k, one can write

$$v_{Fn}(0,t) = \begin{cases} v_o(t), & n = k \\ 0, & n \neq k \end{cases} \tag{13-1}$$

The primary signal in the modal basis is

$$\mathbf{v}'_F(0,t) = \mathbf{u}'_k v_o(t) = \begin{bmatrix} u'_{1k}v_o(t) \\ u'_{2k}v_o(t) \\ \cdot \\ u'_{Nk}v_o(t) \end{bmatrix} \tag{13-2}$$

where $\mathbf{u}'_k$ is the kth column of the matrix $\mathbf{U}^{-1}$. A single mode (row of Equation 13-2) can be written

$$v'_{Fn}(0,t) = u'_{nk} v_o(t) \tag{13-3}$$

This primary mode is approximated as an attenuating signal without any crosstalk as

$$v'_{Fn}(x, t - x/c_n) = u'_{nk} e^{-\alpha'_{nn} x} v_o(t - x/c_n) \tag{13-4}$$

13.2 THE SECONDARY SIGNAL, AN APPROXIMATION OF PROPAGATION CROSSTALK

The secondary signal is approximated as that directly due to the primary signal as implied by the transport Equations (12-24) and (12-25). The crosstalk effects of the secondary signal modes among themselves are neglected; the secondary modes are assumed to propagate as attenuating signals being driven by the distributed effect of the primary signal. The secondary signals will be identified by a tilde; in the modal basis, the secondary forward and backward voltage vectors are denoted $\tilde{\mathbf{v}}'_F(x,t)$ and $\tilde{\mathbf{v}}'_B(x,t)$ with elements $\tilde{v}'_{Fi}(x,t)$ and $\tilde{v}'_{Bi}(x,t)$.

The crosstalk backward signal is approximated using the transport Equation (12-25). The ith row of this equation can be written (neglecting the off-diagonal α'_{ij} terms)

$$\tilde{v}'_{Bi1}(x, t + x/c_i) = \alpha'_{ii}\tilde{v}'_{Bi}(x, t + x/c_i) - \sum_n \beta'_{in} u'_{nk} e^{-\alpha'_{nn} x} v_o(t - x/c_n) \tag{13-5}$$

where the right-most term is the ith row of matrix $\mathbf{B}'$ times the primary signal vector (the nth element of the primary signal vector is given in Equation (13-4)). If $\tau_{Bi} = t + x/c_i$ is held constant, Equation (13-5) is an ordinary differential equation that can be written as

$$\frac{d\tilde{v}'_{Bi}}{dx} - \alpha'_{ii}\tilde{v}'_{Bi} = -\sum_n \beta'_{in} u'_{nk} e^{-\alpha'_{nn} x} v_o(\tau_{Bi} - 2x/c_n) \tag{13-6}$$

It is assumed that there is no backward signal departing from the right end. That is, $\tilde{v}'_{Bi}(l, t + l/c_i) = 0$, where l is the length of the line, i.e., the line is assumed impedance matched at the right end. Applying this as the boundary condition to Equation (13-6), the solution can be written

$$\tilde{v}'_{Bi}(x,\tau_i) = e^{\alpha'_{ii} x} \sum_n \beta'_{in} u'_{nk} \int_x^l e^{-(\alpha'_{ii} + \alpha'_{nn})x'} v_o(\tau_{Bi} - 2x'/c_n)\, dx' \quad (13\text{-}7)$$

At the left end of the line, $x = 0$ and $\tau_{Bi} = t$, the arriving signal is

$$\tilde{v}'_{Bi}(0,t) = \sum_n \beta'_{in} u'_{nk} \int_0^l e^{-(\alpha'_{ii} + \alpha'_{nn})x'} v_o(t - 2x'/c_n)\, dx' \quad (13\text{-}8)$$

This implies that the arriving vector of wire voltages at the left end is

$$\mathbf{v}_B(0,t) = \mathbf{U}\tilde{\mathbf{v}}'_B(0,t) \quad (13\text{-}9)$$

where, $\tilde{\mathbf{v}}'_B(0,t)$ is the column vector of mode voltages given by Equation (13-8). This equation is half of the desired result, i.e., the secondary signal (crosstalk) that arrives at the driven end. We will proceed to find the other half, the secondary signal (crosstalk) that arrives at the load end.

The left end is assumed terminated with zero impedance (the voltage is fixed at all of the wire ends, see Equation (13-1)). Hence, the secondary signal in the modal basis departing from the left end is

$$\tilde{\mathbf{v}}'_F(0,t) = -\tilde{\mathbf{v}}'_B(0,t) \quad (13\text{-}10)$$

with components given by minus Equation (13-8).

Equation (13-10) will be used as the boundary condition for the forward secondary signal. To obtain the forward secondary signal, begin with the transport Equation (12-24); the ith row can be written

$$\tilde{v}'_{Fi1}(x,t - x/c_i) = -\alpha'_{ii}\,\tilde{v}'_{Fi}(x,t - x/c_i) - \sum_n \alpha'_{in} u'_{nk} e^{-\alpha'_{nn} x} v_o(t - x/c_n) \quad (13\text{-}11)$$

Note that there is no backward primary signal, the coupling is from the forward primary signal to the forward secondary signal via the matrix $\mathbf{A}'$. The right-most term is the ith row of matrix $\mathbf{A}'$ times the primary signal vector (the nth element of the primary signal vector is given in Equation (13-4)). If $\tau_{Fi} = t - x/c_i$ is constant, Equation (13-11) is an ordinary differential equation that can be written

$$\frac{d\tilde{v}'_{Fi}}{dx} + \alpha'_{ii}\,\tilde{v}'_{Fi} = -\sum_n \alpha'_{in} u'_{nk} e^{-\alpha'_{nn} x} v_o\left(\tau_{Fi} + \frac{x}{c_i} - \frac{x}{c_n}\right) \quad (13\text{-}12)$$

The solution of Equation (13-12) with the boundary condition at $x = 0$ given by Equation (13-10) can be written

$$\tilde{v}'_{Fi}(x,\tau_{Fi}) = -e^{-\alpha'_{ii}x}\sum_n \beta'_{in}u'_{nk}\int_0^l e^{-(\alpha'_{ii}+\alpha'_{nn})x'}v_o\left(\tau_{Fi}-2\frac{x'}{c_n}\right)dx'$$

$$-e^{-\alpha'_{ii}x}\sum_n \alpha'_{in}u'_{nk}\int_0^x e^{(\alpha'_{ii}-\alpha'_{nn})x'}v_o\left(\tau_{Fi}+\frac{x'}{c_i}-\frac{x'}{c_n}\right)dx'$$

$$(13\text{-}13)$$

At the right (receiving) end of the line, $x = l$ and $\tau_{Fi} = t - l/c_i$. The received secondary (crosstalk) signal is (for the ith mode)

$$\tilde{v}'_{Fi}(l,t-l/c_i) = -e^{-\alpha'_{ii}l}\sum_n \beta'_{in}u'_{nk}\int_0^l e^{-(\alpha'_{ii}+\alpha'_{nn})x'}v_o\left(t-\frac{l}{c_i}-2\frac{x'}{c_n}\right)dx'$$

$$-e^{-\alpha'_{ii}l}\sum_n \alpha'_{in}u'_{nk}\int_0^l e^{(\alpha'_{ii}-\alpha'_{nn})x'}v_o\left(t-\frac{l}{c_i}+\frac{x'}{c_i}-\frac{x'}{c_n}\right)dx'$$

$$(13\text{-}14)$$

In the wire basis we have

$$\tilde{\mathbf{v}}_{Fi}(l,t) = \mathbf{U}\tilde{\mathbf{v}}'_{Fi}(l,t) \tag{13-15}$$

where the elements of column vector $\tilde{\mathbf{v}}'_{Fi}(l,t)$ are given by Equation (13-14) and $\mathbf{U}$ is the transformation matrix from the modal basis to the wire basis.

13.3 PROPAGATION CROSSTALK OF IMPULSE FUNCTION

The secondary signal in the modal basis that arrives at the receiving end of a line, Equation (13-14), will be evaluated for $v_o(t)$ equal to an impulse function, i.e., the Dirac delta function $\delta(t)$. This is of interest both for intuitive understanding of propagation crosstalk and as a computational tool. Once the secondary signal is obtained for $v_o(t) = \delta(t)$, it can be calculated for a driver equal to any function of time by applying the convolution integral theorem. This is easier to implement than a straightforward evaluation of the integrals in Equation (13-14). We proceed to evaluate the two integrals of Equation (13-14) with $v_o(t) = \delta(t)$.

A fundamental integral property of the delta function is used; that is, for $x_1 < x_2$

$$\int_{x_1}^{x_2} f(x)\delta(x-x_o)dx = \begin{cases} f(x_o), & x_1 \leq x_o \leq x_2 \\ 0, & \text{otherwise} \end{cases} \tag{13-16}$$

That is, $\delta(x) = 0$ if $x \neq 0$, but at $x = 0$, $\delta(x)$ is infinite in such a way that its integral equals one and the integral of $\delta(x)$ multiplied by any function is the value of the function at the point where the argument of the delta function is zero. This property will be used to evaluate the integrals in Equation (13-14). By changing variables of integration in Equation (13-16), one can obtain a slightly generalized form of Equation (13-16):

$$\int_{x_1}^{x_2} f(x)\delta(ax - x_o)dx = \begin{cases} \dfrac{1}{|a|}f(x_o/a), & x_1 \le \dfrac{x_o}{a} \le x_2 \\[3mm] 0, & \text{otherwise} \end{cases} \tag{13-17}$$

To use Equation (13-17) to evaluate the first integral in Equation (13-14), we make the following correspondences:

$$x_1 = 0 , \quad x_2 = l$$

$$f(x) = e^{-(\alpha'_{ii} + \alpha'_{nn})x}$$

$$v_o\left(t - \frac{l}{c_i} - 2\frac{x'}{c_n}\right) = \delta\left(t - \frac{l}{c_i} - 2\frac{x}{c_n}\right) = \delta(ax - x_o)$$

Hence,

$$a = -\frac{2}{c_n}$$

$$x_o = \frac{l}{c_i} - t$$

$$\frac{x_o}{a} = \frac{c_n}{2}\left(t - \frac{l}{c_i}\right)$$

Substituting these into Equation (13-17) and simplifying the inequality, one obtains

$$\int_0^l e^{-(\alpha'_{ii} + \alpha'_{nn})x'} v_o\left(t - \frac{l}{c_i} - 2\frac{x'}{c_n}\right)dx'$$

$$= \begin{cases} \dfrac{c_n}{2} e^{-(\alpha'_{ii} + \alpha'_{nn})\frac{c_n}{2}\left(t - \frac{l}{c_i}\right)}, & \dfrac{l}{c_i} \le t \le \dfrac{2l}{c_n} + \dfrac{l}{c_i} \\[3mm] 0, & \text{otherwise} \end{cases} \tag{13-18}$$

The second of the integrals in Equation (13-14) must be evaluated as two cases that depend on whether or not $c_i = c_n$. If $c_i = c_n$, $a = 0$ and a division by zero occurs in Equation (13-17) so that formula can not be used. Assume, first, that $c_i = c_n$, then with v_o set to the delta function,

$$\int_0^l e^{(\alpha'_{ii} - \alpha'_{nn})x'} v_o\left(t - \frac{l}{c_i} + \frac{x'}{c_i} - \frac{x'}{c_n}\right)dx' = \int_0^l e^{(\alpha'_{ii} - \alpha'_{nn})x'} \delta\left(t - \frac{l}{c_i}\right)dx'$$

$$= \delta\left(t - \frac{l}{c_i}\right)\int_0^l e^{(\alpha'_{ii} - \alpha'_{nn})x'} dx'$$

(13-19)

$$= \delta\left(t - \frac{l}{c_i}\right)\begin{cases} \dfrac{e^{(\alpha'_{ii} - \alpha'_{nn})l} - 1}{\alpha'_{ii} - \alpha'_{nn}}, & \alpha'_{ii} \neq \alpha'_{nn} \\[2ex] l, & \alpha'_{ii} = \alpha'_{nn} \end{cases}$$

We see that the delta function comes out of the integral and the result is simply a constant times a delta function. The delta function is extremely easy to use in a convolution integral application of Equation (13-19) because of the integral property of the delta function given in Equation (13-16). But what if c_i is quite close to, but not exactly equal to c_n, i.e., $c_i \cong c_n$? Then, the integral will approximate a delta function but will have a finite width. To accurately use such a very narrow function in a convolution integral, either the small width must be resolved in the numerical integration or the function may be simply approximated by a delta function. In such problems there is a practical limit to the resolution one can use because of both computational speed and the round-off error incurred when using a great number of terms in a summation. Hence, it makes sense to numerically approximate such a function as a delta function even though the equality $c_i = c_n$ is not exact.[1] In the illustrating computer program discussed below (Computer Code 13-1, in file `ch13_1.c`), a subroutine (`Approx-Equal`) is used to make this test. Any value of the tolerance parameter can be used, but as the code is written, it assumes $c_i \cong c_n$ if they differ by one part in 1,000 or less.

If $c_i \neq c_n$, Equation (13-17) is applicable, and the following correspondences can be made:

$$x_1 = 0, \quad x_2 = l$$

$$f(x) = e^{-(\alpha'_{ii} + \alpha'_{nn})x}$$

$$v_o\left(t - \frac{l}{c_i} + \frac{x'}{c_i} - \frac{x'}{c_n}\right) = \delta\left(t - \frac{l}{c_i} + \frac{x}{c_i} - \frac{x}{c_n}\right) = \delta(ax - x_o)$$

Hence,

$$a = \frac{1}{c_i} - \frac{1}{c_n}$$

$$x_o = \frac{l}{c_i} - t$$

---

1. "In nature, no mathematical function exists! All mathematics is an approximation to reality." Dr. Thomas W. Larkin, private communication.

$$\frac{x_o}{a} = \frac{c_i c_n}{c_i - c_n}\left(t - \frac{l}{c_i}\right)$$

With these substitutions, Equation (13-17) yields

$$\int_0^l e^{(\alpha'_{ii} - \alpha'_{nn})x'} v_o\left(t - \frac{l}{c_i} + \frac{x'}{c_i} - \frac{x'}{c_n}\right) dx'$$

$$= \begin{cases} \left|\frac{c_i c_n}{c_i - c_n}\right| e^{-(\alpha'_{ii} + \alpha'_{nn})\frac{c_i c_n}{c_i - c_n}\left(t - \frac{l}{c_i}\right)}, & 0 \le \frac{c_i c_n}{c_i - c_n}\left(t - \frac{l}{c_i}\right) \le l \\ \\ 0, & \text{otherwise} \end{cases} \quad (13\text{-}20)$$

Let $\tilde{v}'_{\delta F i}(l, t - l/c_i)$ be the resulting secondary signal due to a delta-function drive on wire k that is given by Equation (13-14) with Equation (13-18) substituted for the first integral on its right-hand side and Equation (13-19) or (13-20) (whichever is appropriate) substituted for its second integral. Then, for any drive $v_o(t)$, the secondary signal arriving at the right end is given by the convolution integral

$$\tilde{v}'_{Fi}(l, t - l/c_i) = \int_0^t \tilde{v}'_{\delta F i}(l, t' - l/c_i) v_o(t - t')\, dt' \quad (13\text{-}21)$$

These voltage modes can be transformed to wire voltages using the matrix $\mathbf{U}$ as in Equation (13-15).

Computer Code 13-1. Approximate Crosstalk, Delta-Function Drive.

This computer code evaluates Equation (13-15) with the departing voltage function $v_o(t)$ set to the delta function $\delta(t)$. The subroutine that performs the calculation is `PropCrosstalkDelta` and is listed below. It accepts as input the transmission line parameters, the index of the wire that is driven with a delta function and the time t (variable `tout`) of the crosstalk. The output contains both the continuous part at the single time t (vector `vout`) and the entire delta function part of the crosstalk. There is always a delta function part to the secondary signal because of the term coupling each mode of the primary signal to the same mode of the secondary signal, i.e., $i = n$ in Equation (13-18). Subroutine `PropCrosstalkDelta` calls the following subroutines:

`MatrixTransformTMTT` — Transforms matrix $\mathbf{M}$ using $\mathbf{M}' = \mathbf{TMT}^T$ (see transformations of $\mathbf{L}$, $\mathbf{C}$, $\mathbf{R}$, and $\mathbf{G}$ below Equations (12-1) and (12-2))

`ApproxEqual` — Tests two floating point numbers for nearly equal

`ConstByMatrixByColumn` — Multiplies a constant times a matrix times a column vector

The subroutine `PropCrosstalkDelta` evaluates the integrals given in Equations (13-18), (13-19), and (13-20). Test cases and outputs are given below the listing.

```
/* ------ Procedure PropCrosstalkDelta ---%proc%--------------------
   Description: Calculate approximate crosstalk due to delta function
            drive on wire kay.  The output contains, in general, two
            parts (for each wire): A continuous function and, if two
            propagation speeds are equal, a delta function part. The
            continuous part is output in vector vout.(Dielectric
            conductance G is assumed neglegible.)
   Procedures Called
------------------------------------------------------------------*/
// MatrixTransformTMTT -- Transforms matrix using T * M * (T transform)
// ApproxEqual -- Test two floats for nearly equal
// ConstByMatrixByColumn -- constant * matrix * (column vector)
/*-------FunctionDeclaration-----------------------------------------*/
int PropCrosstalkDelta ( // Input parameters
            int     num,      // number of wires/modes
            int     kay,      // index of driven wire (impulse at
                              // time = 0) 0 <= kay < num
            float   tout,     // time of crosstalk output
            float   length,   // length of transmission line
            float   *R,       // wire resistances [ohms/m]
            float   *ci,      // mode prop. speeds (gamma inverse)
            float   *Zo,      // inductance matrix
            float   *U,       // eigenvectors
            float   *Uinv,    // inverse of matrix U
                       // Output parameters
            float   *vdelta,  // Delta function amplitudes
            float   *tdelta,  // Delta function times
                              // (see Eq. 13-19)
            float   *vout)    // Crosstalk vector (at time tout)
{
/* --- Local Variable Declarations ---------------------------------
   Type       Name                  Description & range
   ----       ----         ---------------------------------- */
   int        rcode;                // return code, 0 (false), normal
   float      *Aprime = NULL;       // attenuation matrix
   float      *Bprime = NULL;       // reflection matrix
   float      *Zoprime = NULL;      // Zo prime
   float      *ZopInv = NULL;       // Zo prime inverse
   float      *Rprime = NULL;       // R prime matrix
   int        idx, col;             // for-loop indices
   double     diag;                 // temp. variable
   float      sumc;                 // summation variables
   int        imode, nmode;         // for-loop indices
```

```
   float        arg;                    // argument of exponential
   float        test;                   // temp variable for test
   float        expiiterm;              // temp. variable
   float        coef;                   // coefficient, Eq 13-14
   float        halfcin;                // 0.5 * ci[n]
/*----------------------- Error Conditions Trapped ------------------
 Fatal/Nonfatal     Description
 -------------      -----------
 Fatal      PropCrosstalkDelta = -1, error returned by MatrixTransformTMTT
                                 -2, Zoprime singular
                              -3, error returned by MatrixTransformTMTT
                              -4, invalid value of kay input
----------------ExecutableCode-------------------------------------*/
  //***** Initialize parameters and perform transformations *****
   // Calculate matrix Zoprime (Eq 11-19)
   Zoprime = (float*)malloc(sizeof(float) * num * num);
   rcode = MatrixTransformTMTT (num, Uinv, Zo, Zoprime);
   if (rcode)        // error
   {
      rcode = -1;
      goto AllDone;
   }
   // Note that Zoprime is a diagonal matrix.
   // Just the diagonal elements of its inverse will be calculated.
   // Calculate diagonal elements of inverse of Zoprime
   ZopInv = (float*)malloc(sizeof(float) * num);
   for (idx = 0; idx < num; idx++)
   {
      diag = *(Zoprime + idx * (num+1));
      if (diag == 0.0)        // error, singlular matrix
      {
         rcode = -2;
         goto AllDone;
      }
      *(ZopInv + idx) = (float)(1.0F/diag);
   }
// Calculate Rprime
   Rprime = (float*)malloc(sizeof(float) * num * num);
   for (idx = 0; idx < num*num; idx++)
      *(Rprime + idx) = 0.0F;          // initialize to zero
   for (idx = 0; idx < num; idx++)     // set diagonals to R
      *(Rprime + idx*(num+1)) = R[idx];
   // Transform to R prime (Eq 12-1)
   rcode = MatrixTransformTMTT (num, Uinv, Rprime, Rprime);
   if (rcode)    // error
   {
```

```
        rcode = -3;
        goto AllDone;
    }
    // Calculate matrices Aprime and Bprime (Eq 12-26)
    // (matrix G assumed = 0)
    Aprime = (float*)malloc(sizeof(float) * num*num);
    Bprime = (float*)malloc(sizeof(float) * num*num);
    // multiply  0.5 * Rprime * ZopInv column-by-column
    for (col = 0; col < num; col++)
        for (idx = 0; idx < num; idx++)
        {
            Aprime[idx+num*col] = 0.5F
                    * Rprime[idx+num*col] * ZopInv[col];
            Bprime[idx+num*col] = Aprime[idx+num*col];
        }
    // Check input index kay
    if (kay < 0 || kay > num-1)          // error, kay out of range
    {
        rcode = -4;
        goto AllDone;
    }
    // initialize output arrays
    for (idx = 0; idx < num; idx++)          // row index
    {
        vout[idx] = 0.0F;
        for (col = 0; col < num; col++)   // time-of-delta index
            vdelta[idx + num*col] =   0.0F;
    }
    // Fill delta-function times output
    for (idx = 0; idx < num; idx++)
        tdelta[idx]   = length / ci[idx];
    //***** Calculate modal crosstalk *****
    // loop over output modes (Calc. Eq. 13-14)
    for (imode = 0; imode < num; imode++)
    {
        expiiterm = (float)exp(-Aprime[imode+num*imode]*length);
        sumc = 0.0F;   // initialize sum for continuous function
        // loop over source modes & sum
        for (nmode = 0; nmode < num; nmode++)
        {
            // calculate first term (continuous)
            if (tdelta[imode] <= tout
                    && tout <= 2.0F*tdelta[nmode]+tdelta[imode])
            {
                coef = -expiiterm * Bprime[imode+num*nmode]
                                    * Uinv[nmode+num*kay];
```

```
        halfcin = 0.5F * ci[nmode];
        arg = -(Aprime[imode+num*imode]+Aprime[nmode+num*nmode])
                * halfcin * (tout - tdelta[imode]);
        sumc += (float)(coef * halfcin * exp(arg));
    }
    // second term -- continuous or delta function
    coef = -expiiterm * Aprime[imode+num*nmode]
                            * Uinv[nmode+num*kay];
    if (ApproxEqual (ci[imode], ci[nmode], 1.0e-3F))
    {
        if (ApproxEqual (Aprime[imode+num*imode],
                    Aprime[nmode+num*nmode], 1.0e-3F))
        vdelta[imode + num*nmode] = coef * length;
        else
        {
            arg=(Aprime[imode+num*imode]-Aprime[nmode+num*nmode])
                * length;
            vdelta[imode + num*nmode] =(float)(coef*(exp(arg)-1.0)
              /(Aprime[imode+num*imode]-Aprime[nmode+num*nmode]));
        }
    }
    else        // continuous part of secondary function
    {
        test = ci[imode]*ci[nmode]/(ci[imode]-ci[nmode]) * (tout
                - tdelta[imode]);
        if ( 0.0 <= test && test <= length)
        {
            arg = -(Aprime[imode+num*imode]
                + Aprime[nmode+num*nmode]) * test;
            sumc +=(float)(coef * fabs(ci[imode] * ci[nmode]
            /(ci[imode] -ci[nmode])) * exp(arg));
        }
    }
}
vout[imode] = sumc;    // continuous function mode out
}
//***** Transform crosstalk to wire basis *****
// Transform continuous part
ConstByMatrixByColumn (1.0F, U, vout, num, vout);
// Transform each column of the delta-function amplitudes
for (col = 0; col < num; col++)    // time-of-delta index
{
    ConstByMatrixByColumn (1.0F, U, (vdelta+num*col), num,
                                (vdelta+num*col));
}
rcode = 0;
```

```
AllDone:
    // Free temporary memory
    free(Zoprime);      Zoprime = NULL;
    free(ZopInv);       ZopInv = NULL;
    free(Rprime);       Rprime = NULL;
    free(Aprime);       Aprime = NULL;
    free(Bprime);       Bprime = NULL;
    return (rcode);

}
/* -----End of Procedure PropCrosstalkDelta --------------------- */
```

The output below was generated by the code in file ch13_1.c that calls subroutine PropCrosstalkDelta listed above. The results are for two transmission lines. The first transmission line is that used for illustration in the previous computer code (Computer Code ch10_1.c). Its cross section is given in Figure 10.3. Its modal propagation speeds vary by more than a factor of 2. The second line was chosen to have two propagation speeds close to each other to exercise the delta-function part of the code. Its cross section is shown below in Figure 13.1.

This transmission line has the following parameters:

Capacitance matrix [pF/m]:

	1	2	3
1	23.907	-2.546	-2.546
2	-2.546	222.798	-30.082
3	-2.546	-30.082	222.798

Inductance matrix [micro H/m]:

	1	2	3
1	0.707	0.010	0.010
2	0.010	0.102	0.014
3	0.010	0.014	0.102

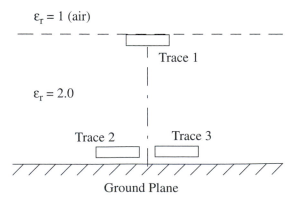

$\varepsilon_r = 1$ (air)

Trace 1

$\varepsilon_r = 2.0$

Trace 2   Trace 3

Ground Plane

Figure 13-1 Second Transmission Line Illustrating Computer Program 13-1.

Propagation speeds:

```
No.  1   2.435516e+008  [m/s]    0.8124c
No.  2   2.120190e+008  [m/s]    0.7072c
No.  3   2.119870e+008  [m/s]    0.7071c
```

U matrix — basis vectors
Transformation from eigenmodes to traces:

```
         1        2         3
1    -1.000    0.038     0.000
2    -0.010    0.707    -0.707
3    -0.010    0.706     0.708
```

U inverse matrix — Transformation to eigenmodes:

```
         1        2         3
1    -1.001    0.027     0.027
2    -0.014    0.708     0.707
3    -0.000   -0.707     0.707
```

The resistance of each wire was taken to be 100 [Ω/m].
Output of Computer Program 13-1:

```
First transmission line
Delta-function drive on wire no. 2
Continuous-function part
      Time          wire 1          wire 2          wire 3
   4.3110e-010   0.0000e+000    0.0000e+000    0.0000e+000
   6.4968e-010   1.9739e+008   -1.0828e+008   -3.3448e+007
   8.6827e-010   1.5312e+008   -1.2805e+008   -1.9309e+007
   1.0869e-009   1.0861e+008   -1.5099e+008   -2.3940e+007
   1.3054e-009  -7.2836e+006   -1.7550e+008    6.3474e+007
   1.5240e-009  -1.8555e+006   -1.5749e+008    9.5661e+007
   1.7426e-009   1.7731e+007   -1.1187e+008    8.9388e+006
   1.9612e-009  -6.6596e+007   -1.1182e+008    7.1611e+006
   2.1798e-009  -6.3141e+007   -1.0661e+008    6.9816e+006
   2.3984e-009  -5.9900e+007   -1.0169e+008    6.8115e+006
   2.6170e-009  -7.3608e+006   -3.7625e+007    2.4845e+007
   2.8356e-009  -8.4477e+006   -3.7428e+007    1.1247e+007
   3.0542e-009  -8.6288e+006   -3.6764e+007    1.0917e+007
   3.2727e-009  -2.8596e+007   -3.7342e+007    1.0239e+007
   3.4913e-009  -2.8100e+007   -3.6660e+007    9.9385e+006
   3.7099e-009  -2.9287e+007   -3.6095e+007    9.6159e+006
   3.9285e-009  -1.8844e+006   -3.1722e+006    1.9210e+007
   4.1471e-009  -2.4688e+006   -3.4443e+006    9.5975e+006
   4.3657e-009   7.0674e+005    3.7974e+005    1.0517e+007
   4.5843e-009   0.0000e+000    0.0000e+000    0.0000e+000
```

```
Delta-function part
     Time          wire 1        wire 2        wire 3
   6.4968e-010    1.8288e-001    1.1361e-002    3.2985e-003
   1.2665e-009   -8.5207e-002   -1.0230e-001   -3.1320e-002
   1.5281e-009    2.9456e-003    1.5827e-003    4.3832e-002

 Second transmission line
 Delta-function drive on wire no. 2
 Continuous-function part
     Time          wire 1        wire 2        wire 3
   5.3217e-010    0.0000e+000    0.0000e+000    0.0000e+000
   6.1589e-010    1.2681e+008    1.0226e+006    1.0230e+006
   6.9960e-010    9.4260e+007    6.0458e+005    6.0505e+005
   7.8332e-010    4.0943e+006   -1.6828e+008    1.4051e+007
   8.6703e-010    4.0864e+006   -1.6129e+008    1.2564e+007
   9.5075e-010    4.0763e+006   -1.5461e+008    1.1176e+007
   1.0345e-009    4.0642e+006   -1.4820e+008    9.8806e+006
   1.1182e-009    4.0502e+006   -1.4207e+008    8.6733e+006
   1.2019e-009    4.0343e+006   -1.3619e+008    7.5487e+006
   1.2856e-009    4.0168e+006   -1.3056e+008    6.5022e+006
   1.3693e-009    3.9977e+006   -1.2517e+008    5.5290e+006
   1.4530e-009    3.9772e+006   -1.2000e+008    4.6251e+006
   1.5368e-009    3.9552e+006   -1.1505e+008    3.7861e+006
   1.6205e-009    3.9319e+006   -1.1031e+008    3.0084e+006
   1.7042e-009    3.9075e+006   -1.0577e+008    2.2883e+006
   1.7879e-009    3.8819e+006   -1.0142e+008    1.6224e+006
   1.8716e-009    3.0170e+006   -9.7255e+007    9.9892e+005
   1.9553e-009    2.9954e+006   -9.3240e+007    4.4988e+005
   2.0391e-009   -2.4111e+006   -8.9466e+007   -1.2666e+005
   2.1228e-009    0.0000e+000    0.0000e+000    0.0000e+000

 Delta-function part
     Time          wire 1        wire 2        wire 3
   6.1589e-010    1.1286e-003    1.1286e-005    1.1286e-005
   7.0748e-010   -6.0686e-003   -1.1298e-001   -1.1267e-001
   7.0759e-010   -5.3623e-006   -1.3454e-001    1.3453e-001
```

Chapter 14

NETWORK SOLUTIONS USING MODAL ANALYSIS

14.1 SEPARATING AND RECOMBINING THE PROPAGATION MODES

For single-speed, multi-wire lines, the basic form of the solution of the transmission line equations was written as Equations (5-4) and (5-5) with the relationship between forward and backward voltage and current given by Equations (5-6) and (5-7). These relationships require a small modification to account for the different propagation speed of each mode. Those equations will be modified to read

$$\mathbf{v}(x,t) = \mathbf{v}_F(x,t) + \mathbf{v}_B(x,t) \tag{14-1}$$

$$\mathbf{i}(x,t) = \mathbf{i}_F(x,t) + \mathbf{i}_B(x,t) \tag{14-2}$$

where the forward and backward currents and voltages are related by

$$\mathbf{v}_F(x,t) = \mathbf{Z}_o\mathbf{i}_F(x,t) \tag{14-3}$$

$$\mathbf{v}_B(x,t) = -\mathbf{Z}_o\mathbf{i}_B(x,t) \tag{14-4}$$

The argument $t - x/c$ (in Equations (5-4) through (5-7)) has been changed to t because each of the forward and backward signals in Equations (14-1) through (14-4) is a mixture of modes that travel at different speeds; they cannot be propagated without separating them into components each of which travels at a discreet speed.

Suppose a network is being solved using the time-stepping method. At the left end of a (typical) transmission line, the line end *wire* voltages will be obtained at time t by solving the network equations; let this voltage (column) vector be $\mathbf{v}_{left}(t)$. By performing a propagation algorithm on the transmission line, the vector of arriving *mode* voltages $\mathbf{v'}_B(0, t)$ is obtained. These are transformed to the vector of arriving *wire* voltages $\mathbf{v}_B(0, t)$ by applying the transformation $\mathbf{U}$, that is

$$\mathbf{v}_B(0,t) = \mathbf{U}\mathbf{v'}_B(0,t) \tag{14-5}$$

Applying Equation (14-1), the departing voltage in the *wire* basis is obtained as

$$\mathbf{v}_F(0,t) = \mathbf{v}_{left}(t) - \mathbf{v}_B(0,t) \tag{14-6}$$

To propagate this signal, it must be separated into discreet-speed *modes*. This is done by applying the transformation $\mathbf{U}^{-1}$. That is,

$$\mathbf{v}'_F(0,t) = \mathbf{U}^{-1}\mathbf{v}_F(0,t) \tag{14-7}$$

The elements of $\mathbf{v}'_F(0,t)$ propagate at the discreet speeds c_i; they can be written

$$\mathbf{v}'_F(x,t) = \begin{bmatrix} v'_{F1}(x,t-x/c_1) \\ v'_{F2}(x,t-x/c_2) \\ . \\ . \\ v'_{FN}(x,t-x/c_N) \end{bmatrix} \tag{14-8}$$

Each element of this vector must be propagated at its appropriate speed. If the line is lossless, the same algorithm (and subroutine) can be used as was used for single-wire lossless lines. If the wire is lossy, the modes couple to each other as they propagate in accordance with the transport Equations (12-24) and (12-25) with elements that (approximately) obey the difference Equation (12-29). An algorithm and code for this coupling is given in the next section. The analogous calculations and transformations are done for each line end (connected to each subnetwork in a large network of lines and discreet elements). After another propagation time step and network solution, this process is repeated.

14.2 SOLUTION OF NETWORKS WITH MULTI-SPEED LINES

Two computer codes are presented in this chapter, they are both general network solvers. The first is for lossless, multi-speed lines. It is a generalization of the network solver for single-speed lines described in Chapter 8 (file ch8_1.c) to multi-speed lossless lines. The second is a further generalization to lossy multi-speed lines; this modification is done by adding to input the wire resistances and calling the routine TransportMultiSpeed just before the propagation routine PropTstep is called. The routine TransportMultiSpeed calculates the propagation cross talk over a single time step.

14.2.1 Lossless Multi-Speed Lines

The network solver for lossless multi-speed lines is a modified version of the lossless single-speed network code in file ch8_1.c; it is in file ch14_1.c. Only the changes will be discussed; see Chapter 8 for the full code description and data flow chart. Additional input is needed that defines the speed of each mode and the transformation matrix $\mathbf{U}$. A new keyword

[MultiSpeed] has been added to head the new data blocks. The new keyword must directly follow the $\mathbf{Z}_O$ matrix in the input data file. The additional data required in the input file is shown below in Table 14-1.

The structure of the linked list for transmission lines has also been modified to include the vector of speeds (rather than a single speed) and the matrices $\mathbf{U}$ and $\mathbf{U}^{-1}$ (for transformations from wire basis to mode basis and mode basis to wire basis). Other changes in this structure include changing the weights, wgt1 and wgt2 and Nhalf to vector pointers rather than scalars (each element of the vector is applicable to a given mode of propagation). The memory circle for propagation PropCircle has been made more complex; instead of consisting of equal parts, one for each wire, it was modified to be comprised of unequal parts, one for each mode. Its sections have unequal lengths because the speed of propagation (and the number of elements needed in each propagation memory circle) are different for each mode. The voltages stored in the propagation memory circle are mode voltages rather than wire voltages.

TABLE 14-1
Transmission Lines Data Block, Multi-Speed Additions (Three-Wire Line).

Line of input file	Explanation	Where stored
[MultiSpeed]	Keyword for line speeds	Not stored
-- list of speeds [m/s]	Comment, not echoed	Not stored
2.308815e+008	Speed c_1 [m/s]	*(TranLineList->Speed)
1.184394e+008	Speed c_2 [m/s]	*(TranLineList->Speed + 1)
9.816166e+007	Speed c_3 [m/s]	*(TranLineList->Speed + 2)
--U matrix -- basis vectors col. by col.	Comment, not echoed	Not stored
0.998	Matrix $\mathbf{U}$ element u_{11}	*(TranLineList->UMatrix)
0.062	Matrix $\mathbf{U}$ element u_{21}	*(TranLineList->UMatrix + 1)
0.018	Matrix $\mathbf{U}$ element u_{31}	*(TranLineList->UMatrix + 2)
0.623	Matrix $\mathbf{U}$ element u_{12}	*(TranLineList->UMatrix + 3)
0.748	Matrix $\mathbf{U}$ element u_{22}	*(TranLineList->UMatrix + 4)
0.229	Matrix $\mathbf{U}$ element u_{32}	*(TranLineList->UMatrix + 5)
0.067	Matrix $\mathbf{U}$ element u_{13}	*(TranLineList->UMatrix + 6)
0.036	Matrix $\mathbf{U}$ element u_{23}	*(TranLineList->UMatrix + 7)
0.997	Matrix $\mathbf{U}$ element u_{33}	*(TranLineList->UMatrix + 8)

The results of an example problem are given. The three-wire transmission line in the problem has the cross section shown in Figure 10.3. The **L** and **C** matrices are given is Section 10.4, Computer Code 10-1. The eigenvalues and eigenvectors are also given in that section as the output of that computer code. A schematic of the sample problem is given in Figure 14.1. The input file is named VERYNON.IN (with computer code file ch14_1.c).

The load resistors (428, 167, and 73.5 [Ω]) are the terminal-to-ground resistors implied by the characteristic impedance matrix; the terminal-to-terminal resistors have been omitted to simulate the crosstalk effect of omitting these resistors. The output files as specified in the input file have been combined and truncated to save space; the output is listed below. The output is shown graphically in Figure 14.2. All the crosstalk in this case is termination crosstalk (because the line is lossless in this model).

	Left End Terminals [V]			Right End Terminals [V]		
Time [s]	Wire 1	Wire 2	Wire 3	Wire 1	Wire 2	Wire 3
1.000000e-010,	0.453166	0.037573	0.002666	0.000000	0.000000	0.000000
2.000000e-010,	0.906333	0.075145	0.005332	0.000000	0.000000	0.000000
3.000000e-010,	1.359499	0.112718	0.007999	0.000000	0.000000	0.000000
4.000000e-010,	1.812666	0.150291	0.010665	0.000000	0.000000	0.000000
5.000000e-010,	2.265832	0.187864	0.013331	0.488542	-0.027062	-0.007371
6.000000e-010,	2.718999	0.225436	0.015997	1.219046	-0.067528	-0.018392
7.000000e-010,	3.172165	0.263009	0.018663	1.949550	-0.107994	-0.029413
8.000000e-010,	3.625331	0.300582	0.021329	2.680054	-0.148459	-0.040434
9.000000e-010,	4.107174	0.316336	0.022329	3.415542	-0.182070	-0.050269
1.000000e-009,	4.617422	0.310477	0.021678	4.154998	-0.210225	-0.059160

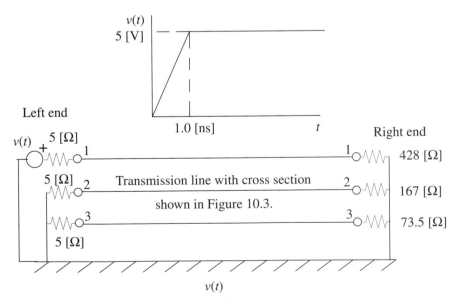

Figure 14-1 Schematic for Multi-Speed Line Example.

```
1.100000e-009,4.681539 0.261694  0.017951 4.895339 -0.236907 -0.075943
1.200000e-009,4.745654 0.212911  0.014225 5.635883 -0.263250 -0.094544
1.300000e-009,4.809600 0.160292  0.010526 6.253450 -0.282780 -0.111289
1.400000e-009,4.873326 0.102722  0.006861 6.688292 -0.292188 -0.125278
1.500000e-009,4.936978 0.043586  0.003422 6.544094 -0.269520 -0.130531
1.600000e-009,5.000625-0.015591  0.000139 6.142990 -0.232623 -0.131908
1.700000e-009,5.057053-0.069202-0.002700 5.733649 -0.212753 -0.136253
1.800000e-009,5.099183-0.111805-0.004708 5.309587 -0.223305 -0.145903
1.900000e-009,5.102019-0.124304-0.004570 4.871543 -0.259203 -0.155052
2.000000e-009,5.072947-0.112366-0.002717 4.427343 -0.304968 -0.159919
2.100000e-009,5.036878-0.094014-0.000712 4.027383 -0.334440 -0.151665
2.200000e-009,5.001878-0.074168 0.001116 3.733894 -0.327012 -0.135291
2.300000e-009,4.967950-0.049240 0.002791 3.659492 -0.302559 -0.115896
2.400000e-009,4.934582-0.018888 0.004316 3.809499 -0.284446 -0.097031
2.500000e-009,4.902898 0.011688 0.005464 4.059131 -0.267885 -0.079074
2.600000e-009,4.875950 0.038113 0.006167 4.324654 -0.240853 -0.060115
2.700000e-009,4.862249 0.053649 0.006413 4.598349 -0.184468 -0.037524
2.800000e-009,4.866968 0.054470 0.006265 4.886533 -0.091095 -0.011268
2.900000e-009,4.883663 0.045643 0.006180 5.173289  0.017668  0.010942
3.000000e-009,4.903467 0.034024 0.006397 5.427487  0.116189  0.025176
3.100000e-009,4.922304 0.021722 0.006664 5.587112  0.179147  0.032045
3.200000e-009,4.939402 0.008189 0.006292 5.601706  0.194778  0.031143
3.300000e-009,4.955021-0.006168 0.005408 5.503684  0.184184  0.024872
3.400000e-009,4.968946-0.019728 0.004546 5.363049  0.169838  0.016592
3.500000e-009,4.978886-0.029729 0.003884 5.218248  0.151387  0.008332
3.600000e-009,4.981608-0.033242 0.003397 5.076939  0.120116 -0.000261
3.700000e-009,4.976722-0.029977 0.002686 4.939326  0.075567 -0.009650
3.800000e-009,4.967276-0.022558 0.001304 4.809701  0.025313 -0.019378
3.900000e-009,4.956754-0.013942-0.000660 4.705939 -0.018619 -0.029082
4.000000e-009,4.947101-0.005670-0.002716 4.656127 -0.043262 -0.037329
4.100000e-009,4.938784 0.001940-0.004126 4.670282 -0.046943 -0.042219
4.200000e-009,4.931652 0.008875-0.004439 4.728265 -0.040428 -0.041740
4.300000e-009,4.925987 0.014631-0.004059 4.801568 -0.032775 -0.037315
4.400000e-009,4.922849 0.018105-0.003637 4.872561 -0.025731 -0.032107
4.500000e-009,4.923133 0.018456-0.003298 4.935783 -0.017785 -0.027365
4.600000e-009,4.926508 0.015988-0.002821 4.991576 -0.007960 -0.022915
4.700000e-009,4.931589 0.011922-0.002082 5.038542  0.002590 -0.017135
4.800000e-009,4.936953 0.007497-0.001169 5.069848  0.011102 -0.008287
4.900000e-009,4.941788 0.003382-0.000358 5.077403  0.014753  0.003017
5.000000e-009,4.945878-0.000293 0.000012 5.059999  0.012629  0.014231
5.100000e-009,4.949184-0.003500-0.000121 5.025907  0.006977  0.021942
5.200000e-009,4.951453-0.005963-0.000430 4.986732  0.001151  0.024033
5.300000e-009,4.952263-0.007248-0.000625 4.950110 -0.003615  0.022065
5.400000e-009,4.951456-0.007161-0.000678 4.918359 -0.008133  0.019130
5.500000e-009,4.949380-0.005967-0.000678 4.891720 -0.013065  0.016282
5.600000e-009,4.946687-0.004201-0.000698 4.871532 -0.018069  0.012747
```

```
5.700000e-009,4.943968-0.002339-0.000707 4.860783 -0.022115  0.007993
5.800000e-009,4.941560-0.000641-0.000601 4.861960 -0.023864  0.002582
5.900000e-009,4.939586 0.000815-0.000304 4.874342 -0.022665 -0.002020
6.000000e-009,4.938129 0.002000 0.000151 4.893749 -0.019321 -0.004150
```

14.2.2 Lossy Multi-Speed Lines

The input change for lossy lines is simply the insertion of the resistance per length [Ω/m] of each wire following the $\mathbf{Z}_O$ matrix (just above the key word [MultiSpeed]). See the input file named VERYNON.IN (with computer code file ch14_2.c). Modifications to the structure for transmission lines (type TRANSMLINES) are: inclusion of the vectors of resistances (Rwires), the matrices $\mathbf{A}'$ and $\mathbf{B}'$ (Aprime and Bprime) and the vector of spatial increments for each mode (deltax). The matrices $\mathbf{A}'$ and $\mathbf{B}'$ are calculated using the wire resistances and the matrices $\mathbf{Z}_O$, $\mathbf{U}$, and $\mathbf{U}^{-1}$.

The major algorithmic change from the preceding lossless case is the inclusion of the subroutine TransportMultiSpeed. This subroutine modifies the voltage signals according to the multispeed transport difference Equation (12-29) prior to propagation (by calling subroutine PropTstep for each mode). This procedure is complicated by the fact that the

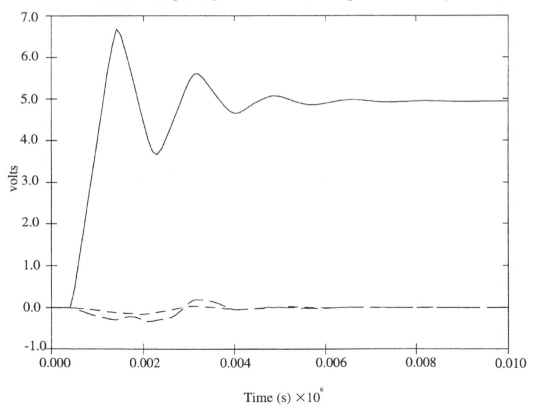

Figure 14-2 Graph of the Terminal Voltages at the Right End.

departing signals must be both modified (they appear on the left side of Equation (12-29)) *and* be used as a source for modifications (they also appear on the right side of Equation (12-29)) and they have not yet been stored in the propagation memory circle. All the other required signals are in the propagation memory circle. Hence, the inputs to `TransportMultiSpeed` are three pointers to: the vectors of departing mode voltages at the left and right ends and the pointer to the appropriate transmission line link (that contains as a member the propagation memory circle). The subroutine `TransportMultiSpeed` calls the subroutine `Interp-PropCircle` to perform the interpolation of mode *i* voltages indicated in Equation (12-30). The subroutine `InterpPropCircle` has the following argument parameters:

Input:

Length	Length of transmission line
xn$x$	Coordinate for interpolated value (mode *n* coordinate)
dxi$x$	Increment in mode *i* (`TStep` * `Speed` of mode *i*)
thi2 * Nhalf, =	Length of memory circle for mode *i*
TStepOff$i$	Time step offset in mode *i* memory circle
offset$i$	Offset to mode *i* memory circle
PropCircleTemp	Pointer to Temporary copy of `PropCircle` for mode *i*
VFLTemp[imode]	Departing left voltage for mode *i*
VBRTemp[imode]	Departing right voltage for mode *i*

Output:

VFinterp	Pointer to forward interpolated voltage (mode *i*)
VBinterp	Pointer to backward interpolated voltage (mode *i*)

A listing of the subroutine `TransportMultiSpeed` is given below.

```
/* ------ Procedure TransportMultiSpeed ----%proc%--------------------
   Description: Modifies mode voltage signal according to multi-speed
           transport equation (12-29) prior to propagation using
           PropTstep for each mode. The departing signals are
           modified as well as the en route signals stored in
           PropCircle.
   Procedures Called
   -----------------------------------------------------------------

   InterpPropCircle              Interpolate propagating voltage on
                                 PropCircleTemp.
   -------FunctionDeclaration-------------------------------------- */
void TransportMultiSpeed (float *VFModesL,
                          float *VBModesR,
                          TRANSMLINES_PTR TranLineList)
{
/* -------- Argument Variables ------------------------------------
   Type        Name                 I/O Description & range
   ----        ----                 --- ------------------------------
   float       *VFModesL            i/o Mode-signal (vector) departing from
                                        left end, but not yet loaded in
                                        propagation memory circle.
```

```
  float      *VBModesR        i/o  Mode-signal (vector) departing from
                                   right end, but not yet loaded in
                                   propagation memory circle.
  TRANSMLINES_PTR TranLineList  i  Pointer to transmission line link
                                   for transmission line whose transport
                                   is being updated.
  --- Local Variable Declarations -----------------------------------
  Type        Name            Description & range
  ----        ----            ------------------------------------ */
  int        memloc;            // number of locations needed for
                                // PropCircleTemp
  int        offsetn, offseti;  // offsets to memory circle
  int        TStepOffn,TStepOffi;// time step offsets
  float      *PropCircleTemp = NULL; // temp. copy of PropCircle
  float      *VFLTemp = NULL;    // Temp. forward signals
  float      *VBRTemp = NULL;    // Temp. backward sigals
  float      VFinterp, VBinterp; // interpolated values
  int        idx;               // for-loop index
  int        num;               // number of modes
  int        nmode;             // index over modes
  int        imode;             // index over modes
  int        Nhalfn;            // half number in PropCirle mode n
  int        thn;               // twice Nhalfn
  float      xn;                // x coord. mode n
  int        Nhalfi;            // half number in PropCirle mode i
  int        thi;               // twice Nhalfi
  int        nx;                // index of x in nmode
  float      dxn;               // delta x, mode n
  float      dxi;               // delta x, mode i
/*----------------------- Error Conditions Trapped ------------------
  Fatal/Nonfatal    Description
  -------------      -----------
  none
  ----------------ExecutableCode----------------------------------------*/
    num = TranLineList->NumWires;  // number of modes
    // Create temporary PropCircle copy so that the r.h. side of
    // Eq. 12-29 can be used immediately to augment values in
    // PropCircle.
    memloc = 0;
    for (nmode = 0; nmode < num; nmode++)
      memloc += 2 * *(TranLineList->Nhalf + nmode);
    PropCircleTemp = (float*)malloc(sizeof(float) * memloc);
    for (idx = 0; idx < memloc; idx++)
      *(PropCircleTemp + idx) = *(TranLineList->PropCircle + idx);
    // Create temporary copy of signal departing left end.
    VFLTemp = (float*)malloc(sizeof(float) * num);
```

```
for (idx = 0; idx < num; idx++)
   VFLTemp[idx] = VFModesL[idx];
// Create temporary copy of signal departing right end.
VBRTemp = (float*)malloc(sizeof(float) * num);
for (idx = 0; idx < num; idx++)
   VBRTemp[idx] = VBModesR[idx];
// Loop over each mode to be augmented by Aprime and Bprime matrices.
// Transport difference Eq. 12-29
offsetn = 0;
for (nmode = 0; nmode < num; nmode++)
{
   Nhalfn = *(TranLineList->Nhalf + nmode);
   thn = 2 * Nhalfn;
   dxn = *(TranLineList->deltax + nmode);
   TStepOffn = *(TranLineList->TStepOffset + nmode);
   // Loop over each modal contribution to r.h. side of 12-29
   offseti = 0;
   for (imode = 0; imode < num; imode++)
   {
      Nhalfi = *(TranLineList->Nhalf + imode);
      thi = 2 * Nhalfi;
      dxi = *(TranLineList->deltax + imode);
      TStepOffi = *(TranLineList->TStepOffset + imode);
      //*** Departing signal, left end ***
      // (propagates next from x = 0 to deltax, mean is dxn/2)
      xn = dxn/2;
      InterpPropCircle ( // Input variables
                  TranLineList->Length, // line length
                  xn,                // x coord. mode n
                  dxi,               // x increment mode i
                  thi,               // 2 * Nhalf, mode i
                  TStepOffi,         // time step offset mode i
                  offseti,           // offset to mode i mem. circ.
                  PropCircleTemp,    // Temp. PropCircle
                  VFLTemp[imode],    // departing left
                  VBRTemp[imode],    // departing right
                  // output variables
                  &VFinterp,         // Forward interpolated
                  &VBinterp);        // Backward interpolated
      *(VFModesL + nmode) -= (*(TranLineList->Aprime + nmode
         + num * imode) * VFinterp - *(TranLineList->Bprime + nmode
         + num * imode) * VBinterp) * dxn;
      //*** Departing signal, right end ***
      // (propagates next from x = length to (length - deltax))
      xn = TranLineList->Length - dxn/2;
      InterpPropCircle ( // Input variables
```

```
                    TranLineList->Length, // line length
                    xn,                // x coord. mode n
                    dxi,               // x increment mode i
                    thi,               // 2 * Nhalf, mode i
                    TStepOffi,         // time step offset mode i
                    offseti,           // offset to mode i mem. circ.
                    PropCircleTemp,    // Temp. PropCircle
                    VFLTemp[imode],    // departing left
                    VBRTemp[imode],    // departing right
                        // output variables
                    &VFinterp,         // Forward interpolated
                    &VBinterp);        // Backward interpolated
        *(VBModesR + nmode) -= (*(TranLineList->Aprime + nmode
          + num * imode) * VBinterp - *(TranLineList->Bprime + nmode
          + num * imode) * VFinterp) * dxn;
        //*** Locations alone the line ***
        // Loop over locations
        for (nx = 1; nx < Nhalfn; nx++)
        {
            // Forward signal
            // propagates next from x = nx * deltax to
            //                      x = (nx + 1)*deltax
            xn = dxn/2 * (2 * nx + 1);
            InterpPropCircle ( // Input variables
                    TranLineList->Length, // line length
                    xn,                // x coord. mode n
                    dxi,               // x increment mode i
                    thi,               // 2 * Nhalf, mode i
                    TStepOffi,         // time step offset mode i
                    offseti,           // offset to mode i mem. circ.
                    PropCircleTemp,    // Temp. PropCircle
                    VFLTemp[imode],    // departing left
                    VBRTemp[imode],    // departing right
                    // output variables
                    &VFinterp,         // Forward interpolated
                    &VBinterp);        // Backward interpolated
            *(TranLineList->PropCircle + offsetn
                + ((TStepOffn - nx + thn) % thn))
              -= (*(TranLineList->Aprime + nmode
            + num * imode) * VFinterp - *(TranLineList->Bprime + nmode
            + num * imode) * VBinterp) * dxn;
            // Backward signal
            xn = TranLineList->Length - xn;
            InterpPropCircle ( // Input variables
                    TranLineList->Length, // line length
                    xn,                // x coord. mode n
```

```
                    dxi,                // x increment mode i
                    thi,                // 2 * Nhalf, mode i
                    TStepOffi,          // time step offset mode i
                    offseti,            // offset to mode i mem. circ.
                    PropCircleTemp,     // Temp. PropCircle
                    VFLTemp[imode],     // departing left
                    VBRTemp[imode],     // departing right
                    // output variables
                    &VFinterp,          // Forward interpolated
                    &VBinterp);         // Backward interpolated
            *(TranLineList->PropCircle + offsetn
               + ((TStepOffn + Nhalfn - nx + thn) % thn))
               -= (*(TranLineList->Aprime + nmode
            + num * imode) * VFinterp - *(TranLineList->Bprime + nmode
               + num * imode) * VBinterp) * dxn;
         }
         offseti += thi;
      }   // end of loop over imodes
      offsetn += thn;
   }   // end of loop over nmodes
   // Free temporary memory
   free (PropCircleTemp);       PropCircleTemp = NULL;
   free (VFLTemp);              VFLTemp = NULL;
   free (VBRTemp);              VBRTemp = NULL;
}
/*-----End of Procedure TransportMultiSpeed------------------------*/
```

The network shown in Figure 14.1 was used again; this time to illustrate use of the lossy algorithm and code. A trace resistance of 100 $[\Omega/m]$ was assumed; all other parameters were the same. Again the output files were combined and truncated to condense the output to that shown below. Figure 14.3 shows the output graphed.

	Left End Terminals [V]			Right End Terminals [V]		
Time [s]	Wire 1	Wire 2	Wire 3	Wire 1	Wire 2	Wire 3
1.000000e-010,	0.453166	0.037573	0.002666	0.000000	0.000000	0.000000
2.000000e-010,	0.904659	0.076905	0.005295	0.000000	0.000000	0.000000
3.000000e-010,	1.354242	0.118022	0.007937	0.000000	0.000000	0.000000
4.000000e-010,	1.802472	0.160689	0.010529	0.000000	0.000000	0.000000
5.000000e-010,	2.248580	0.205469	0.013094	0.448269	-0.024311	-0.006671
6.000000e-010,	2.693273	0.251707	0.015618	1.093505	-0.054332	-0.016136
7.000000e-010,	3.135806	0.300127	0.018154	1.744953	-0.075627	-0.025193
8.000000e-010,	3.577651	0.349605	0.020548	2.421015	-0.089486	-0.034127
9.000000e-010,	4.048785	0.377167	0.021109	3.128982	-0.089820	-0.041908
1.000000e-009,	4.547800	0.383444	0.020082	3.860452	-0.083801	-0.049856
1.100000e-009,	4.601648	0.345413	0.016047	4.612911	-0.080900	-0.067272

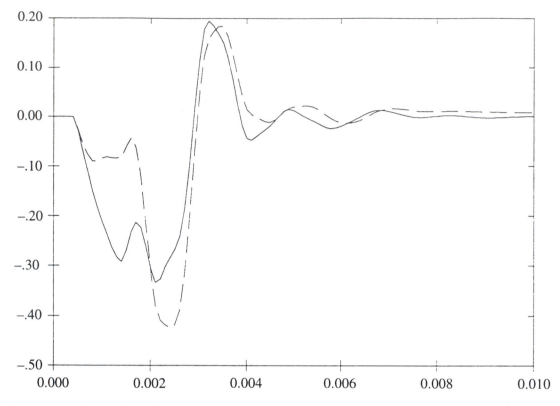

Figure 14-3 Right-End Node 2 Voltages.

```
1.200000e-009,4.659483 0.302967 0.012080 5.378489 -0.083032 -0.086745
1.300000e-009,4.721886 0.251911 0.008186 6.039633 -0.083805 -0.104985
1.400000e-009,4.788244 0.192020 0.004511 6.548242 -0.082188 -0.121361
1.500000e-009,4.859850 0.125987 0.001313 6.530651 -0.060900 -0.130410
1.600000e-009,4.935637 0.056051-0.001804 6.287995 -0.043196 -0.136831
1.700000e-009,5.008033-0.012118-0.004724 6.005012 -0.061817 -0.147273
1.800000e-009,5.068243-0.071696-0.006956 5.658177 -0.123456 -0.162418
1.900000e-009,5.089341-0.101849-0.007173 5.250229 -0.214275 -0.175022
2.000000e-009,5.078481-0.107486-0.005893 4.800619 -0.309905 -0.181195
2.100000e-009,5.058090-0.104049-0.004510 4.360366 -0.380249 -0.172094
2.200000e-009,5.034449-0.095014-0.003223 3.991552 -0.408531 -0.154299
2.300000e-009,5.006667-0.076322-0.002056 3.800813 -0.417098 -0.132837
2.400000e-009,4.974113-0.048103-0.001010 3.796873 -0.423873 -0.111511
2.500000e-009,4.937968-0.015639-0.000287 3.888675 -0.418167 -0.091274
2.600000e-009,4.902192 0.016334 0.000461 4.025537 -0.387429 -0.069233
2.700000e-009,4.876843 0.040037 0.001402 4.214334 -0.314356 -0.041992
2.800000e-009,4.868073 0.050884 0.002577 4.459507 -0.198255 -0.010641
2.900000e-009,4.870893 0.052717 0.004245 4.739002 -0.069088  0.016062
```

```
3.000000e-009,4.878391 0.050741 0.006305 5.017717  0.045805  0.034815
3.100000e-009,4.887874 0.045866 0.008254 5.233426  0.124737  0.046077
3.200000e-009,4.899238 0.037145 0.009366 5.338726  0.160416  0.049272
3.300000e-009,4.913087 0.024820 0.009853 5.357466  0.174282  0.047114
3.400000e-009,4.929093 0.010449 0.010124 5.336545  0.183708  0.043101
3.500000e-009,4.944231-0.002814 0.010115 5.293327  0.183339  0.039008
3.600000e-009,4.954385-0.011480 0.009725 5.225811  0.163936  0.032629
3.700000e-009,4.958225-0.014630 0.008566 5.132170  0.127046  0.022547
3.800000e-009,4.957591-0.013994 0.006351 5.018024  0.081850  0.009201
3.900000e-009,4.954796-0.011565 0.003446 4.903851  0.040607 -0.006406
4.000000e-009,4.950976-0.008184 0.000539 4.819018  0.014738 -0.021671
4.100000e-009,4.946126-0.003716-0.001614 4.775949  0.004124 -0.033829
4.200000e-009,4.939911 0.002063-0.002639 4.763414 -0.000911 -0.040814
4.300000e-009,4.932782 0.008603-0.002975 4.766088 -0.005974 -0.044352
4.400000e-009,4.926241 0.014535-0.003177 4.777067 -0.010139 -0.046766
4.500000e-009,4.921760 0.018628-0.003234 4.796905 -0.010714 -0.047828
4.600000e-009,4.919700 0.020678-0.002933 4.828222 -0.006326 -0.046598
4.700000e-009,4.919435 0.021304-0.002254 4.869212  0.001710 -0.041318
4.800000e-009,4.920202 0.021170-0.001389 4.911335  0.010351 -0.030877
4.900000e-009,4.921647 0.020488-0.000668 4.944185  0.016692 -0.017054
5.000000e-009,4.923806 0.019093-0.000403 4.962744  0.019586 -0.003195
5.100000e-009,4.926691 0.016898-0.000580 4.969432  0.020437  0.007308
5.200000e-009,4.929917 0.014223-0.000856 4.969213  0.021151  0.012784
5.300000e-009,4.932800 0.011672-0.001014 4.964286  0.021501  0.014833
5.400000e-009,4.934816 0.009711-0.001082 4.953853  0.019880  0.015773
5.500000e-009,4.935876 0.008417-0.001143 4.937185  0.015521  0.015704
5.600000e-009,4.936201 0.007620-0.001243 4.916055  0.008973  0.013674
5.700000e-009,4.936024 0.007165-0.001316 4.894725  0.001630  0.009606
5.800000e-009,4.935446 0.007032-0.001240 4.877681 -0.004755  0.004568
5.900000e-009,4.934470 0.007297-0.000957 4.866996 -0.009134  0.000271
6.000000e-009,4.933174 0.007981-0.000523 4.861946 -0.011746 -0.001767
```

PART IV

TRANSMISSION LINE PARAMETER DETERMINATION

Chapter **15**

INTRODUCTION TO TRANSMISSION LINE PARAMETER DETERMINATION

The algorithms developed so far in this book have little value unless the transmission line capacitance and inductance matrices can be obtained for the transmission lines of interest. The focus of this part of the book is on a method to calculate these two parameters for transmission lines consisting of parallel traces embedded in or on the surface of multi-dielectric-layer circuit boards. Methods will be developed that are applicable to layers over a single ground plane and between two ground planes.

Calculation of the capacitance matrix of parallel traces embedded in and/or on the surface of a multi-dielectric-layer circuit board is a challenging task. The capacitance matrix is defined in terms of the charges on the conductors. Its evaluation requires that the charge on each trace be calculated under conditions to be defined below (i.e., the potentials of the traces are to be set in certain ways). The charge on a given trace is the surface integral of the surface charge density on the (surface of the) trace. Hence, the required fundamental algorithm must be one that determines (or approximates) the surface charge density at each point on the surface of each trace with the potentials of the traces held at various values.

Mathematically, there are several ways one can express the surface charge density as a function of position on the trace surface. One method is to express the charge density as an infinite series. (Using this method, the constant coefficients in the series are the unknowns that must be determined to obtain the charge density.) In this book, a more direct approach is taken. The surface of each trace is divided into segments. The charge density on each segment is assumed uniform over the segment. A set of linear equations is derived where the vector of unknowns consists of the list of charges on the segments. Solution of the linear equations yields the vector of segment charges. The sum of the charges on the segments of a given trace yields the charge on that trace.

The accuracy of the calculation depends both on the number of segments used and their relative sizes on different parts of the trace. The charge density is not uniformly distributed on the trace surface. To accurately describe (or resolve) the variation of the charge density, many segments are needed in regions where the charge density varies rapidly. In regions where the charge density varies slowly, few segments are needed. The solution algorithm developed will allow complete flexibility in the relative sizes of the segments. Defining clever ways to define relative segment sizes to improve calculational efficiency is left to the reader. (Using many very small segments will yield high accuracy, but at a large computational cost.) The example code given simply divides the horizontal and vertical surfaces of each trace into equal segments; the numbers of segments in each of the directions may be input by the user.

The charge density at any point on the surface of a trace is given by the normal component of the electric displacement vector **D** at the given point. The vector **D** is related to the electric field vector **E** by the equation

$$\mathbf{D} = \varepsilon\mathbf{E} \quad [\text{C/m}^2] \tag{15-1}$$

where ε is the permittivity of the (homogeneous) dielectric [F/m].

The determination of the vector **D** (and, hence, the surface charge density on the trace) therefore becomes one of determining the electric field vector **E** at the surface of the trace. The (static) electric field **E** can be written as minus the gradient of a scalar potential function v defined throughout the space.

$$\mathbf{E} = -\nabla v \quad [\text{V/m}] \tag{15-2}$$

where ∇ is the gradient operator. In two dimensions it is $\nabla \equiv \mathbf{i}\dfrac{\partial}{\partial x} + \mathbf{j}\dfrac{\partial}{\partial y}$. The unit vectors **i** and **j** are in the x and y directions, respectively.

There are many methods of determining the potential function v. This function must satisfy the Laplace equation throughout the problem domain. That is, it must satisfy the partial differential equation (Laplace's equation)

$$\frac{\partial^2 v}{\partial x^2} + \frac{\partial^2 v}{\partial y^2} = 0 \tag{15-3}$$

The potential v must have specified values on the surfaces of the traces and the value zero on the ground plane(s). Methods of solution of this type of problem are discussed by many authors (see Jackson, 1975, Chapter 2, Morse and Feshbach, 1953, Chapter 10, and Smythe, 1950, Chapter IV). The method used in this book avoids calculating v throughout the (two-dimensional) space surrounding the traces; it uses the potential of a point charge as a Green's function to calculate the potential (or its gradient when needed) at trace surfaces or dielectric boundaries. The method is derived in the following chapters.

This part of the book proceeds as follows. First, the algorithm for a single trace in a homogeneous medium over a ground plane is derived. A computer code is presented and explained that does this single-trace problem. Then the method is generalized to multiple traces in a homogeneous medium. A computer code is used to illustrate and implement this algorithm. A method to determine the effect of layered dielectrics is then developed and the homogeneous medium algorithm is extended to a layered dielectric circuit board over a ground plane. Again, a computer code is given to implement the more general algorithm. Finally, the method is modified to include traces embedded in dielectric layers between two ground planes.

Simulation of the inductance matrix is also defined. Because the dielectric layers do not affect the inductance, it can always be calculated from the capacitance as found for the homogeneous case. If the circuit board is layered, a special calculation of capacitance is performed as if the dielectric were not layered. The inductance is then calculated from this capacitance matrix.

Chapter 16
CAPACITANCE AND INDUCTANCE IN A HOMOGENEOUS MEDIUM

16.1 SINGLE-TRACE CAPACITANCE AND INDUCTANCE SIMULATION

The algorithm for the capacitance of a single trace is very similar to that for multiple traces. This is not obvious because the single-trace capacitance is, of course, a scalar while the multi-trace capacitance is a matrix. The similarity is in calculating the charges on the trace surfaces (discussed below); the linear equation matrix for this calculation is the same in both cases. The single trace algorithm is developed first.

16.1.1 Definition of Capacitance

The capacitance (in farads [F]) of any conducting body is the ratio of the electrical charge on the body (in coulombs [C]) to the electrical potential of the body (in volts [V]) for the static field case. This can be written as

$$C = \frac{q_b}{v_b} \quad [\text{F}] \tag{16-1}$$

Because the body is a *conducting* body, the potential of every point on its surface is the same; i.e., its surface is an *equi-potential* surface. Formally, if we let the set of points on the surface of the body be Γ, we can write

$$v(\mathbf{x}) = v_b, \qquad \mathbf{x} \in \Gamma \tag{16-2}$$

where $\mathbf{x} = (x, y, z)$, the coordinate vector of any point; if $\mathbf{x} \in \Gamma$, then the point (x, y, z) is defined to be on the surface of the body. For transmission lines, we need the capacitance *per meter length* of a wire or trace; the potentials are calculated in the plane perpendicular to the transmission line's axis or direction of propagation (i.e., in the *x-y* plane). This is what's needed for TEM (transverse electromagnetic) or QS (quasi-static) propagation.[1] To calculate

---

1. Other modes of propagation can exist on a transmission line. However, when signal wavelengths are long compared to the separation of the conductors, all the other modes attenuate rapidly as they propagate due to the waveguide "cut-off frequency" effect. TEM is the dominant propagation mode on a transmission line. See Jackson, 1975, Chapter 8, especially Section 8.2.

the capacitance, the electrostatic field problem will be solved with the surface of the trace at 1 [V] potential; under this condition, the capacitance is simply the calculated charge (see Figure 16.1).

From each tiny element of charge, an electric field line emerges. Each electric field line follows an (in general, curved) path from an element of positive charge to a corresponding element of negative charge on which the electric field line ends. For charges on the surface of a rectangular trace (near a ground plane) the electric field lines resemble those shown in Figure 16.2. In this case, there is a negative charge distributed over the surface of the ground plane that is exactly equal, but of opposite sign, to the positive charge on the trace.

The charge is not uniformly distributed on either the trace or the ground plane. It distributes in such a way that the potential resulting from the combined effect of all the charges is a uniform potential on the surface of the trace and on the surface of the ground plane. A method will be given to simulate the charge distribution that results in the uniform potential.

PROBLEM 16.1   A trace is 10 [cm] long. It's measured capacitance is 10 [pF]. If the trace is held at a potential of 1 [V], what is the charge per length [C/m]? What is its capacitance in MKS units [C/m]?

16.1.2 The Capacitance of a Single Trace Over a Ground Plane

Suppose the surface charge density on the trace is $\rho_s(\mathbf{x})$ [C/m²], where, $\mathbf{x} = (x,y)$, are points on the surface of the trace. Then, the total charge on the trace can be written as the line integral

$$q = \int_\Gamma \rho_s(\mathbf{x})ds \ \ [\text{C/m}] \tag{16-3}$$

where the path of the integral is on Γ, the surface of the trace (see Figure 16.3).

To perform a numerical simulation of the charge on the trace, the charge will be defined on small, but finite, line segments on the surface of the trace. An illustration of such segmentation is shown in Figure 16.4 where a trace has been divided into 10 segments. For this trace $j_\Gamma = 1, 2, \ldots 10$. Note that the segments do not have equal lengths. This is intentional. An efficient calculation will have long segments in regions where the surface charge density is relatively uniform; this normally occurs near the middle of the top and base of a trace. Near the corners, the charge density varies rapidly; to resolve this behavior small segments are needed. The number of segments to be used depends on the required accuracy of the simulation.

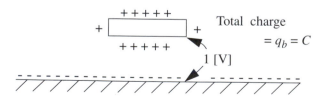

Total charge

$= q_b = C$

1 [V]

Figure 16.1   Charge on a Single Trace at 1 [V] Potential.

Figure 16.2 Schematic of Electric Field near a Trace over a Ground Plane.

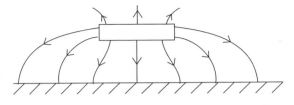

The surface charge density is assumed to be uniform over each small segment. The charge on a typical segment j_Γ is

$$q_{j_\Gamma} = \delta l_{j_\Gamma} \rho_{s\,j_\Gamma} \tag{16-4}$$

where δl_{j_Γ} is the length of the segment identified by the index j_Γ and $\rho_{s\,j_\Gamma}$ is the surface charge density on that segment. Using this finite segment (or finite element) approximation, the total charge on the trace is approximated by

$$q = \sum_{j_\Gamma} q_{j_\Gamma} \tag{16-5}$$

In the next Section, the potential at an arbitrary point due to a unit charge (1 [C]) uniformly distributed over the j_Γ-th segment of Γ) will be derived. Let this potential at the arbitrary point $\mathbf{x}_i = (x_i, y_i)$, due to the charged segment j_Γ, be given by $v_{i\,j_\Gamma} \equiv v(\mathbf{x}_i, \mathbf{x}_{j_\Gamma})$. Then, the potential at point i due to the combined effect of all the segments that comprise Γ (i.e., that lie on the surface of the trace) is[2]

$$w_i = \sum_{j_\Gamma} v_{i\,j_\Gamma} q_{j_\Gamma} \tag{16-6}$$

The algorithm for simulating the capacitance of the trace can now be defined. The first step is to find the charge q_{j_Γ} on each segment when the potential of all of the segments is 1 [V]. Then, the capacitance C is given by the sum of these charges. Let the number of seg-

Figure 16.3 Path of Integral over the Surface of a Trace.

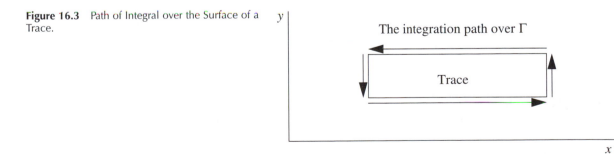

The integration path over Γ

Trace

---

2. The symbol w will be used for potentials due to the combined effect of charged segments (to avoid confusion with v_{ij}, the potential at a point due to the charge on a single segment).

Figure 16.4 Illustration of Segmented Trace (Segments Numbered).

ments on the trace be N_Γ. Then there are N_Γ unknown segment charges. Suppose Equation (16-6) is written N_Γ times where the points $\mathbf{x}_i$ are all chosen to be on Γ; e.g., at the centers of the segments. Then, let $w_i = 1$ [V] for each of the equations. This assures that the potential of all the segments will be near 1 [V]. The resulting set of simultaneous linear equations can be written

$$1 = \mathbf{V}\mathbf{q}_\Gamma \qquad (16\text{-}7)$$

where $\mathbf{1}$ is the column vector of all 1s, $\mathbf{V}$ is the matrix with elements v_{ij_Γ}, and $\mathbf{q}_\Gamma$ is the column vector whose elements are the charges q_{j_Γ}. The vector of segment charges is then given by the solution of the linear equations defined in Equation (16-7). The solution can be written

$$\mathbf{q}_\Gamma = \mathbf{V}^{-1}\mathbf{1} \qquad (16\text{-}8)$$

This solution can be evaluated using `SimpleGauss` or another simultaneous equation solver (see discussion of matrix software in Section 4.1).

The method derived above is numerically efficient. It depends on the solution of an N_Γ by N_Γ set of linear equations, where N_Γ is the number of segments on the (one-dimensional) *surface* of the conductor. The result is the charge distribution on the surface of the trace needed to calculate the capacitance; this is exactly the information needed and no more. Other methods depend on the solution of Laplace's equation throughout the two-dimensional *area* surrounding the trace. This usually requires a segmentation or "gridding" of the two-dimensional area and solution of Laplace's equation throughout the area; hence, many unneeded potentials are calculated. The efficiency attained is through the use of the known analytic solution of Laplace's equation for a point charge ($v(\mathbf{x},\mathbf{x}')$ in Equation (16-10)) or for a charged segment (v_{ij_Γ} in Equation (16-6), to be derived below).

PROBLEM 16.2 A trace is divided into 10 segments (as in Figure 16.4). How many potential functions must be evaluated to calculate its capacitance? (That is, how many elements are in the matrix V?) If it were divided into 100 segments, how many potential functions would be required?

PROBLEM 16.3 Suppose the capacitance is approximated by using only one segment per side, for a total of four segments. Write out explicitly the matrix V and the four linear equations that the matrix Equation (16-7) represents.

16.1.3 Integral Equation for the Charge

If the length of each of the segments that comprise Γ is allowed to approach zero while the number of them increases to infinity, the summation in Equation (16-6) becomes an integral; Equation (16-6) becomes

$$w(\mathbf{x}) = \int_{\Gamma} v(\mathbf{x},\mathbf{x}')\rho_s(\mathbf{x}')\,dl' \qquad (16\text{-}9)$$

where $v(\mathbf{x},\mathbf{x}')$ is the potential at point $\mathbf{x}$ due to a unit point charge at point $\mathbf{x}'$. The integration is a line integral on the surface of the trace. If the points $\mathbf{x}$ are on Γ and the potential of Γ is set to 1 [V], then $w(\mathbf{x}) = 1$, and Equation (16-9) becomes the integral equation

$$1 = \int_{\Gamma} v(\mathbf{x},\mathbf{x}')\rho_s(\mathbf{x}')\,dl', \qquad \mathbf{x} \in \Gamma \qquad (16\text{-}10)$$

A solution of this equation yields the exact surface charge density on the trace, whereas the solution of Equation (16-8) is a finite-segment numerical approximation. However, there is no known general analytic solution of Equation (16-10). There are various ways of approximating solutions of this integral equation; one of them is the finite-segment approximation derived above.

16.1.4 Potential and Electric Field of a Uniformly Charged Segment

The potential due to a unit point charge (1 [C/m] in two-dimensional space) can be written as[3]

$$v(\mathbf{x},\mathbf{x}') = \frac{-1}{2\pi\varepsilon_o}\ln r \qquad (16\text{-}11)$$

where, $r = \sqrt{(x-x')^2 + (y-y')^2}$ is the distance between the unit charge and the point where the potential is observed and ε_o is the permittivity of free space. This is illustrated in Figure 16.5. Note that a "point charge" in a two-dimensional geometry is physically a linear charge density in the third direction (the z direction using the coordinate system discussed here). Hence, its dimensions are [C/m].

The potential given in Equation (16-11) can be applied directly to geometries with circular cylindrical symmetry (such as a coaxial cable geometry) or it can be integrated over a charge distribution as in Equation (16-9). When used in an integral of the form of Equation (16-9), $v(\mathbf{x}, \mathbf{x}')$ is called the *kernel* of the integral. In the present application, a solution of Laplace's equation is required, where the source of the field is the charge distribution on the boundary of the domain of the solution. In this situation, the function $v(\mathbf{x}, \mathbf{x}')$ is also called a *Green's function* (see Morse and Feshbach, 1953, Chapter 7 for a discussion of this method of solution of partial differential equations). Before performing the integration, it is beneficial to discuss reference points or surfaces.

---

3. See Smythe, 1950, p. 63.

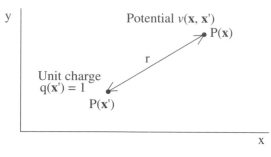

Figure 16.5 Potential Due to Point Charge in Two Dimensions.

The potential given in Equation (16-11) is a formal solution of Laplace's equation (for a point charge) but has little usefulness unless it is algebraically manipulated so as to define a *reference point* or *reference surface;* the reference point or surface is defined to have potential zero. Note that $v(\mathbf{x},\mathbf{x}')$ equals zero on the surface $r = 1$, a circular cylinder with center at the point charge. But in what units of r is this valid? Is zero potential automatically assigned to $r = 1$ meter? Not so! One might simply assert, "I have an equi-potential cylinder whose potential I wish to assign the value zero. I'll measure radii in units of my potential-equals-zero radius." Smart fellow — that will work, of course. Usually, though, the problem is approached algebraically. Any constant can be added to a potential without changing its validity; the constant is a trivial solution of Laplace's equation and any two (or more) solutions can be added to obtain another solution of Laplace's equation. Hence, consider the constant $\dfrac{1}{2\pi\varepsilon_o}\ln R$, where R is a radius that is to have the reference potential zero. Adding this to both sides of Equation (16-11), one obtains the new potential $v'(\mathbf{x},\mathbf{x}')$.

$$v'(\mathbf{x},\mathbf{x}') \;=\; v(\mathbf{x},\mathbf{x}') + \frac{1}{2\pi\varepsilon_o}\ln R \;=\; \frac{-1}{2\pi\varepsilon_o}\ln\frac{r}{R} \qquad (16\text{-}12)$$

Clearly, the new potential is zero when $r = R$. This formula is useful in calculating the capacitance of two concentric cylinders, the geometry of a coaxial cable.

The present concern is with the potential of charged segments over a ground plane. Image theory provides a convenient tool to obtain a potential whose value is zero on the ground plane. Therefore, the function $v(\mathbf{x},\mathbf{x}')$ of Equation (16-11) will be integrated over the charge on a segment of the trace and then the method of images will be applied to obtain the potential of the segment including the effect of the ground plane.

Consider the segment shown in Figure 16.6. It is assumed to have a uniform surface charge density ρ_s. The formal potential due to this charge distribution is given by Equation (16-9) with Equation (16-11) substituted for $v(\mathbf{x},\mathbf{x}')$ and ρ_s given by the *constant* surface charge on this segment (Γ defined as the segment of interest). Because ρ_s is constant, it can be brought out from under the integral sign. A second argument has been added to v to indicate that the center of the charged segment is at the point $\mathbf{x}_s = (x_s, y_s)$ and the subscript $s \parallel x$ has been added to indicate that the charged segment is parallel to the x axis. The result is Equation (16-13).

$$v_{s \parallel x}(\mathbf{x}, \mathbf{x}_s) = \frac{-\rho_s}{2\pi\varepsilon_o} \int_{x_s - l/2}^{x_s + l/2} \ln \sqrt{(x' - x)^2 + (y_s - y)^2} \, dx' \qquad (16\text{-}13)$$

If the segment of width l has unit charge (per length z **direction**) on it, the surface charge density is $\rho_s = 1/l$ [C/m^2] and Equation (16-13) becomes

$$v_{s \parallel x}(\mathbf{x}, \mathbf{x}_s) = \frac{-1}{2\pi\varepsilon_o l} \int_{x_s - l/2}^{x_s + l/2} \ln \sqrt{(x' - x)^2 + (y_s - y)^2} \, dx' \qquad (16\text{-}14)$$

The potential due to this segment can be conveniently expressed in terms of the indefinite integral

$$U(X, Y) = \frac{-1}{2\pi\varepsilon_o l} \int \ln r \, dX \qquad (16\text{-}15)$$

where

$$r = \sqrt{X^2 + Y^2}$$

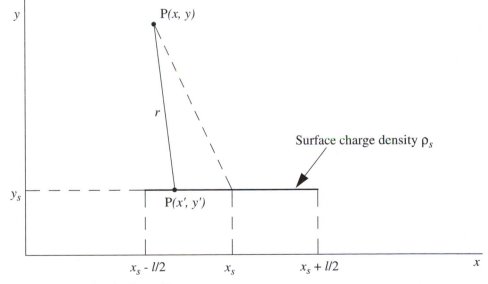

Figure 16.6   Uniformly Charged Segment.

The integral in Equation (16-15) can be evaluated analytically to yield

$$U(X,Y) = \frac{-1}{4\pi\varepsilon_o l}\left[X\ln(X^2 + Y^2) - 2X + 2Y\operatorname{atan}\frac{X}{Y}\right] \tag{16-16}$$

In terms of the indefinite integral $U(X,Y)$, $v_{s\parallel x}(\mathbf{x},\mathbf{x}_s)$ can be written

$$v_{s\parallel x}(\mathbf{x},\mathbf{x}_s) = U(x_s + \frac{l}{2} - x, y_s - y) - U(x_s - \frac{l}{2} - x, y_s - y) \tag{16-17}$$

If the charged segment is parallel to the y axis, the roles of x and y are reversed; the potential becomes

$$v_{s\parallel y}(\mathbf{x},\mathbf{x}_s) = U(y_s + \frac{l}{2} - y, x_s - x) - U(y_s - \frac{l}{2} - y, x_s - x) \tag{16-18}$$

16.1.5 Charged Segment Near a Conducting Plane

If the segment of charge is near an infinite ground plane, charges are induced on the surface of the ground plane such that every point of the ground plane has the same potential. The ground plane is arbitrarily assigned the potential zero, making it the reference surface. The effect of the charges induced on the ground plane can be represented above the ground plane by an "image" of the distribution of charges at the image location of the charged segment. The image location is below the ground plane the same distance as the segment is above the ground plane.

Referring to Figure 16.7, note that the ground plane is a plane of symmetry between the charged segment and the image segment. The image segment (by assumption) has minus the

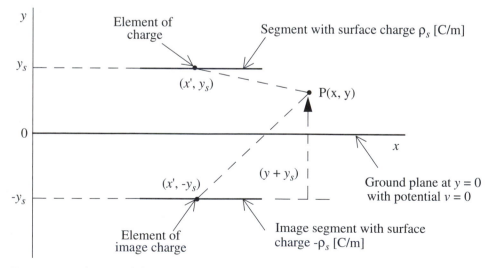

Figure 16.7   Schematic of Physical and Image Charges of a Charged Segment.

charge distribution of the physical segment. Because of the geometrical symmetry and the behavior of the potential function, the potential *on the ground plane* due to the image segment is exactly minus the potential due to the physical segment. The potential of the sum of the two is exactly zero on the ground plane.

The combined potential is

$$v_{s \parallel x}(\mathbf{x,x}_s) = U(x_s + \frac{l}{2} - x, y_s - y) - U(x_s - \frac{l}{2} - x, y_s - y)$$

$$-U(x_s + \frac{l}{2} - x, y_s + y) + U(x_s - \frac{l}{2} - x, y_s + y)$$

(16-19)

EXERCISE 16-1.   Show that at points on the ground plane where $y = 0$ that $v_{s \parallel x} = 0$.

The potential due to the image charged segment is minus the potential due to the charged segment with the distance in the y direction from the physical segment to the observer point $(y_s - y)$ replaced with the distance in the y direction from the image segment to the observer point $(y_s + y)$.

If the charged segment is parallel to the y axis and the ground plane is again at $y = 0$ (the x axis), the combined potential including the image segment is

$$v_{s \parallel y}(\mathbf{x,x}_s) = U(y_s + \frac{l}{2} - y, x_s - x) - U(y_s - \frac{l}{2} - y, x_s - x)$$

$$-U(y_s + \frac{l}{2} + y, x_s - x) + U(y_s - \frac{l}{2} + y, x_s - x)$$

(16-20)

EXERCISE 16-2.   Show that at points on the ground plane where $y = 0$ that $v_{s \parallel y} = 0$.

PROBLEM 16.4   Consider a segment parallel to the x axis as shown in Figure 16.7. Let the length of the segment be 1.25×10^{-4} [m] (about 5 mils) and let the center of the segment be located at the point $(0, 2.5 \times 10^{-5})$ [m], about 1 mil above the ground plane. If the charge on the segment is unity (1 [C/m]), what is the potential at the center of the segment due to this charge?

16.1.6 Inductance Simulation

In a homogeneous medium, the capacitance and inductance of a single trace are related by

$$LC = \mu\varepsilon$$

(16-21)

where in free space, $\mu\varepsilon = \mu_o \varepsilon_o = 1/(299{,}792{,}458)^2$ [s^2/m^2]. The speed of light in free space is (exactly, by definition) 299,792,458 [m/s]. Therefore, if C has been calculated, L can be calculated from it by dividing both sides of Equation (16-21) by C.

Computer Code 16-1. Single-Trace Capacitance and Inductance.

A computer code (file `ch16_1.c`) has been written to illustrate and demonstrate the calculation of C and L for a single trace. The code is listed below (excluding `SimpleGauss` that is listed in Chapter 3 in Computer Code 3-1). The trace position and size is input in mils (thousandths of an inch); this is a common unit in circuit board layout. The conversion constant to meters is defined in the header; it is applied near the top of program `main` and the calculation is carried out in MKS units. The constant $2\pi\varepsilon_o = 5.5632502\times10^{-11}$ is also defined in the header for use in calculating the function $U(X,Y)$; it is calculated in the C function `FunctU`, the last function in the code listing below.

Notice the structure `GammaSeg` defined in the header of the file. It contains parameters pertinent to each segment of the trace. The parameters `epsratio` ($\varepsilon_o/\varepsilon = 1/\varepsilon_r$ for the dielectric adjacent to the segment) and `TraceNo` (the index of the trace) are not needed for the single trace calculation in a homogeneous dielectric, but are needed for the generalized case of multiple traces in a layered dielectric. Hence, they are included from the start. The variables (`xs`, `ys`) are the coordinates of the center of the segment, `length` is the segment length and `parallel` equals the ASCII value `'x'` or `'y'` to indicate whether the trace is parallel to the x or y axes, respectively. The variable `next` is a pointer to the next `GammaSeg` structure in a linked list. The linked list approach allows the segmenting parameters to be calculated (or input) at run time; memory is allocated as needed. Note how the linked list is built and then used in the program below.

The main program begins by defining the positions of the left and right sides and top and bottom of the trace. These inputs could easily be modified by the user to become interactive inputs; or program `main` could be converted to a subroutine and they could be input through the argument list. Interactive input is provided for the user to input the numbers of segments in the x and y directions of the trace. This allows the user to quickly observe the dependence of the resulting values of C and L on the size of the segments when he runs the executable program. The effect of such size variations is discussed below the program listing.

The linked list of segments is then built by stepping horizontally and creating links for segments across the top and bottom of the trace. Then vertical steps are taken to create links for segments on the left- and right-hand sides of the trace. Each time a new link is needed, the memory allocation routine `malloc` is called. It returns a pointer to the newly allocated memory block (link). The pointer to the previously obtained memory block is saved in the `->next` variable of the new block. When these data are used, the process moves in the opposite direction starting with the last link allocated. The `->next` variable then points to next link to be accessed, and so on until `->next` contains a `NULL` pointer that identifies the last link (i.e., the first link allocated).

Memory is then allocated for the voltage matrix $v_{ij\Gamma}$ (see Equation 16-6); it is named `vijgamma`. The matrix is filled by calling function `vspygp` if the charged segment (source) is parallel to the y axis, or function `vspxgp` if the segment is parallel to the x axis. The observer locations where the potential is calculated are the centers of the segments (where the potential will be forced to be 1 [V]). The `int` variable `->parallel` contains either `'x'` or `'y'` to identify the axis to which the charged segment is parallel. In a more general code, the segments might have arbitrary orientation; the integer `->parallel` could be replaced with `float` or `double` direction cosines to define the orientation of the segment.

The forcing function side of the linear equations (see Equation (16-7)) is then loaded into column $(n + 1)$ of the matrix `vijgamma` so that after calling `SimpleGauss`, it will contain the solution vector, i.e., the segment charges that yield 1 [V] at the centers of the segments. Function `SimpleGauss` is then called. The charges on the segments are summed to obtain the capacitance and the inductance is then calculated using the relationship (true only in a homogeneous dielectric)

$$L = \frac{\varepsilon_r}{Cc^2} \tag{16-22}$$

where c is the speed of light. Note that the speed of light, $c = 1/(\sqrt{\varepsilon_o \mu_o})$ and C/ε_r is the capacitance as if the dielectric were in free space. The input trace parameters, the relative permittivity and the calculated capacitance and inductance are output to the screen. The capacitance and inductance are calculated in MKS units ([F] and [H]). For more readable output, they are printed in [pF] and [μH]. The units conversion is done in the `printf` statement.

The C computer code for the functions `vspygp` (the potential due to a segment parallel to the y axis with a ground plane), `vspxgp` (the potential due to a segment parallel to the x axis with a ground plane) and `FunctU` (the indefinite integral) are all listed below for examination and study.

```
/* ------ File ch16_1.c ------------------------------------------------
   Description:  Capacitance and inductance of a single trace over a
                 ground plane.  Homogeneous dielectric.

   Include files and comp. directives     Description
   ---------------------------------      -----------                 */
#include <math.h>
#include <stdio.h>
#include <malloc.h>
#define TwoPiEps0      5.56325028e-11F   // [F/m]
#define SMALL          1.0e-30F
#define MILS_TO_METERS 2.54E-5F          // [m/mil]
#define C_LIGHT        299792458.0F      // Speed of light
struct GammaSeg
{
   float    epsratio;          // eps0/eps at segment surface [-]
   int      TraceNo;           // Trace number segment is on
   float    xs;                // x coordinate of segment center [m]
   float    ys;                // y coordinate of segment center [m]
   float    length;            // length of segment [m]
   int      parallel;          // parallel to 'x' or 'y'
   struct   GammaSeg  *next;   // pointer to next link in linked list
};
typedef struct GammaSeg GAMMA_SEG; // gamma segment type
/*   Procedures Called (all are in this file)
```

```
    ------------------------------------------------------------------*/
    float vspxgp (float xp, float yp, float xs, float ys, float length);
    float FunctU (float length, float X, float Y);
    float vspygp (float xp, float yp, float xs, float ys, float length);
    int   SimpleGauss (int nrows, int ncols, float *matrix);
    int   IndexMatrix (int irow, int jcol, int nrows);
    /*-----FunctionDeclaration-------------------------------------------*/
    void main ()
    {
    /* --- Local Variable Declarations ----------------------------------
      Type         Name                 Description & range
      ----         ----                 -------------------------------- */
      float xmin, xmax, ymin, ymax;  // trace sides, top & bottom
      GAMMA_SEG  *gammaList = NULL,  // pointer to list of gamma segments
                 *gammaTmp  = NULL,  // temp. pointer to gamma segment
                 *gammaTmp2 = NULL;  // temp. pointer to gamma segment
      float      *vijgamma  = NULL;  // pointer for voltage matrix
      float       Cap;               // capacitance of trace
      float       Inductance;        // inductance of trace
      int         xsegs,             // number of segments, top & bottom
                  ysegs,             // number of segments each side
                  numgammas;         // number of gamma segments
      float       lx;                // length of a segment
      int         idx;               // for-loop index
      int         row, col;          // row, column indices
      float       eps;               // relative permittivity
      float       potential,         // potential of a point
                  xi, yi,            // coordinates of observer point
                  xs, ys;            // coordinates of center of segment
    /*--------------ExecutableCode---------------------------------------*/
      // Define trace position and dimensions
      xmin = 0.0F;             // enter [mils]
      xmax = 1.0F;
      ymin = 1.0F;
      ymax = 5.0F;
      // Define dielectric constant
      eps = 3.0F;
      xmin *= MILS_TO_METERS;      // convert to meters
      xmax *= MILS_TO_METERS;
      ymin *= MILS_TO_METERS;
      ymax *= MILS_TO_METERS;
      // Input segmenting parameters
      printf("\nSingle trace capacitance\n");
      printf("Enter number of x-direction segments: ");
      scanf("%i", &xsegs);
      printf("Enter number of y-direction segments: ");
```

```c
scanf("%i", &ysegs);
// Build linked list of segments
numgammas = 0;                 // gamma segment counter
lx = (xmax - xmin)/xsegs;   // length of segments in x direction
for (idx = 0; idx < xsegs; idx++)
{
   // top of trace
   gammaTmp = gammaList;   // save pointer to previous link
                           // first time this is NULL
   gammaList = (GAMMA_SEG*)malloc(sizeof(GAMMA_SEG));
   gammaList->next      = gammaTmp;
   gammaList->epsratio = 1.0F/eps;
   gammaList->TraceNo   = 1;
   gammaList->length    = lx;
   gammaList->xs        = xmin + lx * (idx + 0.5F);
   gammaList->ys        = ymax;
   gammaList->parallel = 'x';
   // bottom of trace
   gammaTmp = gammaList;   // save pointer to previous link
   gammaList = (GAMMA_SEG*)malloc(sizeof(GAMMA_SEG));
   gammaList->next      = gammaTmp;
   gammaList->epsratio = 1.0F/eps;
   gammaList->TraceNo   = 1;
   gammaList->length    = lx;
   gammaList->xs        = xmin + lx * (idx + 0.5F);
   gammaList->ys        = ymin;
   gammaList->parallel = 'x';
   numgammas += 2;
}
lx = (ymax - ymin)/ysegs;   // length of segments in y direction
for (idx = 0; idx < ysegs; idx++)
{
   // left side of trace
   gammaTmp = gammaList;   // save pointer to previous link
   gammaList = (GAMMA_SEG*)malloc(sizeof(GAMMA_SEG));
   gammaList->next      = gammaTmp;
   gammaList->epsratio  = 1.0F/eps;
   gammaList->TraceNo   = 1;
   gammaList->length    = lx;
   gammaList->xs        = xmin;
   gammaList->ys        = ymin + lx * (idx + 0.5F);
   gammaList->parallel  = 'y';
   // right side of trace
   gammaTmp = gammaList;   // save pointer to previous link
   gammaList = (GAMMA_SEG*)malloc(sizeof(GAMMA_SEG));
   gammaList->next      = gammaTmp;
```

```
         gammaList->epsratio  = 1.0F/eps;
         gammaList->TraceNo    = 1;
         gammaList->length     = lx;
         gammaList->xs         = xmax;
         gammaList->ys         = ymin + lx * (idx + 0.5F);
         gammaList->parallel   = 'y';
         numgammas += 2;
      }
      // Load potential matrix   vijgamma
      // Allocate memory for voltage matrix
      vijgamma = (float*)malloc(sizeof(float) * numgammas
              * (numgammas+1));     // n x (n+1) matrix for SimpleGauss
      // load voltage matrix (left n x n part)
      gammaTmp = gammaList;
      for (row = 0; row < numgammas; row++)
      {
         gammaTmp2 = gammaList;
         for (col = 0; col < numgammas; col++)
         {
            xi = gammaTmp->xs;       // point to evaluate potential
            yi = gammaTmp->ys;
            xs = gammaTmp2->xs;      // center of charged segment
            ys = gammaTmp2->ys;
            if (gammaTmp2->parallel == 'y')
               potential = vspygp (xi, yi, xs, ys, gammaTmp2->length);
            else if   (gammaTmp2->parallel == 'x')
               potential = vspxgp (xi, yi, xs, ys, gammaTmp2->length);
            else
            {
               printf ("Error -- parallel not correct\n");
               goto AllDone;
            }
            *(vijgamma + row + numgammas * col) = potential
               * gammaTmp2->epsratio;
            gammaTmp2 = gammaTmp2->next;
         }
         gammaTmp = gammaTmp->next;
      }
      // Solve for capacitance
      // Load column (numgammas + 1) with 1's   (equation right-hand side)
      for (row = 0; row < numgammas; row++)
         *(vijgamma + row + numgammas * numgammas) = 1.0F;
      // Solve using SimpleGauss
      // (After this call, the trace charges are in column numgammas + 1)
      if (SimpleGauss (numgammas, numgammas+1, vijgamma))
```

```
   {
      printf ("Error -- SimpleGauss\n");
      goto AllDone;
   }
   // Calculate capacitance -- the sum of the charges on the segments
   Cap = 0.0F;
   for (row = 0; row < numgammas; row++)
      Cap += *(vijgamma + row + numgammas * numgammas);
   // Calculate inductance
   Inductance = eps/(Cap * C_LIGHT * C_LIGHT);
   // Printf output to the screen
   printf("xmin = %.2f    xmax = %.2f [mils]\n", xmin/MILS_TO_METERS,
      xmax/MILS_TO_METERS);
   printf("ymin = %.2f    ymas = %.2f [mils]\n", ymin/MILS_TO_METERS,
      ymax/MILS_TO_METERS);
   printf("relative permittivity = %.2f\n", eps);
   printf("capacitance = % .4f [pF/m]\n", Cap * 1.e12F);
   printf("inductance = % .6f [microH/m]\n", Inductance * 1.e6F);
AllDone:
   // Free memory
   free (vijgamma);       vijgamma = NULL;
   while (gammaList != NULL)
   {
      gammaTmp = gammaList->next;
      free(gammaList);
      gammaList = gammaTmp;
   }
}
/*-----EndofProceduremain----------------------------------------*/
/*-------FunctionDeclaration--------------------------------------*/
/*   Description: Potential of segment parallel to the y axis that has a
              unit charge uniformly distributed on it.
              A ground plane is at the x axis.
                                    ------------------*/
  float vspygp (float xp, float yp, float xs, float ys, float length)
{
/*-------- Argument Variables -------------------------------------
  Type        Name                I/O Description & range
  ----        ----                --- --------------------------------
  float       vspygp               o  Potential due to the segment [V]
  float       xp                   i  x coordinate of observer point [m]
  float       yp                   i  y coordinate of observer point [m]
  float       xs                   i  x coordinate of center of segment [m]
  float       ys                   i  x coordinate of center of segment [m]
  float       length               i  length of the segment [m]
```

```
    --- Local Variable Declarations --------------------------------
   Type         Name                    Description & range
   ----         ----                    ------------------------------ */
   float        potential,          // potential of point (xp, yp)
                xsmp,               // xs - xp
                yspp,               // ys + yp
                ysmp,               // ys - yp
                l2;                 // half the length of the segment
/* ------------- Executable Code ------------------------------- */
   xsmp = xs - xp;
   yspp = ys + yp;
   ysmp = ys - yp;
   l2 = length * 0.5F;
   potential =   FunctU ( length, ysmp + l2, xsmp)
               - FunctU ( length, ysmp - l2, xsmp)
               - FunctU ( length, yspp + l2, xsmp)
               + FunctU ( length, yspp - l2, xsmp);
   return (potential);
}
/* -----End of Procedure vspygp  ------------------------------- */
/*------- Function Declaration ------------------------------- */
/*   Description: Potential of segment parallel to the x axis that has
a
               unit charge uniformly distributed on it.
               A ground plane is at the x axis.
                              ----------------*/
float vspxgp (float xp, float yp, float xs, float ys, float length)
{
/* -------- Argument Variables ----------------------------------
   Type         Name            I/O Description & range
   ----         ----            --- --------------------------------
   float        vspxgp           o  Potential due to the segment [V]
   float        xp               i  x coordinate of observer point [m]
   float        yp               i  y coordinate of observer point [m]
   float        xs               i  x coordinate of center of segment [m]
   float        ys               i  x coordinate of center of segment [m]
   float        length           i  length of the segment [m]
    --- Local Variable Declarations -------------------------------
   Type         Name                    Description & range
   ----         ----                    ------------------------------ */
   float        potential,          // potential of point (xp, yp)
                xsmp,               // xs - xp
                yspp,               // ys + yp
                ysmp,               // ys - yp
                l2;                 // half the length of the segment
/* ------------- Executable Code ------------------------------- */
```

```
   xsmp = xs - xp;
   yspp = ys + yp;
   ysmp = ys - yp;
   l2 = length * 0.5F;
   potential =   FunctU ( length, xsmp + l2, ysmp)
               - FunctU ( length, xsmp - l2, ysmp)
               - FunctU ( length, xsmp + l2, yspp)
               + FunctU ( length, xsmp - l2, yspp);
   return (potential);
}
/* -----End of Procedure vspxgp ------------------------------------ */
/*-------FunctionDeclaration----------------------------------------*/
// Indefinite integral Eq IV.2-16
float FunctU (float length, float X, float Y)
{
/* -------- Argument Variables ------------------------------------
   Type          Name                   I/O Description & range
   ----          ----                   --- -----------------------------

   float         FunctU                 o   Value of the indefinte integral
   float         length                 i   length of the charged segment
   float         X                      i   X argument
   float         Y                      i   Y argument
   --- Local Variable Declarations ---------------------------------
   Type          Name                   Description & range
   ----          ----                   ------------------------------- */
   float         t1, t2, t3;            // temporary variables
/*--------------ExecutableCode--------------------------------------*/
   t1 = X * (float)log(X * X + Y * Y);
   t2 = 2.0F * X;
   if (fabs(X) < SMALL || fabs(Y) < SMALL)    // X or Y very near zero
      t3 = 0.0F;
   else
      t3 = 2.0F * Y * (float)atan(X/Y);
   return (-(t1 - t2 + t3) / (2.0F * TwoPiEps0 * length));
}
/* -----End of Procedure FunctU ------------------------------------ */
```

16.1.7 Calculated Results and their Accuracy

The trace geometry defined in the above computer code to illustrate the capacitance and inductance algorithms is shown in Figure 16.8. The trace orientation was deliberately chosen with the long dimension perpendicular to the ground plane so that the charge distribution on it would be very nonuniform along its sides. (The surface charge density will be much larger near the bottom than near the top.) With such a nonuniform charge distribution, the result will depend quite strongly on the sizes of the segments. The results of the five calculations given below illustrate this. A series of several calculations with decreasing segment size can allow one to obtain accuracy greater than that of the calculation with the smallest segment sizes.

Running the above code with one segment on each side (by entering 1 and 1 for the number of x-direction and y-direction segments, respectively), the following output is printed to the screen:

```
Single trace capacitance
Enter number of x-direction segments: 1
Enter number of y-direction segments: 1
xmin = 0.00      xmax = 1.00 [mils]
ymin = 1.00      ymas = 5.00 [mils]
relative permittivity = 3.00
capacitance =   125.8463 [pF/m]
inductance =   0.265240 [microH/m]
```

We shall find, by increasing the number of segments, that the correct capacitance is very close to 134.1 [pF/m]. The simple model using only one segment on each side yields a capacitance of 125.8 [pF/m] (see output above), about 6 percent smaller than the actual capacitance. (This error is considered too large for most engineering purposes.) Let us explore the convergence properties of the algorithm.

The algorithmic error for the method given is approximately proportional to the segment size if the segments are all scaled by the same amount. Under this condition, let δx be any of the segment sizes. (For the series of calculations given for illustration, all the segment sizes are equal.) The situation is illustrated in Figure 16.9. It was pointed out above, that as the segment sizes approach zero (and the number of segments become infinite) the finite-segment approximation approaches the integral equation that defines the exact capacitance. We seek to calculate a value close to this limit. As a practical matter, the algorithm cannot be carried out to this limit. Inversion of an infinite matrix cannot be performed numerically. As the matrix becomes very large, not only does the computational time and memory required become impractical, but the round-off error in the matrix inversion becomes dominant (rather than the error due to the finite-segment approximation of the integral equation). Therefore, using direct application of the capacitance algorithm, one can expect increasing

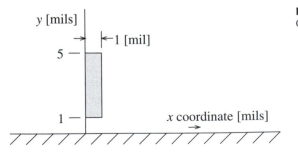

Figure 16.8 Trace Geometry Defined in Computer Code 16-1.

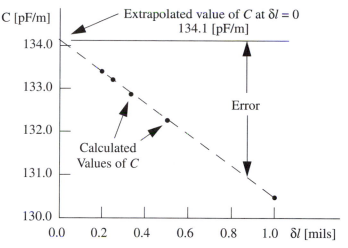

Figure 16.9 Straight Line Fit of C(δl).

accuracy with decreasing segment sizes only up to some (generally unknown) limit. As illustrated above, an *extrapolation* of a series of calculations can be carried out to obtain results more accurate than that obtained directly with the smallest segment size directly used.[4]

The limit is expressed mathematically as

$$C(0) = \lim_{\delta l \to 0} C(\delta l) \tag{16-23}$$

If the dependence of C on δl is assumed linear, we write

$$C(\delta l) = C(0) + b\delta l \tag{16-24}$$

The computer program above was run for keyboard entries (numbers of x- and y-direction segments): (1, 4), (2, 8), (3, 12), (4, 16), and (5, 20). This sequence results in the segment sizes shown in column three of Table 16-1, below. The capacitance corresponding to each of those segment sizes is shown in column four. A linear regression (least squares fit) was performed[5]. That is, the capacitance $C(\delta l)$ was assumed to have the form of Equation (16-24). Then, the value of $C(0)$ and b were found that minimize

$$\chi^2 = \sum_{i=1}^{5} [C(\delta l_i) - C(0) - b\delta l]^2 \tag{16-25}$$

where $C(\delta l_i)$, $i = 1, \ldots, 5$ are the calculated values of C in column four of Table 16-1. The value of χ is the r.m.s. error associated with the fit. Small values indicate a good fit to the straight-line behavior. For the five calculations performed, the best fit value of $C(0)$ was found to be $C(0) = 134.125$ [pF/m] with $\chi = 0.0460$ [pF/m].

---

4. Methods of extrapolation to zero more sophisticated than the one given here exist. See, for instance, *Richardson's deferred approach to the limit*, as applied to Bulirsch-Stoer integration, Press, Flannery, et al. 1986, Section 15.4.

5. The details of performing a linear regression will not be given here. See Press, Flannery, *et al.*,1986, Section 14.2.

TABLE 16-1
Sequence of Capacitance Calculations.

Index i of Calculation	Numbers of Segments	Segment Lengths δl [mils]	Capacitance $C(\delta l)$ [pF/m]
1	1, 4	1.0000	130.3763
2	2, 8	0.5000	132.2749
3	3, 12	0.3333	132.9050
4	4, 16	0.2500	133.1887
5	5, 20	0.2000	133.3463

How accurate is this extrapolated value of C? One estimate of the error might be the value of $\chi = 0.0460$ [pF/m]. Another estimate (preferred by many) is the difference between the value of $C(0)$ obtained using calculations 1 through 4 (Table 16-1) and the value obtained using calculations 1 through 5, that is, the change in the estimate of $C(0)$ that is due to the addition of the most accurate calculation (index 5) to the fit. Using the first four points, the value 134.148 [pF/m] is obtained. This suggested an estimated error of 0.023 [pF/m]. Hence, the rounded value of 134.1 [pF/m] is significant to all four decimals.

PROBLEM 16.5 If only two points are used in a fit to Equation (16-24), an exact fit can be made (because two points determine a straight line). Using calculations indexed 1 and 2 from Table 16-1, what value of $C(0)$ is obtained? How accurate is that value?

Without the extrapolation, the last value in Table 16-1 has about 1 percent error; the final value using the extrapolation has about 0.02 percent error. The extrapolation method improves the accuracy dramatically for a given smallest segment size at the expense of a more complex algorithm and an increase in computation time. Its use is a judgment determination that depends on the accuracy needed and the trade-offs between software development time and computer run time. The extrapolation method will not be applied or pursued further in this book.

16.2 MULTI-TRACE CAPACITANCE AND INDUCTANCE SIMULATION

The calculation of the capacitance matrix for a multi-trace transmission line is very much like that of the single-trace capacitance. The segment parameters are defined for all of the traces just as they were for a single trace. Saving the trace number (->TraceNo) for each segment is the only change in that part of the algorithm. The voltage matrix is calculated in exactly the same way. The solution for the segment charges is different and depends on the definition of the capacitance matrix to be discussed below. The modified C computer code for the multi-trace case is discussed below.

16.2.1 Definition of Capacitance Matrix

The capacitance matrix is defined algebraically as a generalization of the scalar definition given in Equation (16-1). Because the voltage in the generalized definition is a vector (whose elements are the potentials of the traces), it does not make sense to write it in the denominator as in Equation (16-1). Hence, the generalization of Equation (16-1) is written

$$\mathbf{q} = \mathbf{C}\mathbf{v} \tag{16-26}$$

where $\mathbf{q}$ is a column vector of charges on the conductors of the transmission line, $\mathbf{C}$ is the capacitance matrix, and $\mathbf{v}$ is the vector of potentials of the conductors.

The capacitance matrix yields the conductor charges when it multiplies a column of potentials that can have any combination of values. To understand the capacitance matrix and develop an algorithm to calculate its elements, a special vector of potentials is considered. Consider the vector of potentials whose entries are all zero except one, and the one conductor that's not at zero potential has the potential 1 volt. This corresponds physically to grounding all but one of the conductors and attaching a voltage source of 1 volt between ground and the remaining (nongrounded) conductor. Of course, this is a conceptual grounding, a physical ground wire could distort the fields (i.e., it could attract field lines to itself and attain a charge of its own) thus causing an error in the capacitance determination. Assume that conductor j has potential 1 volt and all the rest of the conductors have zero potential. The situation is illustrated in Figure 16.10.

Algebraically, the voltage column vector can be written

$$\mathbf{v} = \begin{bmatrix} 0 \ \dots \ 0 \ 1 \ 0 \ \dots \ 0 \end{bmatrix}^{\mathrm{T}} \tag{16-27}$$

where the number 1 is in row j. In the matrix product $\mathbf{C}\mathbf{v}$, each element (row) of $\mathbf{v}$ multiplies a column of the $\mathbf{C}$ matrix, i.e., the kth row of $\mathbf{v}$ multiplies the kth column of $\mathbf{C}$. Substituting the column vector given by Equation (16-27) into Equation (16-26), one obtains

Figure 16.10 Schematic of Conductor j at One Volt, Others at Zero Potential.

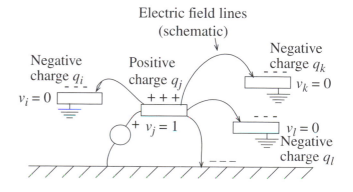

$$
\begin{bmatrix} q_1 \\ q_2 \\ \cdot \\ \cdot \\ \cdot \end{bmatrix} = \begin{bmatrix} \ldots + 0 + c_{1j} + 0 + \ldots \\ \ldots + 0 + c_{2j} + 0 + \ldots \\ \cdot \\ \cdot \\ \cdot \end{bmatrix} \tag{16-28}
$$

All of the columns of the $\mathbf{C}$ matrix have been multiplied by zero except the jth column, it has been multiplied by one to yield the result shown in Equation (16-28). Therefore, determining the charges on the traces for this special potential combination yields the jth column of the $\mathbf{C}$ matrix. Because the $\mathbf{C}$ matrix is symmetric, this also yields the jth row of the $\mathbf{C}$ matrix. This is the procedure that will be developed below.

Two properties of the capacitance matrix can be observed at this point. Referring to Figure 16.10, it is evident from the physical situation that the charge on each of the conductors that have zero potential is negative. This means that all of the off-diagonal elements of the $\mathbf{C}$ matrix are negative. One can summarize the first property as follows:

Capacitance matrices always have positive diagonal elements and negative off-diagonal elements.

We note that because charge is conserved, the following equation must hold

$$
q_j = -q_{ground} - \sum_{k \ne j} q_k \tag{16-29}
$$

where q_j is positive and q_{ground} and the q_k, $k \ne j$, are negative. Substituting c_{kj} for q_k (where c_{kj} is the kth element of the jth column of matrix $\mathbf{C}$), one obtains

$$
c_{jj} = -q_{ground} - \sum_{k \ne j} c_{kj} \tag{16-30}
$$

Equation (16-30) can be written

$$
\sum_k c_{kj} = -q_{ground} \tag{16-31}
$$

Because $-q_{ground} > 0$, one can conclude that

$$
\sum_k c_{kj} > 0 \tag{16-32}
$$

That is, the sum of any column (or any row) of any $\mathbf{C}$ matrix must be greater than zero (it could be very close to zero if the conductors are very far from the reference conductor). One can summarize this second property by stating:

The sum of every row and the sum of every column of a capacitance matrix is greater than zero.

If a numerical procedure that calculates the matrix $\mathbf{C}$ yields results that contradict either of these two properties, there must be an error or inaccuracy in the calculation. Any such result is nonphysical, and should not be used for transmission line analysis.

16.2.2 Capacitance Matrix Simulation

Calculation of the capacitance matrix begins in much the same way as calculation of the capacitance of a single trace. Let all of the traces be divided into segments; points on the surface of the jth trace will be denoted Γ_j. The set of points Γ_j is defined such that if $\mathbf{x} \in \Gamma_j$, then the point $\mathbf{x}$ lies on the surface of trace j. Let the set of points Γ be all the points on the surfaces of all the traces. That is,

$$\Gamma = \bigcup_{j=1}^{N} \Gamma_j \tag{16-33}$$

where N is the number of traces.

Equation (16-6) is repeated below. When first written, it represented the potential at an arbitrary point due to all the segments on a single trace. With Γ now defined as the combined surfaces of all the traces, the same equation yields the potential at an arbitrary point $\mathbf{x}_i$ due to the charges on all the traces.

$$w_i = \sum_{j_\Gamma = 1}^{J_\Gamma} v_{ij_\Gamma} q_{j_\Gamma} \tag{16-6}$$

where J_Γ is the total number of segments (on all N traces).

For the case of a segment parallel to the x axis, v_{ij_Γ} was evaluated and is given by $v_{ij_\Gamma} = v_{s \parallel x}\left(\mathbf{x}_i, \mathbf{x}_{sj_\Gamma}\right)$ in Equation (16-19). And for the case of a segment parallel to the y axis, v_{ij_Γ} was evaluated and is given by $v_{ij_\Gamma} = v_{s \parallel y}\left(\mathbf{x}_i, \mathbf{x}_{sj_\Gamma}\right)$ in Equation (16-20). The same substitutions are valid for the multi-trace situation.

To define the capacitance of a single trace, the points $\mathbf{x}_i$ are defined to be the center points of the segments on the trace and the potentials at those points are set to one. For the multi-trace problem, the points $\mathbf{x}_i$ are again defined to be the center points of the segments on (all) the traces. To obtain the charges required for the kth column of the capacitance matrix, we set w_i as follows:

$$w_i = \begin{cases} 1, & \mathbf{x}_i \in \Gamma_k \\ 0, & \mathbf{x}_i \in \Gamma_{i \neq k} \end{cases} \tag{16-34}$$

This is simply Equation (16-27) written for each segment on each trace (with potential 1 [V] on trace k rather than on trace j). A subscript k can be added to this potential to identify it with trace k. A subscript k will also be added to q_{j_Γ} in Equation (16-6) to identify it with the segment charges due to 1 [V] on segment k with all the other segments at zero potential. Substituting these into Equation (16-6), one obtains

$$w_{ik} = \sum_{j_\Gamma = 1}^{J_\Gamma} v_{ij_\Gamma} q_{j_\Gamma k} \tag{16-35}$$

Letting i assume the values 1 to J_Γ, where J_Γ (the total number of segments on all of the traces) a set of J_Γ linear equations is obtained in the same number of unknown charges $q_{j_\Gamma k}$ (for each trace index k). The solution of this set of linear equations yields the charges on the segments when 1 [V] is applied to trace k and all the other traces have zero potential. Summation of the charges on (the segments of) each trace yields the elements of the kth column of the capacitance matrix. We can write the mth element of the kth column of the matrix $\mathbf{C}$ as

$$c_{mk} = \sum_{x_{j_\Gamma} \in \Gamma_m} q_{j_\Gamma k} \tag{16-36}$$

where $x_{j_\Gamma} \in \Gamma_m$ means that the sum includes only those values of j_Γ that represent segments on trace number m.

Before putting aside this formulation, we observe that the quantities $q_{j_\Gamma k}$ can be regarded as the kth column of a $J_\Gamma \times N$ matrix $\mathbf{Q}$ and w_{ik} can be regarded as the kth column of a $J_\Gamma \times N$ matrix $\mathbf{W}$. Hence, Equation (16-35) for all the rows $1 \le i \le J_\Gamma$ and the columns $1 \le k \le N$ of $\mathbf{W}$ and $\mathbf{Q}$ can be combined into the single matrix equation

$$\mathbf{W} = \mathbf{VQ} \tag{16-37}$$

where $\mathbf{V}$ is the square $J_\Gamma \times J_\Gamma$ matrix whose elements are v_{ij_Γ}. The solution for the segment charges under all the required potential combinations can be written simply as the matrix equation

$$\mathbf{Q} = \mathbf{V}^{-1}\mathbf{W} \tag{16-38}$$

16.2.3 Inductance Simulation

For lines in a homogeneous medium (i.e., single-speed lines), the $\mathbf{L}$ and $\mathbf{C}$ matrices are related as given by Equation (5-1); it is repeated here

$$\mathbf{LC} = \mathbf{CL} = \frac{1}{c^2}\mathbf{I} \tag{5-1}$$

Multiplication of this equation by the inverse of the matrix $\mathbf{C}$ yields the matrix $\mathbf{L}$. One obtains

$$\mathbf{L} = \frac{\mathbf{C}^{-1}}{c^2} = \mu\epsilon\mathbf{C}^{-1} \qquad (16\text{-}39)$$

Equation (16-39) can be applied to obtain the inductance matrix after the capacitance matrix has been calculated.

Computer Code 16-2. Multiple Trace Capacitance and Inductance.

The algorithms for calculating the capacitance and inductance matrices are implemented in the computer code given below. Only the main program is listed below; the header statements and the subroutines (`vspxgp`, `FunctU`, `vspygp`, `SimpleGauss` and `IndexMatrix`) are identical to those listed in computer code 16-1. This homogeneous medium, multi-trace code is in file `ch16_2.c`; all of the required routines are in this file.

The changes to code 16-1 to generalize it to perform the multi-trace algorithm are surprisingly few. The local variable declarations have several additions and modifications (from scalar to vector or matrix arrays) to store the trace description and the new capacitance matrix. A new working variable (matrix) named `work` has been added to use in the call to `SimpleGauss` in performing the inverse of the matrix **C** needed in Equation (16-39). The matrix `work` is declared as a linear array (of $2N^2$ elements). It is used as an $N \times 2N$ matrix array.

The algorithm changes are:

1. The input of the trace descriptions was modified to accommodate multiple traces.
2. The trace numbers (1 to N) are saved in each of the segment structure links.
3. The voltage matrix `vijgamma` was expanded from an $I_\Gamma \times (I_\Gamma + 1)$ matrix to an $I_\Gamma \times (I_\Gamma + N)$ to include a column for each additional trace. This is to perform the linear equation solution required by Equation (16-38) using `SimpleGauss`. In that solution, The left $I_\Gamma \times I_\Gamma$ block of the matrix `vijgamma` is filled just as before. The right $I_\Gamma \times N$ block is filled column-by-column in accordance with Equation (16-34).
4. A call to `SimpleGauss` then solves Equation (16-37) to obtain the matrix **Q** as shown in Equation (16-38).
5. The capacitance matrix is then initialized to zero and the sums of the charges expressed in Equation (16-36) is performed for each column of the matrix **C**.
6. Because **C** is a symmetrical matrix, its elements must obey $c_{ij} = c_{ji}$. Each element of **C** is calculated independently and because of the approximations in the model, c_{ij} is close to c_{ji}, but they are not equal. To force the appropriate symmetry, the program calculates the mean of these two elements and defines that mean as the output for both matrix elements. This completes the calculation of the matrix **C**.
7. The inverse of matrix L, $\mathbf{C}/\mu\epsilon$, is loaded into the left $N \times N$ half of the matrix `work` and the unit vector is loaded into the right $N \times N$ half. A call to `SimpleGauss` then calculates the inverse of $\mathbf{C}/\mu\epsilon = \mathbf{L}^{-1}$ using elementary operators and operates on the unit vector with this inverse. On exit, the right $N \times N$ half of matrix `work` contains the matrix **L**.

The listing of the main program follows.

```c
/* ------ File ch16_2.c --------------------------------------------
--
  Description:  Capacitance and inductance of multiple traces over a
              ground plane.  Homogeneous dielectric.
  Include files and comp. directives       Description
  ---------------------------------        ----------
*/
#include <math.h>
#include <stdio.h>
#include <malloc.h>
#define TwoPiEps0      5.56325028e-11F   // [F/m]
#define SMALL          1.0e-30
#define MILS_TO_METERS 2.54E-5F          // [m/mil]
#define C_LIGHT        299792458.0F      // Speed of light
struct GammaSeg
{
    float    epsratio;          // eps0/eps at segment surface [-]
    int      TraceNo;           // Trace number segment is on
    float    xs;                // x coordinate of segment center [m]
    float    ys;                // y coordinate of segment center [m]
    float    length;            // length of segment [m]
    int      parallel;          // parallel to 'x' or 'y'
    struct   GammaSeg  *next;   // pointer to next link in linked list
};
typedef struct GammaSeg GAMMA_SEG; // gamma segment type
/*   Procedures Called (all are in this file)
     ---------------------------------------------------------------
-- */
float vspxgp (float xp, float yp, float xs, float ys, float length);
float FunctU (float length, float X, float Y);
float vspygp (float xp, float yp, float xs, float ys, float length);
int   SimpleGauss (int nrows, int ncols, float *matrix);
int   IndexMatrix (int irow, int jcol, int nrows);
/* ----- Function Declaration --------------------------------------
-- */
void main ()
{
/*  --- Local Variable Declarations ---------------------------------
---
  Type         Name                 Description & range
  ----         ----                 ---------------------------------
-- */
  const int numtraces = 3;        // defines the number of traces
   float      xmin[numtraces],     // trace sides, top & bottom of
traces
             xmax[numtraces],
             ymin[numtraces],
             ymax[numtraces];
  int        tracenum;            // trace number in for loop
  GAMMA_SEG *gammaList = NULL,     // pointer to list of gamma segments
```

```
             *gammaTmp   = NULL,    // temp. pointer to gamma segment
             *gammaTmp2 = NULL;     // temp. pointer to gamma segment
    float    *vijgamma  = NULL;     // pointer for voltage matrix
    float    Cap[numtraces*numtraces];     // capacitance matrix
    float    work[2*numtraces*numtraces];  // working matrix
                                   // (inductance will be in right half)
    int      xsegs,                // number of segments, top & bottom
             ysegs,                // number of segments each side
             numgammas;            // number of gamma segments
    float    lx;                   // length of a segment
    int      idx;                  // for-loop index
    int      row, col;             // row, column indices
    int      colCap;               // column index of capacitance matrix
    float    eps;                  // relative permittivity
    float    potential,            // potential of a point
             xi, yi,               // coordinates of observer point
             xs, ys;               // coordinates of center of segment
    float    sum;                  // sum of matrix element and its
                                   // transpose element
    float    c2;                   // speed of light squared
    int      num2;                 // number of traces squared
/*--------------ExecutableCode----------------------------------*/
    // Define trace positions and dimensions
    // Trace 1
    xmin[0] = 0.0F * MILS_TO_METERS;          // enter [mils]
    xmax[0] = 4.0F * MILS_TO_METERS;
    ymin[0] = 4.0F * MILS_TO_METERS;
    ymax[0] = 5.0F * MILS_TO_METERS;
    // Trace 2
    xmin[1] = 0.0F * MILS_TO_METERS;          // enter [mils]
    xmax[1] = 4.0F * MILS_TO_METERS;
    ymin[1] = 1.0F * MILS_TO_METERS;
    ymax[1] = 2.0F * MILS_TO_METERS;
    // Trace 3
    xmin[2] = 5.0F * MILS_TO_METERS;          // enter [mils]
    xmax[2] = 9.0F * MILS_TO_METERS;
    ymin[2] = 1.0F * MILS_TO_METERS;
    ymax[2] = 2.0F * MILS_TO_METERS;
    // Define dielectric constant
    eps = 3.0F;
    // Input segmenting parameters
    printf("\nCapacitance of %i traces\n", numtraces);
    printf("Enter number of x-direction segments: ");
    scanf("%i", &xsegs);
    printf("Enter number of y-direction segments: ");
    scanf("%i", &ysegs);
```

```c
                // Build linked list of segments
numgammas = 0;                    // gamma segment counter
for (tracenum = 0; tracenum < numtraces; tracenum++)
{
    lx = (xmax[tracenum] - xmin[tracenum])/xsegs;  // length of segs
    for (idx = 0; idx < xsegs; idx++)
    {
        // top of trace
        gammaTmp = gammaList;   // save pointer to previous link
                                // first time it's NULL
        gammaList = (GAMMA_SEG*)malloc(sizeof(GAMMA_SEG));
        gammaList->next      = gammaTmp;
        gammaList->epsratio  = 1.0F/eps;
        gammaList->TraceNo    = tracenum+1;
        gammaList->length    = lx;
        gammaList->xs        = xmin[tracenum] + lx * (idx + 0.5F);
        gammaList->ys        = ymax[tracenum];
        gammaList->parallel  = 'x';
        // bottom of trace
        gammaTmp = gammaList;  // save pointer to previous link
        gammaList = (GAMMA_SEG*)malloc(sizeof(GAMMA_SEG));
        gammaList->next      = gammaTmp;
        gammaList->epsratio  = 1.0F/eps;
        gammaList->TraceNo    = tracenum+1;
        gammaList->length    = lx;
        gammaList->xs        = xmin[tracenum] + lx * (idx + 0.5F);
        gammaList->ys        = ymin[tracenum];
        gammaList->parallel  = 'x';
        numgammas += 2;
    }
    lx = (ymax[tracenum] - ymin[tracenum])/ysegs;  // length of segs
    for (idx = 0; idx < ysegs; idx++)
    {
        // left side of trace
        gammaTmp = gammaList;  // save pointer to previous link
        gammaList = (GAMMA_SEG*)malloc(sizeof(GAMMA_SEG));
        gammaList->next      = gammaTmp;
        gammaList->epsratio  = 1.0F/eps;
        gammaList->TraceNo    = tracenum+1;
        gammaList->length    = lx;
        gammaList->xs        = xmin[tracenum];
        gammaList->ys        = ymin[tracenum] + lx * (idx + 0.5F);
        gammaList->parallel  = 'y';
        // right side of trace
        gammaTmp = gammaList;  // save pointer to previous link
        gammaList = (GAMMA_SEG*)malloc(sizeof(GAMMA_SEG));
```

```c
        gammaList->next        = gammaTmp;
        gammaList->epsratio    = 1.0F/eps;
        gammaList->TraceNo      = tracenum+1;
        gammaList->length       = lx;
        gammaList->xs           = xmax[tracenum];
        gammaList->ys           = ymin[tracenum] + lx * (idx + 0.5F);
        gammaList->parallel     = 'y';
        numgammas += 2;
    }
}
// Load potential matrix vijgamma
// Allocate memory for voltage matrix
// n x (n+numtraces) matrix for SimpleGauss
vijgamma = (float*)malloc(sizeof(float) * numgammas
         * (numgammas+numtraces));
// load voltage matrix (left n x n part of vijgamma)
gammaTmp = gammaList;
for (row = 0; row < numgammas; row++)
{
    gammaTmp2 = gammaList;
    for (col = 0; col < numgammas; col++)
    {
        xi = gammaTmp->xs;        // point to evaluate potential
        yi = gammaTmp->ys;
        xs = gammaTmp2->xs;       // center of charged segment
        ys = gammaTmp2->ys;
        if (gammaTmp2->parallel == 'y')
            potential = vspygp (xi, yi, xs, ys, gammaTmp2->length);
        else if  (gammaTmp2->parallel == 'x')
            potential = vspxgp (xi, yi, xs, ys, gammaTmp2->length);
        else
        {
            printf ("Error -- parallel not correct\n");
            goto AllDone;
        }
        *(vijgamma + row + numgammas * col) = potential
            * gammaTmp2->epsratio;
        gammaTmp2 = gammaTmp2->next;
    }
    gammaTmp = gammaTmp->next;
}
// Solve for capacitance
// Load columns (numgammas + 1) to (numgammas + numtraces)
// Elements are (potentials) 1 or 0 [V]
for (col = numgammas; col < numgammas + numtraces; col++)
{
```

216 CAPACITANCE AND INDUCTANCE IN A HOMOGENEOUS MEDIUM

```
      gammaTmp = gammaList;  // loop through gamma list
      for (row = 0; row < numgammas; row++)
      {
         if (gammaTmp->TraceNo == col - numgammas + 1)
            *(vijgamma + row + numgammas * col) = 1.0F;
         else
            *(vijgamma + row + numgammas * col) = 0.0F;
         gammaTmp = gammaTmp->next;
      }
   }
   // Solve for charges on segments using SimpleGauss
   // (After this call, the segment charges are in columns
   // (numgammas + 1) to (numgammas + numtraces)).
   if (SimpleGauss (numgammas, numgammas+numtraces, vijgamma))
   {
      printf ("Error -- SimpleGauss -- charge calc.\n");
      goto AllDone;
   }
   // Calculate capacitance -- the sum of the charges on the segments
   // for each trace
   for (col = numgammas; col < numgammas + numtraces; col++)
   {
      // Initialize this column of capacitance to 0
      colCap = col - numgammas; // column index for capacitance matrix
      for (row = 0; row < numtraces; row++)
         Cap[row + numtraces * colCap] = 0.0F;
      gammaTmp = gammaList;  // loop through segment list
      for (row = 0; row < numgammas; row++)
      {
         if (gammaTmp->TraceNo > 0 && gammaTmp->TraceNo <= numtraces)
            Cap[gammaTmp->TraceNo - 1 + numtraces * colCap]
               += *(vijgamma + row + numgammas * col); //add segment
                                              //charge [C]
         gammaTmp = gammaTmp->next;
      }
   }
   // Force symmetry -- set each element to the mean of the transposed
   // capacitance matrix elements
   for (row = 0; row < numtraces; row++)
      for (col = row + 1; col < numtraces; col++)   // upper triangle
      {
         sum =Cap[row + col * numtraces] + Cap[col + row * numtraces];
         Cap[row + col * numtraces] = 0.5F * sum;
         Cap[col + row * numtraces] = 0.5F * sum;
      }
   // Calculate the inductance matrix
```

```
c2 = C_LIGHT * C_LIGHT;          // speed of light squared
num2 = numtraces*numtraces;    // number of elements in L matrix
// Load inverse of inductance matrix in left half of working matrix
for (idx = 0; idx < num2; idx++)
   work[idx] = c2 * Cap[idx]/eps;
// Load unit matrix in right half of working matrix
for (idx = num2; idx < 2 * num2; idx++)
   if ((idx - num2) % (numtraces + 1) == 0)  // diagonal element
      work[idx] = 1.0F;                       // set diagonals to 1
   else
      work[idx] = 0.0F;
// Call SimpleGauss
if (SimpleGauss (numtraces, 2*numtraces, work))
{
   printf ("Error -- SimpleGauss -- inductance calc.\n");
   goto AllDone;
}
// Printf output to the screen
for (idx = 0; idx < numtraces; idx++)
{
   printf ("Trace number: %i\n",idx+1);
   printf("xmin = %.2f    xmax = %.2f [mils]\n",
      xmin[idx]/MILS_TO_METERS, xmax[idx]/MILS_TO_METERS);
   printf("ymin = %.2f    ymas = %.2f [mils]\n",
      ymin[idx]/MILS_TO_METERS, ymax[idx]/MILS_TO_METERS);
}
printf("relative permittivity = %.2f (homogeneous)\n", eps);
printf("\ncapacitance matrix [pF/m]\n");
for (row = 0; row < numtraces; row++)
{
   for (col = 0; col < numtraces; col++)
      printf("%.4f ", Cap[row + col * numtraces]*1.0e12F);
   printf("\n");
}
printf("\ninductance matrix [microH/m]\n");
for (row = 0; row < numtraces; row++)
{
   for (col = 0; col < numtraces; col++)
      printf("%.6f ", work[num2 + row + col * numtraces]*1.0e6F);
   printf("\n");
}
printf("\nNORMAL EXIT\n");
AllDone:
// Free memory
free (vijgamma);       vijgamma = NULL;
```

```
        while (gammaList != NULL)
        {
            gammaTmp = gammaList->next;
            free(gammaList);
            gammaList = gammaTmp;
        }
    }
}
/*-----Endof Proceduremain--------------------------------------------*/
```

The trace configuration defined in the main program (for test and demonstration) is shown below in Figure 16.11.

The output of the program (as directed to the screen) is listed below.

```
Capacitance of 3 traces
Enter number of x-direction segments: 8
Enter number of y-direction segments: 3
Trace number: 1
xmin = 0.00    xmax = 4.00 [mils]
ymin = 4.00    ymas = 5.00 [mils]
Trace number: 2
xmin = 0.00    xmax = 4.00 [mils]
ymin = 1.00    ymas = 2.00 [mils]
Trace number: 3
xmin = 5.00    xmax = 9.00 [mils]
ymin = 1.00    ymas = 2.00 [mils]
relative permittivity = 3.00 (homogeneous)
capacitance matrix [pF/m]
121.1585 -63.0900 -22.9002
-63.0900 250.5547 -39.4041
-22.9002 -39.4041 221.5802
```

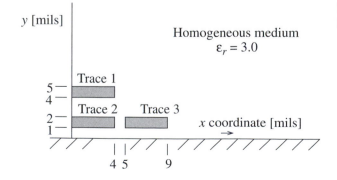

Figure 16.11 Multiple-Trace Geometry Defined in Computer Code 16-2.

```
inductance matrix [microH/m]
0.332894 0.091801 0.050730
0.091801 0.162371 0.038362
0.050730 0.038362 0.162708
NORMAL EXIT
```

Chapter 17
ELECTRIC FIELDS IN A LAYERED CIRCUIT BOARD

A digital circuit board often has several layers of dielectric over a ground plane as shown schematically in Figure 17.1. The layers have different dielectric constants. Traces may be placed on top of the layers or embedded in them. The top or bottom of a trace may lie tangent to a dielectric boundary or the boundary may intersect the trace. This is a challenging geometry for capacitance calculations. The method developed for this situation is an extension of that for the homogeneous case; the extension is the addition of two additional surface charge densities that contribute to the potential throughout the domain of the problem. The surface charge densities are located at the boundaries of the dielectrics (the surfaces where two media of different dielectric constants touch each other) and the surfaces where the conducting traces contact dielectrics with relative permittivity greater than one. With the potential function properly modified, the capacitance can be calculated as it was using Equations (16-35) and (16-36).

The development of the new algorithm focuses on the derivations of the new terms to be included in the potential. The two new terms can be written

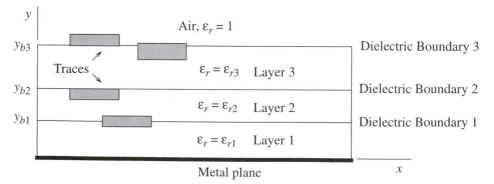

Figure 17.1 Schematic of Multi-Layered Circuit Board with Single Metal Ground Plane.

$$w_{\beta i} = \sum_{j_\beta} v_{i,j_\beta} \tilde{q}_{j_\beta} \qquad (17\text{-}1)$$

and

$$w_{\Gamma i} = \sum_{j_\Gamma} v_{i,j_\Gamma} \tilde{q}_{j_\Gamma} \qquad (17\text{-}2)$$

where β is the set of points on the dielectric boundaries, $\tilde{q}_{j_\beta}$ is the charge (or more correctly, the equivalent charge) on segment j_β of a dielectric boundary and $\tilde{q}_{j_\Gamma}$ is the equivalent charge on the dielectric adjacent to the trace surface segment j_Γ. The potential throughout the region of the transmission line will be expressed as the sum of these two new terms and the original potential (due to the physical charges on the trace segments) given by Equation (16-6). Including these two new terms, the potential at point $\mathbf{x}_i$ is

$$w_i = \sum_{j_\beta} v_{i,j_\beta} \tilde{q}_{j_\beta} + \sum_{j_\Gamma} v_{i,j_\Gamma} \tilde{q}_{j_\Gamma} + \sum_{j_\Gamma} v_{i,j_\Gamma} q_{j_\Gamma} \qquad (17\text{-}3)$$

Each of these terms will be discussed and evaluated below.

17.1 BOUNDARY CONDITIONS AT A DIELECTRIC–DIELECTRIC BOUNDARY

When an electric field passes from one dielectric to another, its normal vector component changes discontinuously at the boundary. This can be visualized as due to the electric field's relationship with the electric displacement vector $\mathbf{D}$. The electric field vector $\mathbf{E}$ and the electric displacement vector $\mathbf{D}$ in an (ordinary noncrystalline) dielectric are related by the (constitutive) equation

$$\mathbf{D} = \varepsilon \mathbf{E} \qquad (17\text{-}4)$$

where ε is the permittivity of the dielectric. At a boundary between two dielectrics that have no physical charge residing at the boundary, the normal component of the electric displacement vector $\mathbf{D}$ is continuous. That is $D_{n1} = D_{n2}$ (see Figure 17.2).[1]

The fact that the normal component of $\mathbf{D}$ is unchanged across the boundary combined with Equation (17-4) implies that the normal component of the electric field $\mathbf{E}$ *does* change across the boundary. The change is implied by the equation

$$\varepsilon_1 E_{n1} = \varepsilon_2 E_{n2} \qquad (17\text{-}5)$$

To calculate the electric field with layers of discontinuity, an equivalent, but somewhat different situation can be analyzed. The electric field configuration can be regarded as existing entirely in a homogeneous medium with the permittivity ε_o of free space. The discontinuity in

---

1. For further explanation of these boundary conditions, see Smythe, 1950, pp. 18–20.

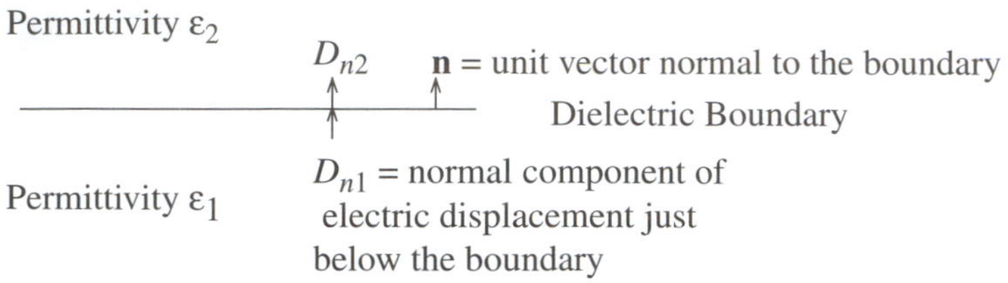

Permittivity ε_2

D_{n2} **n** = unit vector normal to the boundary

Dielectric Boundary

Permittivity ε_1

D_{n1} = normal component of
  electric displacement just
  below the boundary

Figure 17.2   Electric Displacement Normal to a Dielectric Boundary.

the normal component of **E** at each point on the dielectric boundaries can be regarded as due to an *equivalent* surface charge density (the physical surface charge density at these boundaries is zero). Hence, the situation shown above in Figure 17.2 is viewed as the equivalent situation shown below in Figure 17.3. The appropriate value of the equivalent surface charge density $\tilde{\rho}_s$ will be derived below (see Stratton, 1941, pp. 183, 184 for a further discussion of this equivalence).

PROBLEM 17-1   Referring to Figure 17.2 and Equations (17-4) and (17-5), let $\varepsilon_1 = \varepsilon_0$, the permittivity of free space, $\varepsilon_2 = 5\varepsilon_0$, and $E_{n1} = 5{,}000$ [V/m]. What is E_{n2}?

17.2 EQUIVALENT CHARGE AT DIELECTRIC–DIELECTRIC BOUNDARIES

Let $\tilde{E}_n$ be the normal component of the electric field *caused by* a layer of surface charge density. Because of symmetry, this field must be equal, but of opposite sign, on each side of the layer of surface charge. The electric field on each side must account for half of the divergence of the electric displacement from the surface charge. Referring to Figure 17.3, one has

$$\tilde{E}_{n2} = \frac{\tilde{\rho}_s}{2\varepsilon_o} \qquad (17\text{-}6)$$

$$\tilde{E}_{n1} = \frac{-\tilde{\rho}_s}{2\varepsilon_o} \qquad (17\text{-}7)$$

Let the normal electric field at the point on the dielectric boundary due to *all other charges* be $\bar{E}_n$. Then, below the charged layer, the normal electric field is

$$E_{n1} = \bar{E}_n + \tilde{E}_{n1} = \bar{E}_n - \frac{\tilde{\rho}_s}{2\varepsilon_o} \qquad (17\text{-}8)$$

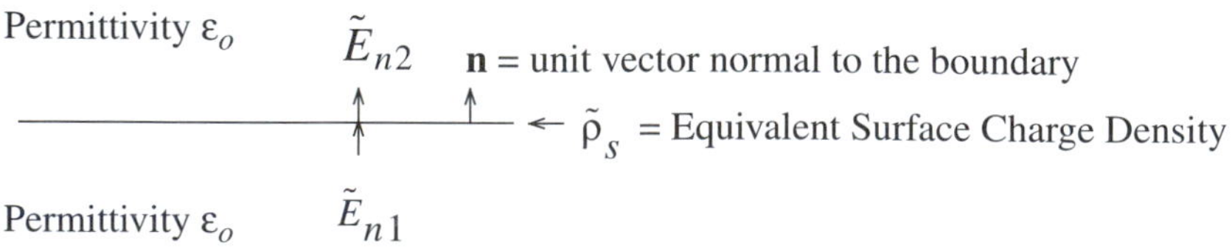

Permittivity ε_o $\tilde{E}_{n2}$ **n** = unit vector normal to the boundary

$\leftarrow \tilde{\rho}_s$ = Equivalent Surface Charge Density

Permittivity ε_o $\tilde{E}_{n1}$

Figure 17.3   Electric Field Diverging from a Layer of Surface Charge.

and above the charged layer, the normal electric field is

$$E_{n2} = \bar{E}_n + \tilde{E}_{n2} = \bar{E}_n + \frac{\tilde{\rho}_s}{2\varepsilon_o} \tag{17-9}$$

Substituting these into Equation (17-5), we obtain

$$\varepsilon_1\left(\bar{E}_n - \frac{\tilde{\rho}_s}{2\varepsilon_o}\right) = \varepsilon_2\left(\bar{E}_n + \frac{\tilde{\rho}_s}{2\varepsilon_o}\right) \tag{17-10}$$

Equation (17-10) can be solved for $\tilde{\rho}_s$ to obtain

$$\tilde{\rho}_s = 2\varepsilon_o\bar{E}_n\frac{\varepsilon_1 - \varepsilon_2}{\varepsilon_1 + \varepsilon_2} \tag{17-11}$$

Therefore, the equivalent surface charge density at any point (on any segment) of a dielectric boundary can be found in terms of the normal electric field (at that point) due to all the other charges. The other charges include both the physical charges on the surfaces of the traces and the other equivalent charges on the dielectric boundaries. The charge on any segment affects the charges on all the other segments. The solution for the charges must involve simultaneously all of the charges present.

PROBLEM 17-2 Assume that the normal electric field due to all charges present at a point on a dielectric boundary is $\bar{E}_n = 5,000$ [V/m]. Let $\varepsilon_1 = \varepsilon_o$, the permittivity of free space and $\varepsilon_2 = 5\varepsilon_o$ (as in the preceding problem). Use Equation (17-11) to obtain the equivalent surface charge density in [C/m²]. Use $\varepsilon_o = 8.854\times10^{-12}$ [F/m].

Let the points on the dielectric boundaries be defined as the set β. These surfaces can be divided into segments just as the surfaces of the traces were divided into segments. And, as in the case of the segments on the traces, the equivalent surface charge density on the segments of β will be approximated as constant on each segment. If the length of β segment j_β is l_{j_β}, then the charge on the segment is

$$\tilde{q}_{j_\beta} = l_{j_\beta}\tilde{\rho}_{s j_\beta} \tag{17-12}$$

Substituting Equation (17-11) into Equation (17-12), we obtain

$$\tilde{q}_{j_\beta} = 2\varepsilon_o\frac{\varepsilon_{1j_\beta} - \varepsilon_{2j_\beta}}{\varepsilon_{1j_\beta} + \varepsilon_{2j_\beta}}l_{j_\beta}\bar{E}_{n j_\beta} \tag{17-13}$$

where subscripts have been added to identify $\bar{E}_n$, ε_1, and ε_2 with segment j_β.

The electric field is given as minus the gradient of the potential; therefore, the previously defined potentials for the uniformly charged trace surface segments can be differentiated to obtain functions for the required normal component of the electric field at the required points, the centers of the dielectric segments.

17.3 DIELECTRICS ADJACENT TO TRACE SURFACES

In considering the field configuration very close to a conducting surface, it is instructive to allow a very small (infinitesimal) slab of free space between the conducting surface and the surface of the dielectric. The geometry is illustrated below in Figure 17.4. Using this concept, the physical surface charge density on the surface of the trace and the *equivalent* surface charge density on the dielectric boundary (that allows us to calculate electric fields and potentials as if the permittivity were that of free space) can be separated and the physics of the situation better understood. In reality, the infinitesimal slab has zero thickness though it is illustrated with a finite thickness.

To find the equivalent surface charge density $\tilde{\rho}_s$, first apply Equation (17-5) to the situation shown in Figure 17.4. In Equation (17-5), set ε_1 to ε_o and ε_2 to ε_1, and E_{n1} to $E_n = \rho_s/\varepsilon_o$; we obtain

$$\varepsilon_o \frac{\rho_s}{\varepsilon_o} = \varepsilon_1 E_{n2} \tag{17-14}$$

Rearranging the terms of Equation (17-14), we have

$$E_{n2} = \frac{\rho_s}{\varepsilon_1} \tag{17-15}$$

By adding Equations (17-8) and (17-9), we note that

$$\bar{E}_n = \frac{1}{2}(E_{n1} + E_{n2}) \tag{17-16}$$

And substituting ρ_s/ε_o for E_{n1} and $E_{n2} = \rho_s/\varepsilon_1$ from Equation (17-15) we obtain

$$\bar{E}_n = \frac{\rho_s}{2}\left(\frac{1}{\varepsilon_o} + \frac{1}{\varepsilon_1}\right) \tag{17-17}$$

Before proceeding further, note that Equation (17-17) is a rather interesting result. That is, $\bar{E}_n$ was *defined* to be the normal electric field due to *all of the charges* other than the sur-

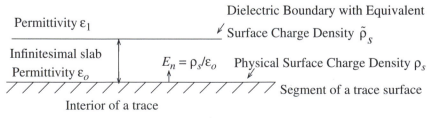

Permittivity ε_1

Dielectric Boundary with Equivalent
Surface Charge Density $\tilde{\rho}_s$

Infinitesimal slab
Permittivity ε_o

$E_n = \rho_s/\varepsilon_o$ Physical Surface Charge Density ρ_s

Segment of a trace surface

Interior of a trace

Figure 17.4 Dielectric Adjacent to Conducting Trace Surface.

face charge density at the given point on the dielectric-to-dielectric boundary. At dielectric–dielectric boundaries, it will be expressed exactly that way, as a sum of the normal electric field components due to each of the charges on all of the segments (both Γ and β). However, when this boundary is infinitesimally separated from a conducting surface, its value is obtained in terms of the surface charge density on the adjacent metal surface. The capacitance can be calculated by expressing $\bar{E}_n$ in either way, but the method given here, Equation (17-17), is the more numerically efficient of the two.

Equation (17-11) applied to the model shown in Figure 17.4 becomes

$$\tilde{\rho}_s = 2\varepsilon_o \bar{E}_n \frac{\varepsilon_o - \varepsilon_1}{\varepsilon_o + \varepsilon_1} \tag{17-18}$$

We then substitute Equation (17-17) into Equation (17-18) and simplify to obtain

$$\tilde{\rho}_s = \rho_s \frac{\varepsilon_o - \varepsilon_1}{\varepsilon_1} \tag{17-19}$$

Equation (17-19) can be multiplied by the length of segment j_Γ, to obtain the equivalent charge on the segment in terms of the physical charge

$$\tilde{q}_{j_\Gamma} = q_{j_\Gamma} \frac{\varepsilon_o - \varepsilon_{1 j_\Gamma}}{\varepsilon_{1 j_\Gamma}} \tag{17-20}$$

where $\tilde{q}_{j_\Gamma} = l_{j_\Gamma} \tilde{\rho}_{s j_\Gamma}$ and $q_{j_\Gamma} = l_{j_\Gamma} \rho_{s j_\Gamma}$.

Equation (17-20) can be used to write a simple expression for the combined effect of the physical charge and the equivalent charge on the adjacent dielectric at any Γ segment. Using Equation (17-20), we find the sum of the two charges to be

$$\tilde{q}_{j_\Gamma} + q_{j_\Gamma} = q_{j_\Gamma} \frac{\varepsilon_o}{\varepsilon_{1 j_\Gamma}} = \frac{q_{j_\Gamma}}{\varepsilon_{r j_\Gamma}} \tag{17-21}$$

where $\varepsilon_{r j_\Gamma} = \varepsilon_{1 j_\Gamma} / \varepsilon_o$ is the relative permittivity of the dielectric adjacent to trace segment j_Γ. Hence, the single value $q_{j_\Gamma} / \varepsilon_{r j_\Gamma}$ includes both the physical and equivalent surface charge densities at the interface of a dielectric and a conductor surface (where q_{j_Γ} is the *physical* charge). In using Equation (17-3) to calculate potentials, the last two terms combine using Equation (17-21) to yield

$$w_i = \sum_{j_\beta} v_{i,j_\beta} \tilde{q}_{j_\beta} + \sum_{j_\Gamma} v_{i,j_\Gamma} \frac{q_{j_\Gamma}}{\varepsilon_{r j_\Gamma}} \tag{17-22}$$

The combined charge will be used in the next section as the source for electric fields due to the charge on Γ segments.

PROBLEM 17-3 Assume the normal electric field adjacent to a conductor (in the dielectric) is $E_n = 5,000$ [V/m]. Let the permittivity of the dielectric be $5\varepsilon_o$. What part of the electric field is due to the physical charge and what part is due to the equivalent surface charge on the dielectric. *Hint*: The total electric field (5,000 [V/m]) is given by Equation (17-15); the part due to the physical charge is given by ρ_s/ε_o.

17.4 EQUIVALENT CHARGES INDUCED BY PHYSICAL CHARGES

The electric field at any point can be obtained by taking the gradient of Equation (17-22). Symbolically, we write the normal electric field at point $\mathbf{x}_i$ due to a segment of charge at point $\mathbf{x}_s$ as

$$\bar{E}_n(\mathbf{x}_i) = -\nabla_n v(\mathbf{x},\mathbf{x}_s)\Big|_{\mathbf{x}=\mathbf{x}_i} \qquad (17\text{-}23)$$

where $\nabla_n = \dfrac{\partial}{\partial x_n}$ is the derivative in the direction of the normal. For dielectric boundaries parallel to the x-z plane (as shown in Figure 17.1) the normal is defined in the y direction (as shown in Figure 17.2) and $\nabla_n = \dfrac{\partial}{\partial y}$. At the edge of a circuit board the normal may be in the x, or the minus x direction; the normal gradient operator is $\nabla_n = \pm\dfrac{\partial}{\partial x}$.

To write the normal gradient of Equation (17-22) at an arbitrary point $\mathbf{x}_i$, we need the normal gradients of v_{i,j_β} and v_{i,j_Γ}. For normal vectors in the x and y directions, expressions for their partial derivatives with respect to x and y are needed. The potential function $v(\mathbf{x},\mathbf{x}_s)$ is given explicitly in Equations (16-19) and (16-20) for the cases of a segment parallel to the x and y axis, respectively (with a ground plane at $y = 0$). The normal gradients needed are derivatives of these expressions. There are four cases of interest; the results are given below.

Segment Parallel to the x Axis, Derivatives of Equation (16-19):

$$E_x(\mathbf{x},\mathbf{x}_s) = U_1\left(x_s + \frac{l}{2} - x, y_s - y\right) - U_1\left(x_s - \frac{l}{2} - x, y_s - y\right)$$

$$-U_1\left(x_s + \frac{l}{2} - x, y_s + y\right) + U_1\left(x_s - \frac{l}{2} - x, y_s + y\right) \qquad (17\text{-}24)$$

$$E_y(\mathbf{x},\mathbf{x}_s) = U_2\left(x_s + \frac{l}{2} - x, y_s - y\right) - U_2\left(x_s - \frac{l}{2} - x, y_s - y\right)$$

$$+ U_2\left(x_s + \frac{l}{2} - x, y_s + y\right) - U_2\left(x_s - \frac{l}{2} - x, y_s + y\right) \qquad (17\text{-}25)$$

Segment Parallel to the y Axis, Derivatives of Equation (16-20):

$$E_x(\mathbf{x},\mathbf{x}_s) = U_2\left(y_s + \frac{l}{2} - y, x_s - x\right) - U_2\left(y_s - \frac{l}{2} - y, x_s - x\right)$$

$$-U_2\left(y_s + \frac{l}{2} + y, x_s - x\right) + U_2\left(y_s - \frac{l}{2} + y, x_s - x\right) \qquad (17\text{-}26)$$

$$E_y(\mathbf{x},\mathbf{x}_s) = U_1\left(y_s + \frac{l}{2} - y, x_s - x\right) - U_1\left(y_s - \frac{l}{2} - y, x_s - x\right)$$

$$+ U_1\left(y_s + \frac{l}{2} + y, x_s - x\right) - U_1\left(y_s - \frac{l}{2} + y, x_s - x\right) \tag{17-27}$$

where the electric fields in Equations (17-24) to (17-27) are fields due to a unit charge and

$$U_1(X,Y) \equiv \frac{\partial U}{\partial X} = \frac{-\ln(X^2 + Y^2)}{4\pi\varepsilon_o l} \tag{17-28}$$

$$U_2(X,Y) \equiv \frac{\partial U}{\partial Y} = \frac{-\mathrm{atan}(X/Y)}{2\pi\varepsilon_o l} \tag{17-29}$$

The electric field at an arbitrary point $\mathbf{x}_i$ is minus the gradient of Equation (17-22). In terms of the electric fields given in Equations (17-24) to (17-27), it can be written

$$\bar{E}_n(\mathbf{x}_i) = \sum_{k_\beta} E_n(\mathbf{x}_i, \mathbf{x}_{k_\beta}) \tilde{q}_{k_\beta} + \sum_{j_\Gamma} E_n(\mathbf{x}_i, \mathbf{x}_{j_\Gamma}) \frac{q_{j_\Gamma}}{\varepsilon_{r j_\Gamma}} \tag{17-30}$$

where the summation index j_β has been changed to k_β. This expression is now evaluated at point j_β and substituted into Equation (17-13) to obtain an expression for the charge on β segment j_β.

$$\tilde{q}_{j_\beta} = \sum_{k_\beta} \varepsilon_o l_{j_\beta} \tilde{b}_{j_\beta} E_n(\mathbf{x}_{j_\beta}, \mathbf{x}_{k_\beta}) \tilde{q}_{k_\beta} + \sum_{j_\Gamma} \varepsilon_o l_{j_\beta} \tilde{b}_{j_\beta} E_n(\mathbf{x}_{j_\beta}, \mathbf{x}_{j_\Gamma}) \frac{q_{j_\Gamma}}{\varepsilon_{r j_\Gamma}} \tag{17-31}$$

where

$$\tilde{b}_{j_\beta} = 2 \frac{\varepsilon_{1 j_\beta} - \varepsilon_{2 j_\beta}}{\varepsilon_{1 j_\beta} + \varepsilon_{2 j_\beta}}$$

and the direction of the normal electric field is determined by the direction of the segment j_β.

Equation (17-31) can be solved for the equivalent charges at the dielectric boundaries $\tilde{q}_{j_\beta}$ as a function of the physical charges q_{j_Γ}; the solution is simplified by writing the equation in matrix form. Let $\tilde{\mathbf{T}}$ be the (square) $J_\beta \times J_\beta$ matrix whose elements are

$$\tilde{T}_{j_\beta k_\beta} = \varepsilon_o l_{j_\beta} \tilde{b}_{j_\beta} E_n(\mathbf{x}_{j_\beta}, \mathbf{x}_{k_\beta}) \tag{17-32}$$

And let $\mathbf{T}$ be the $J_\beta \times J_\Gamma$ matrix whose elements are

$$T_{j_\beta j_\Gamma} = \varepsilon_o l_{j_\beta} \tilde{b}_{j_\beta} E_n(\mathbf{x}_{j_\beta}, \mathbf{x}_{j_\Gamma})/\varepsilon_{r j_\Gamma} \tag{17-33}$$

where J_β is the number of segments that comprise the dielectric boundaries and J_Γ is the number of segments that comprise trace surfaces. Equation (17-31) can be written as the matrix equation

$$\tilde{\mathbf{q}} = \tilde{\mathbf{T}}\tilde{\mathbf{q}} + \mathbf{Tq} \tag{17-34}$$

where $\tilde{\mathbf{q}}$ is the column vector with J_β elements of equivalent charge $\tilde{q}_{j_\beta}$ and $\mathbf{q}$ is the column vector with J_Γ elements of physical charge q_{j_Γ}. Subtracting the term $\tilde{\mathbf{T}}\tilde{\mathbf{q}}$ from both sides of Equation (17-34) and multiplying both sides by $(\mathbf{I} - \tilde{\mathbf{T}})^{-1}$, one obtains

$$\tilde{\mathbf{q}} = (\mathbf{I} - \tilde{\mathbf{T}})^{-1}\mathbf{Tq} \tag{17-35}$$

where $\mathbf{I}$ is the identity matrix. Hence, the equivalent charges at the dielectric boundaries have been expressed as a function of the physical charges on the trace surfaces. This equation will be combined with Equation (17-22) to obtain an expression for the electric potential at any point as a function of the physical charge only (but including the induced, equivalent charges, on the dielectric boundaries). The new expression for the electric potential will be used in place of Equation (16-6) to simulate the capacitance matrix for traces in a layered medium.

17.5 DIELECTRIC BOUNDARY INTERSECTING A CONDUCTOR SURFACE

In a preceding section (17.3), the equivalent charge on a dielectric surface adjacent to a conductor boundary was analyzed. It was found that the sum of the physical charge and the equivalent charge is given by q/ε_r ; see Equation (17-21). In this Section, we consider the situation where different dielectrics are adjacent to different parts of the trace segment. That is, a dielectric boundary intersects the trace surface at right angles to it. The geometry is illustrated in Figure 17.5. For this situation, an effective value of the relative permittivity will be derived. The effective value is a weighted average of the two permittivities that can be used as a single permittivity at such a trace segment.

The length of the segment is $l = l_1 + l_2$ and the physical charge on the segment is

$$q = l_1 \rho_{s1} + l_2 \rho_{s2} \tag{17-36}$$

where ρ_{s1} and ρ_{s2} are the physical surface charge densities on parts 1 and 2 of the segment, respectively. The electric field parallel to such a dielectric boundary is continuous across it. Hence, the electric fields normal to each part of the segment are (approximately) equal.

$$E_n = E_{n1} = E_{n2} \tag{17-37}$$

Using the results of section 17.3, E_n is given by (including both the physical and the equivalent charges)

$$\varepsilon_o E_n = \frac{\rho_{s1}}{\varepsilon_{r1}} = \frac{\rho_{s2}}{\varepsilon_{r2}} \tag{17-38}$$

Equations (17-36) and (17-38) can be solved simultaneously to yield

$$\rho_{s1} = \frac{q\varepsilon_{r1}}{l_1\varepsilon_{r1} + l_2\varepsilon_{r2}} \tag{17-39}$$

$$\rho_{s2} = \frac{q\varepsilon_{r2}}{l_1\varepsilon_{r1} + l_2\varepsilon_{r2}} \tag{17-40}$$

We desire an effective value of relative permittivity $\bar{\varepsilon}_r$ such that

$$\frac{q}{\bar{\varepsilon}_r} = \frac{l_1\rho_{s1}}{\varepsilon_{r1}} + \frac{l_2\rho_{s2}}{\varepsilon_{r2}} \tag{17-41}$$

Substituting Equations (17-39) and (17-40) into Equation (17-41) yield

$$\frac{q}{\bar{\varepsilon}_r} = \frac{q(l_1 + l_2)}{l_1\varepsilon_{r1} + l_2\varepsilon_{r2}} \tag{17-42}$$

and

$$\bar{\varepsilon}_r = \frac{l_1\varepsilon_{r1} + l_2\varepsilon_{r2}}{(l_1 + l_2)} \tag{17-43}$$

Equation (17-43) is the weighted average that was sought. The total charge (physical plus equivalent) is therefore given by $q/\bar{\varepsilon}_r$ at a trace segment such as that shown in Figure 17.5.

PROBLEM 17-4   If $\varepsilon_{r1} = \varepsilon_{r2}$, what is $\bar{\varepsilon}_r$? Let $l_1 = l_2$, $\varepsilon_{r1} = 2\varepsilon_o$ and $\varepsilon_{r2} = 5\varepsilon_o$, What is $\bar{\varepsilon}_r$?

Figure 17.5   Dielectric Boundary Intersecting a Trace Segment.

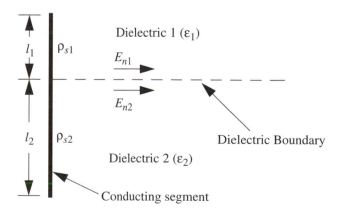

Chapter 18

CALCULATION OF CAPACITANCE IN A LAYERED MEDIUM

The capacitance is calculated using the same procedure that was defined in Section 16.2, but with the potentials modified to take into account the equivalent charges at dielectric boundaries and at the dielectrics in contact with the trace surfaces. The potentials w_i (given as a function of the physical charges only in Equation (16-6)) will be replaced with the potential that includes the equivalent charges given by Equation (17-22). Equation (17-22) can be written in matrix form as

$$\mathbf{w} = \tilde{\mathbf{V}}\tilde{\mathbf{q}} + \mathbf{V}\mathbf{q} \tag{18-1}$$

where $\tilde{\mathbf{V}}$ is defined as the $J_\Gamma \times J_\beta$ matrix with elements v_{ij_β} (that is, the potential at the ith Γ segment due to the equivalent charge at the j_βth dielectric boundary segment) and $\mathbf{V}$ is defined as the $J_\Gamma \times J_\Gamma$ matrix with elements $v_{ij_\Gamma}/\varepsilon_{rj_\Gamma}$ (that is, the potential at the ith Γ segment due to the physical charge plus the equivalent charge at the j_Γth trace surface segment). Substituting the expression for the equivalent charges given by Equation (17-35) into Equation (18-1), one obtains

$$\mathbf{w} = [\tilde{\mathbf{V}}(\mathbf{I} - \tilde{\mathbf{T}})^{-1}\mathbf{T} + \mathbf{V}]\mathbf{q} \tag{18-2}$$

This equation expresses the potentials at the trace segment centers (the elements of $\mathbf{w}$) as a function of the physical charges on the trace segments (the elements of $\mathbf{q}$), including the effect of all of the equivalent charges at dielectric boundaries and at the surfaces of the traces. Each matrix term in Equation (18-2) has a purpose and its purpose can be described as follows:

1. The matrix $\mathbf{V}$ transforms the physical charges (plus the equivalent charges adjacent to them) to the potentials at the trace surfaces caused directly by them (without regard to any equivalent charges at dielectric boundaries).

2. The matrix $\mathbf{T}$ transforms the physical charges (plus the equivalent charges adjacent to them) to the equivalent charges that they directly cause on the dielectric boundary seg-

ments without regard to the effect of the charges on the dielectric boundary segments on each other.

3. The matrix $(\mathbf{I} - \tilde{\mathbf{T}})^{-1}$ transforms the directly caused equivalent charges on the dielectric boundary segments to the actual charges on the dielectric boundary segments including their effect on each other, that is, including the effect of the electric field caused by each dielectric boundary segment on every other dielectric boundary segment.

4. The matrix $\tilde{\mathbf{V}}$ transforms the actual equivalent charges on the dielectric boundary segments to the resulting potentials at the centers of the trace segments.

As in Section 16.2 (capacitance calculation in a homogeneous medium), each column of the matrix $\mathbf{C}$ is determined by the charges resulting from a given potential distribution. The kth column of $\mathbf{C}$ is the column vector of trace charges that result when 1 [V] is impressed on trace k while the other traces are held at zero (ground) potential. To calculate the kth column of $\mathbf{C}$, we set the potentials of the segments on trace k (on the left side of Equation (18-2)) to 1 [V] and all the other potentials, elements of $\mathbf{w}$, to zero. Let $\mathbf{w}_k$ be the column vector of potentials set to these values. The resulting physical charges on the trace segments are given by the solution of Equation (18-2) for $\mathbf{q}$. Let $\mathbf{q}_k$ be the solution; it is given by

$$\mathbf{q}_k = [\tilde{\mathbf{V}}(\mathbf{I} - \tilde{\mathbf{T}})^{-1}\mathbf{T} + \mathbf{V}]^{-1}\mathbf{w}_k \tag{18-3}$$

Each element of the kth column of the matrix $\mathbf{C}$ is then given by the sum of the charges on the segments of the corresponding trace. Element c_{1k} is given by the sum of the charges on the segments of trace 1, and so on, for all the traces.

One can write a single matrix expression for the segment charges that includes all of the potential combinations. Define the $J_\Gamma \times N$ matrix $\mathbf{W}$ as the matrix with columns $\mathbf{w}_k$, $k = 1, \ldots N$, where N is the number of traces. Also define the $J_\Gamma \times N$ matrix $\mathbf{Q}$ as the matrix with columns $\mathbf{q}_k$. Then, Equation (18-3) for $k = 1, \ldots N$ can be written as the single matrix equation

$$\mathbf{Q} = [\tilde{\mathbf{V}}(\mathbf{I} - \tilde{\mathbf{T}})^{-1}\mathbf{T} + \mathbf{V}]^{-1}\mathbf{W} \tag{18-4}$$

In terms of the elements of $\mathbf{Q}$, the elements of the capacitance matrix are then given by Equation (16-36) (repeated here)

$$c_{mk} = \sum_{x_{j_\Gamma} \in \Gamma_m} q_{j_\Gamma k} \tag{16-36}$$

where $x_{j_\Gamma} \in \Gamma_m$ means that the sum includes only those values of j_Γ that represent segments on trace number m.

Computer Code 18-1. Capacitance in a Layered Circuit Board.

A computer code is included that illustrates the calculation of the capacitance matrix on a layered circuit board with a single ground plane. The example problem has only one layer of dielectric. Boundaries between solid dielectric layers can be treated identically to the air–dielectric boundaries in the example problem. In the computer code solution previously given for multiple traces in a homogeneous medium, linked lists were created for the trace parameters that included a link for each trace segment. The method is extended in this code to include an additional linked list for the dielectric boundaries; it includes a link for each dielectric boundary segment. The solution algorithm is completely general with regard to the number of dielectric boundaries and traces once the linked lists have been created. The example problem creates linked lists only for the circuit board geometry shown in Figure 18.1 (shown following the computer code listing).

The method of calculating the capacitance including the effect of dielectric boundaries is the same as that without dielectric boundaries except that Equation (18-4) replaces Equation (16-38). These matrix equations each express the charges on the trace segments under the various combinations of applied trace potential; Equation (18-4) includes the effect of the dielectric boundaries, Equation (16-38) does not. Suppose several dielectric boundaries existed but their dielectric constants are the same; then, the matrix **T** is zero (see Equation (17-33)) and the two equations are the same.

To clarify the method the computer code employs, it's helpful to define the matrix $\tilde{\mathbf{Q}}$, where

$$\tilde{\mathbf{Q}} = (\mathbf{I} - \tilde{\mathbf{T}})^{-1}\mathbf{T} \tag{18-5}$$

Equation (18-4) can then be writen

$$\mathbf{Q} = [\tilde{\mathbf{V}}\tilde{\mathbf{Q}} + \mathbf{V}]^{-1}\mathbf{W} \tag{18-6}$$

The subroutine `CalculateQtilda` has been provided to evaluate Equation (18-5). The routine performs the following computational steps.

1. It allocates memory for a $J_\beta \times (J_\beta + J_\Gamma)$ matrix (`work`) for matrix inversion (solution for $\tilde{\mathbf{Q}}$) using `SimpleGauss`.

2. It loads the matrix $(\mathbf{I} - \tilde{\mathbf{T}})$ into the left $J_\beta \times J_\beta$ part of the matrix `work`. Equation (17-32) is evaluated to provide the elements of $\tilde{\mathbf{T}}$. The electric field functions `Eypx`, `Expx`, `Eypy` and `Expy` are called to evaluate the electric field needed for these. These electric field functions evaluate Equations (17-24) to (17-27).

3. The matrix **T** is calculated using Equation (17-33) and it is loaded into the right $J_\beta \times J_\Gamma$ part of matrix `work`.

4. Finally, `SimpleGauss` is called. On exit, the $J_\beta \times J_\Gamma$ matrix $\tilde{\mathbf{Q}}$ is in the right part of matrix `work`. It is copied to the output matrix `Qtilda`.

A listing of subroutine `CalculateQtilda` is given below.

```
/* ------ Procedure CalculateQtilda ---%proc%--------------------
Description: Calculates the matrix Qtilda (see text for definition)
   Procedures Called
```

```
                ------------------------------------------------------------ */
// SimpleGauss matrix inversion
// IndexMatrix matrix offset for given indices
// EypxEy due to segment parallel to x axis
// ExpxEx due to segment parallel to x axis
// EypyEy due to segment parallel to y axis
// ExpyEx due to segment parallel to y axis
 /*------- Function Declaration -------------------------------- */
int CalculateQtilda (// ---- Input parameters ----
     int numbetas,         // number of beta segments
     BETA_SEG *betaList,   // pointer to beta segment linked list
     int numgammas,        // number of gamma segments
     GAMMA_SEG *gammaList, // pointer to gamma segment linked list
                 // ---- Output ----
       float *Qtilda)// Qtilda matrix (pointer)
{
 /*--- Local Variable Declarations ------------------------------
   Type            Name                  Description & range
   ----            ----                  ----------------------- */
   float           *work = NULL;         // working matrix
   float           term;                 // temp. variable
   BETA_SEG        *betaCol = NULL,
                   *betaRow = NULL;      // temp. pointers to beta segments
   GAMMA_SEG       *gammaCol = NULL;     // temp. pointer to gamma segment
   int             row, col;             // matrix row, column index
   float           En;                   // normal electric field
   int             numbetasq;            // number of beta segments squared
   int             rcode;                // return code
/* ---------Constants----------------------------------------
   Type            Name                  Description           Source
   ----            ----                  -----------           ------
    float          Eps0                  permittivity of space  header
 Fatal/Nonfatal     Description
 --------------     -----------
  Fatal             bad input or singular matrix input to SimpleGauss
                    value returned = 0, normal exit, < 0, on error
---------------- Executable Code ---------------------------- */
// Check input
if (numgammas < 4 || gammaList == NULL)// error -- no traces
     return -1;
if (numbetas < 1 || betaList == NULL)// Qtilda has zero rows
     return 0;
// Allocate Jbeta X (Jbeta + Jgamma) working matrix for solution
// using SimpleGauss
work = (float*)malloc(sizeof(float)
```

```
                          * numbetas * (numbetas + numgammas));
// Build matrix (I - Ttilda) in left (Jbeta X Jbeta) part of work
// (column is the source charge, row is electric field location)
betaCol = betaList;     // initialize column pointer
col = 0;                // initialize column index
while (betaCol != NULL) // loop through all the beta segments
{
  col++;                      // increment column index
  betaRow = betaList;   // initialize row pointer
  row = 0;              // initialize row index
  while (betaRow != NULL) // loop through all the beta segments
  {
    row++;                    // increment row index
    if (row == col)
          term = 1.0F;    // identity matrix
    else
          term = 0.0F;
    if (betaRow->parallel == 'x')// normal electric field
                                  // y direction
      {
        if (betaCol->parallel == 'x')// source segment || to x
          En = Eypx (betaRow->xs, betaRow->ys, // Obs. coordinates
                betaCol->xs, betaCol->ys,     // segment center
                betaCol->length);
        else    // source segment || to y
          En = Eypy (betaRow->xs, betaRow->ys,// Obs. coordinates
                betaCol->xs, betaCol->ys,     // segment center
                betaCol->length);
      }
    else                              // elect. field x direction
      {
        if (betaCol->parallel == 'x') // source segment || to x
          En = Expx (betaRow->xs, betaRow->ys,// Obs. coordinates
                betaCol->xs, betaCol->ys,     // segment center
                betaCol->length);
        else                          // source segment || to y
          En = Expy (betaRow->xs, betaRow->ys,// Obs. coordinates
                betaCol->xs, betaCol->ys,     // segment center
                betaCol->length);
      }
        term -= Eps0 * betaRow->length * betaRow->bjbeta * En;
        *(work + IndexMatrix (row, col, numbetas)) = term;
        betaRow = betaRow->next;   // advance row pointer
  }
  betaCol = betaCol->next;      // advance column pointer
}
```

```
// Build matrix T in right Jbeta X Jgamma part of work
numbetasq = numbetas * numbetas; // offset to right part of work
gammaCol = gammaList;
col = 0;
while (gammaCol != NULL)        // loop over gamma segments
{
  col++;                             // column index
  betaRow = betaList;
  row = 0;
  while (betaRow != NULL)      // loop over beta segments
  {
    row++;                           // row index
    if (betaRow->parallel == 'x')    // normal electric field
                                     // y direction
    {
      if (gammaCol->parallel == 'x')   // source segment || to x
        En = Eypx (betaRow->xs, betaRow->ys, // Obs. coordinates
               gammaCol->xs, gammaCol->ys,     // segment center
               gammaCol->length);
      else                                // source segment || to y
        En = Eypy (betaRow->xs, betaRow->ys, // Obs. coordinates
               gammaCol->xs, gammaCol->ys,     // segment center
               gammaCol->length);
    }
    else                                      // elect. field x direction
    {
      if (gammaCol->parallel == 'x')    // source segment || to x
        En = Expx (betaRow->xs, betaRow->ys, // Obs. coordinates
               gammaCol->xs, gammaCol->ys,     // segment center
               gammaCol->length);
      else                                // source segment || to y
        En = Expy (betaRow->xs, betaRow->ys, // Obs. coordinates
               gammaCol->xs, gammaCol->ys,     // segment center
               gammaCol->length);
    }
    *(work + numbetasq + IndexMatrix (row, col, numbetas))
        = Eps0 * betaRow->length * betaRow->bjbeta
        * En * gammaCol->epsratio;
    betaRow = betaRow->next;
  }
  gammaCol = gammaCol->next;
}
// Calculate Qtilda (call SimpleGauss)
rcode = SimpleGauss (numbetas, numbetas + numgammas, work);
// Load resulting matrix into Qtilda
for (col = 0; col < numbetas*numgammas; col++)
```

```
      *(Qtilda + col) = *(work + numbetasq + col);
free (work);       work = NULL;
return rcode;      // = 0, normal exit
}
/* -----End of Procedure CalculateQtilda ----------------- */
```

The computer code `ch18_1.c` was written to perform the layered circuit board capacitance simulation. The code is a modified version of the multi-trace homogeneous medium code (`ch16_2.c`). It was modified to include input of the corners of the example circuit board as shown in Figure 18.1. Two vertical dielectric boundaries are included (the two sides of the circuit board) and two horizontal dielectric boundaries are included, the boundary from the upper-left corner to the top-left corner of trace 1 and the boundary from the top-right corner of trace 1 to the upper-right corner of the circuit board. The new code creates linked lists containing the parameters of each segment of dielectric boundary. It uses this linked list and that of the trace surface segments to construct the new matrices required and solve for the matrix $\tilde{\mathbf{Q}}$ as described above.

The user may input the number of segments to be used in each of the horizontal dielectric boundaries and the number of segments to be used in each of the vertical dielectric boundaries. Near any corner of a trace, the surface charge density varies rapidly with position along the trace surface; these are areas that require small segment sizes to resolve this variation in order to obtain acceptable accuracy. Likewise, the electric field and equivalent surface charge density will vary rapidly at dielectric boundaries near corners of a trace. In the example calculation, this occurs in the two horizontal dielectric boundaries near the corners of trace 1. Therefore, in the example calculation (results shown below) more segments were used to represent the horizontal boundaries than were used in the vertical boundaries. Along the vertical boundaries the electric field is small and varies slowly because they are relatively far from the traces; the results are not very sensitive to the number of segments used along them. The code output (shown on the screen) of an example calculation follows.

```
Capacitance of 3 traces
Enter number of x-direction trace segments: 8
Enter number of y-direction trace segments: 3
```

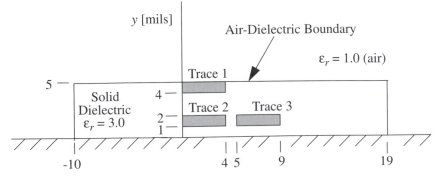

Figure 18.1 Circuit Board Example Problem.

```
Enter number of x-direction dielectric segments: 15
Enter number of y-direction dielectric segments: 5
Trace number: 1
xmin = 0.00    xmax = 4.00 [mils]
ymin = 4.00    ymas = 5.00 [mils]
Trace number: 2
xmin = 0.00    xmax = 4.00 [mils]
ymin = 1.00    ymas = 2.00 [mils]
Trace number: 3
xmin = 5.00    xmax = 9.00 [mils]
ymin = 1.00    ymas = 2.00 [mils]
relative permittivity = 3.00
Dielectric upper left corner (x, y) -10.0 5.0 [mils]
Dielectric upper right corner (x, y) 19.0 5.0 [mils]
capacitance matrix [pF/m]
104.4800 -62.8667 -20.5555
-62.8667 250.2768 -39.6233
-20.5555 -39.6233 216.9735
NORMAL EXIT
```

We note that the trace geometry for this calculation is the same as that used to illustrated the simulation of multi-trace capacitance and inductance in a homogeneous medium (computer code ch16_2.c). The addition of the dielectric boundaries is the only change. Therefore, the inductance for this example is the same as that output by the previous code. However, the new capacitance matrix with the dielectric boundaries included does not commute with the inductance matrix. The circuit board has multiple propagation speeds. These propagation speeds and their basis vectors (columns of the matrix **U**) are shown below.

```
Propagation speeds:
No.  1  1.892752e+008 [m/s]    0.6314c
No.  2  1.749189e+008 [m/s]    0.5835c
No.  3  1.731714e+008 [m/s]    0.5776c
U matrix -- basis vectors
Transformation from eigenmodes to traces
        1        2        3
 1   -0.963    0.117   -0.007
 2   -0.252    0.228   -0.998
 3   -0.092    0.967    0.055
```

Chapter 19

CAPACITANCE AND INDUCTANCE BETWEEN TWO GROUND PLANES

A line consisting of parallel traces in a layered dielectric between two ground planes is analyzed in this Section. The algorithm is the same as that for traces over a single ground plane except that the potential for a point charge between two ground planes is substituted for that of a point charge over a single ground plane. That leads to a quite different evaluation of the potential due to a charged segment. The integral leading to this potential cannot be evaluated analytically for this case (as it was for the single ground plane case); therefore, a simple numerical integral will be used instead.

The potential of a point charge between two ground planes is given by Smythe (1950, 4.20(5), p. 85). If the permittivity is that of free space, it can be written

$$w(x,y,b) = \frac{q}{2\pi\varepsilon_o} \text{atanh} \left[\frac{\sin\frac{\pi b}{a}\sin\frac{\pi y}{a}}{\cosh\frac{\pi x}{a} - \cos\frac{\pi b}{a}\cos\frac{\pi y}{a}} \right] \qquad (19\text{-}1)$$

where the planes are located at $y = 0$ and $y = a$ and the charge q is located at $x = 0$, $y = b$. The potential w is given at the point (x, y); see Figure 19.1. An analytic integral of this function is unknown to the author; therefore, the integral over a segment of charge will be calculated numerically using Gaussian integration.

The hyperbolic arctangent function atanh(x) can be expressed as

$$\text{atanh}(x) = \frac{1}{2}\log\frac{1+x}{1-x} \qquad (19\text{-}2)$$

with the derivative

$$\frac{d}{dx}\text{atanh}(x) = \frac{1}{1-x^2} \qquad (19\text{-}3)$$

To calculate the electric field, the partial derivatives of w with respect to x and y are needed. By differentiating Equation (19-1), one obtains

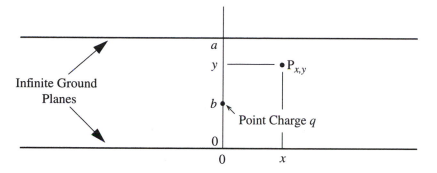

Figure 19.1   Geometry of Unit Charge between Two Infinite Ground Planes.

$$\frac{\partial}{\partial x}w(x,y,b) = \frac{q}{2\pi\varepsilon_o}\frac{1}{1-u^2}\frac{\partial u}{\partial x} \tag{19-4}$$

and

$$\frac{\partial}{\partial y}w(x,y,b) = \frac{q}{2\pi\varepsilon_o}\frac{1}{1-u^2}\frac{\partial u}{\partial y} \tag{19-5}$$

where

$$u(x,y,b) = \frac{\sin\frac{\pi b}{a}\sin\frac{\pi y}{a}}{\cosh\frac{\pi x}{a} - \cos\frac{\pi b}{a}\cos\frac{\pi y}{a}}$$

$$\frac{\partial u(x,y,b)}{\partial x} = \frac{\pi}{a}\frac{-\sin\frac{\pi b}{a}\sin\frac{\pi y}{a}\sinh\frac{\pi x}{a}}{\left[\cosh\frac{\pi x}{a} - \cos\frac{\pi b}{a}\cos\frac{\pi y}{a}\right]^2}$$

$$\frac{\partial u(x,y,b)}{\partial y} = \frac{\pi}{a}\sin\frac{\pi b}{a}\left[\frac{\cos\frac{\pi y}{a}}{\cosh\frac{\pi x}{a} - \cos\frac{\pi b}{a}\cos\frac{\pi y}{a}} - \frac{\sin^2\frac{\pi y}{a}\cos\frac{\pi b}{a}}{\left[\cosh\frac{\pi x}{a} - \cos\frac{\pi b}{a}\cos\frac{\pi y}{a}\right]^2}\right]$$

EXERCISE 19-1.   Show that as the observer point (x,y) approaches the location of the point charge q (i.e., $(x,y) \rightarrow (0,b)$), the potential w approaches infinity. *Hint:* Show that $u(0,b,b) = 1$.

19.1 POTENTIAL DUE TO A UNIFORMLY CHARGED SEGMENT

Analogous to the analysis of trace and dielectric-boundary segments over a single ground plane, it is assumed that a unit charge exits uniformly distributed on a segment (j) of length

l_j. The surface charge density on the segment is, therefore, $\rho_{sj} = 1/l_j$. If the position of the center of the segment is (x_{sj}, y_{sj}), the potential at point (x, y) due to the segment is:

For a segment parallel to the x axis,

$$v_{s \parallel x}(\mathbf{x}, \mathbf{x}_{sj}) = \frac{1}{2\pi\varepsilon_o l_j} \int_{-l/2}^{l/2} \mathrm{atanh}\, u(x - x_{sj} - l', y, y_{sj}) dl' \tag{19-6}$$

The geometry for this potential is shown in Figure 19.2.

For a segment parallel to the y axis,

$$v_{s \parallel y}(\mathbf{x}, \mathbf{x}_{sj}) = \frac{1}{2\pi\varepsilon_o l_j} \int_{-l/2}^{l/2} \mathrm{atanh}\, u(x - x_{sj}, y, y_{sj} + l') dl' \tag{19-7}$$

where, $\mathbf{x} = (x, y)$ and $\mathbf{x}_{sj} = (x_{sj}, y_{sj})$. The geometry for the potential of Equation (19.7) is shown in Figure 19.3.

The integrals in Equations (19.6) and (19.7) cannot (to the author's knowledge) be integrated analytically. Numerical integration is appropriate. A numerical integral is given, in general, by a sum of the following form:

$$\int_a^b f(x)dx \cong \sum_i w_i f(x_i) \tag{19-8}$$

where w_i are the "weights" and x_i are the "abscissas" of the approximation. The number of points used and the values of the weights and abscissas are chosen in various ways depending on the behavior of the function $f(x)$, the accuracy required and the willingness to expend computer resources. Two considerations are immediately apparent for the problem at hand.

1. The use of uniformly charged segments to represent the charge distribution along a surface is already a physical approximation that is adjusted by varying the number (and sizes) of

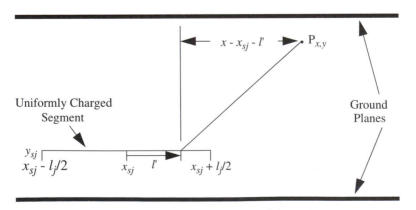

Figure 19.2   Integration Geometry over Segment Parallel to X Axis.

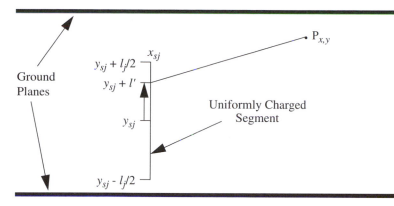

Figure 19.3   Integration Geometry over Segment Parallel to Y Axis.

segments along the surface. Therefore, little is gained by using a highly accurate integration method for each segment. When the number of segments is increased and the width of each segment is decreased, the accuracy of the numerical integral will also increase automatically because the segment is shorter. Therefore, a small number of points should be used in the numerical integral.

2. The method should not use an evaluation of the integrand at the center point of the domain of integration (i.e., at $l' = 0$). This is because the potentials $v_{s \parallel y}(\mathbf{x}, \mathbf{x}_{sj})$ and $v_{s \parallel x}(\mathbf{x}, \mathbf{x}_{sj})$ will be evaluated with the (observer) point $\mathbf{x}$ at the center of the segment. If the integral approximation has a source point there, the potential will be singular at that point and cannot be evaluated there.

Experience has shown that a two-point integration works well for this application. To obtain the highest possible accuracy from a two-point integration, the two abscissas and weights are chosen using the Gauss-Legendre numerical integration method. This method optimizes both the values of the abscissas and the weights for the highest order accuracy. For a discussion of this method, see Press, et al., 1986, Section 4.5. The abscissas and weights can be found in Abramowitz and Stegun, 1970, p. 916 for various numbers of points in the sum.

For evaluation of the integrals in Equations (19-6) and (19-7) the (two) weights are $l_j/2$ and the abscissas are $l' = \pm l_{abs}$, where

$$l_{abs} = \frac{l_j}{2\sqrt{3}} = l_j \times 0.28867513 \ldots \qquad (19\text{-}9)$$

Using these values, the potentials of Equations (19-6) and (19-7) are approximated by the following:

For a segment parallel to the x axis,

$$v_{s \parallel x}(\mathbf{x}, \mathbf{x}_{sj}) \cong \frac{1}{4\pi\varepsilon_o} \left[\operatorname{atanh} u(x - x_{sj} - l_{abs}, y, y_{sj}) + \operatorname{atanh} u(x - x_{sj} + l_{abs}, y, y_{sj}) \right] \qquad (19\text{-}10)$$

For a segment parallel to the y axis,

$$v_{s \parallel y}(\mathbf{x}, \mathbf{x}_{sj}) \cong \frac{1}{4\pi\varepsilon_o} [\operatorname{atanh} u(x - x_{sj}, y, y_{sj} - l_{abs})$$

$$+ \operatorname{atanh} u(x - x_{sj} + l', y, y_{sj} + l_{abs})] \tag{19-11}$$

19.2 ELECTRIC FIELD DUE TO SEGMENT PARALLEL TO THE X AXIS

For a unit point charge, the electric fields E_x and E_y are given by minus the gradient components shown in Equations (19-4) and (19-5) (with $q = 1$ in the expression for w, Equation (19-1)). For a segment parallel to the x axis carrying a unit charge, the x component of the electric field is

$$E_x(\mathbf{x}, \mathbf{x}_{sj}) = \frac{-1}{l_j} \int_{-l_j/2}^{l_j/2} \frac{\partial}{\partial x} w(x - x_{sj} - l', y, y_{sj}) dl' \tag{19-12}$$

The integration can be performed exactly, to obtain

$$E_x(\mathbf{x}, \mathbf{x}_{sj}) = \frac{1}{l_j} [w(x - x_{sj} - l_j/2, y, y_{sj}) - w(x - x_{sj} + l_j/2, y, y_{sj})] \tag{19-13}$$

$$E_x(\mathbf{x}, \mathbf{x}_{sj}) = \frac{1}{2\pi\varepsilon_o l_j} [\operatorname{atanh} u(x - x_{sj} - l_j/2, y, y_{sj}) - \operatorname{atanh} u(x - x_{sj} + l_j/2, y, y_{sj})] \tag{19-14}$$

However, for very small segments, evaluation of Equation (19-14) requires the subtraction of two numbers that are very nearly equal. This always causes significant round-off error. More consistent accuracy can be obtained by applying the two-point Gauss-Legendre expression to obtain

$$E_x(\mathbf{x}, \mathbf{x}_{sj}) \cong \frac{-1}{4\pi\varepsilon_o} \left[\frac{1}{1 - u^2(x - x_{sj} - l_{abs}, y, y_{sj})} \frac{\partial u(x - x_{sj} - l_{abs}, y, y_{sj})}{\partial x} \right.$$

$$\left. + \frac{1}{1 - u^2(x - x_{sj} + l_{abs}, y, y_{sj})} \frac{\partial u(x - x_{sj} + l_{abs}, y, y_{sj})}{\partial x} \right] \tag{19-15}$$

where u and $\partial u/\partial x$ are given below Equation (19-5). Equation (19-15) will be used in the illustrating computer code.

The y component of the electric field due to a segment with a unit charge is

$$E_y(\mathbf{x}, \mathbf{x}_{sj}) = \frac{-1}{l_j} \int_{-l_j/2}^{l_j/2} \frac{\partial}{\partial y} w(x - x_{sj} - l', y, y_{sj}) dl' \tag{19-16}$$

This integral is approximated using the two-point Gauss-Legendre expression to obtain

$$E_y(\mathbf{x},\mathbf{x}_{sj}) \cong \frac{-1}{4\pi\varepsilon_o}\left[\frac{1}{1-u^2(x-x_{sj}-l_{abs},y,y_{sj})}\frac{\partial u(x-x_{sj}-l_{abs},y,y_{sj})}{\partial y}\right.$$

$$\left.+\frac{1}{1-u^2(x-x_{sj}+l_{abs},y,y_{sj})}\frac{\partial u(x-x_{sj}+l_{abs},y,y_{sj})}{\partial y}\right]$$

(19-17)

where u and $\partial u/\partial y$ are given below Equation (19-5).

19.3 ELECTRIC FIELD DUE TO SEGMENT PARALLEL TO THE *Y* AXIS

For a segment parallel to the *y* axis carrying a unit charge, the electric field in the *x* direction is

$$E_x(\mathbf{x},\mathbf{x}_{sj}) = \frac{-1}{l_j}\int_{-l_j/2}^{l_j/2}\frac{\partial}{\partial x}w(x-x_{sj},y,y_{sj}+l')dl'$$

(19-18)

This integral is approximated using the two-point Gauss-Legendre expression to obtain

$$E_x(\mathbf{x},\mathbf{x}_{sj}) \cong \frac{-1}{4\pi\varepsilon_o}\left[\frac{1}{1-u^2(x-x_{sj},y,y_{sj}-l_{abs})}\frac{\partial u(x-x_{sj},y,y_{sj}-l_{abs})}{\partial x}\right.$$

$$\left.+\frac{1}{1-u^2(x-x_{sj},y,y_{sj}+l_{abs})}\frac{\partial u(x-x_{sj},y,y_{sj}+l_{abs})}{\partial x}\right]$$

(19-19)

where u and $\partial u/\partial x$ are given below Equation (19-5).

The *y* component of the electric field is

$$E_y(\mathbf{x},\mathbf{x}_{sj}) = \frac{-1}{l_j}\int_{-l_j/2}^{l_j/2}\frac{\partial}{\partial y}w(x-x_{sj},y,y_{sj}+l')dl'$$

(19-20)

Again, using the Gauss-Legendre approximation, we obtain

$$E_y(\mathbf{x},\mathbf{x}_{sj}) \cong \frac{-1}{4\pi\varepsilon_o}\left[\frac{1}{1-u^2(x-x_{sj},y,y_{sj}-l_{abs})}\frac{\partial u(x-x_{sj},y,y_{sj}-l_{abs})}{\partial y}\right.$$

$$\left.+\frac{1}{1-u^2(x-x_{sj},y,y_{sj}+l_{abs})}\frac{\partial u(x-x_{sj},y,y_{sj}+l_{abs})}{\partial y}\right]$$

(19-21)

where u and $\partial u/\partial y$ are given below Equation (19-5).

19.4 CALCULATING THE CAPACITANCE AND INDUCTANCE MATRICES

The potentials given above in Equations (19-10) and (19-11) were substituted for those applicable to a single ground plane (Equations (16-19) and (16-20)) to result in a code to simulate the capacitance and inductance matrices for traces in a homogeneous medium between two ground planes. The potentials were substituted in computer code ch16_2.c to result in computer code ch19_1.c. The code was listed and discussed in Chapter 16; it will not be repeated here.

The illustrating example is the same as that used to illustrate the original code except that an additional ground plane was added at $y = 10$ [mils]. The geometry is shown in Figure 19.4.

Output from the new code is shown below. This output appears on the screen when the code is run.

```
Capacitance of 3 traces
Enter number of x-direction segments: 8
Enter number of y-direction segments: 3
Trace number: 1
xmin = 0.00    xmax = 4.00 [mils]
ymin = 4.00    ymas = 5.00 [mils]
Trace number: 2
xmin = 0.00    xmax = 4.00 [mils]
ymin = 1.00    ymas = 2.00 [mils]
Trace number: 3
xmin = 5.00    xmax = 9.00 [mils]
ymin = 1.00    ymas = 2.00 [mils]
relative permittivity = 3.00 (homogeneous)
second ground plane at 10.0 [mils]
capacitance matrix [pF/m]
 142.4653  -64.7038  -19.2961
 -64.7038  261.5160  -41.9080
 -19.2961  -41.9080  232.8026
inductance matrix [microH/m]
 0.272343  0.073109  0.035734
```

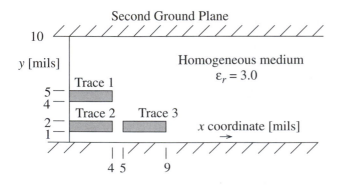

Figure 19.4 Homogeneous Medium between Two Infinite Ground Planes.

```
0.073109   0.151056   0.033252
0.035734   0.033252   0.152329
NORMAL EXIT
```

The potentials and also the electric field expressions for two ground planes, Equations (19-14) to (19-21), were substituted into computer code ch18_1.c to create computer code ch19_2.c for a layered medium between two ground planes. The same illustrating example was used except that a second ground plane was included at $y = 10$ [mils]. The geometry is shown in Figure 19.5.

The output of a sample calculation for this geometry is given below.

```
Capacitance of 3 traces
Enter number of x-direction trace segments: 8
Enter number of y-direction trace segments: 3
Enter number of x-direction dielectric segments: 15
Enter number of y-direction dielectric segments: 5
Trace number: 1
xmin = 0.00    xmax = 4.00 [mils]
ymin = 4.00    ymas = 5.00 [mils]
Trace number: 2
xmin = 0.00    xmax = 4.00 [mils]
ymin = 1.00    ymas = 2.00 [mils]
Trace number: 3
xmin = 5.00    xmax = 9.00 [mils]
ymin = 1.00    ymas = 2.00 [mils]
relative permittivity = 3.00
second ground plane at 10.0 [mils]
Dielectric upper left corner (x, y) -10.0 5.0 [mils]
Dielectric upper right corner (x, y) 19.0 5.0 [mils]
capacitance matrix [pF/m]
113.8329 -65.6804 -19.4045
-65.6804 261.2103 -42.2116
-19.4045 -42.2116 227.1683
NORMAL EXIT
```

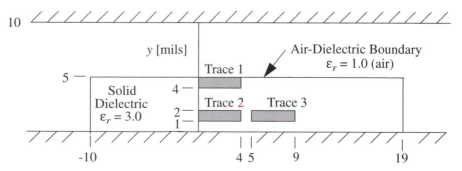

Figure 19.5 Integration Geometry over Segment Parallel to Y Axis.

PART V

SIMULATION OF SKIN EFFECT

Chapter 20
PHYSICS OF THE SKIN EFFECT

The skin effect arises from the inability of electric currents and magnetic fields to penetrate into the bulk of a conductor instantaneously. If a magnetic field could be established instantaneously (or very quickly) in a conductor, then an infinite (or very large) electric field would result in the conductor in accordance with Faraday's law of induction. But if an infinite electric field occurred in a conductor, then an infinite current would flow in the conductor (in accordance with Ohm's law) and that would require an infinite amount of energy. Hence, the field can't be established instantaneously. As it turns out, the field must be established in accordance with a partial differential equation (pde) known as the diffusion equation. This pde is sometimes called the heat equation because heat diffuses into a thermal conductor in accordance with the same pde (but with a different diffusion constant).

When the magnetic field of interest is oscillating sinusoidally, the field amplitude (and the current density amplitude) beneath the surface of the conductor decays exponentially with depth. The exponential decay length is called the skin depth and is traditionally denoted δ. Because of the behavior of the integral of the exponential, the total current flowing beneath the surface of the conductor is given by the product of the current density at the surface times the skin depth. Hence, conceptually, the current density can be regarded as flowing uniformly in a surface layer, or skin, of depth δ. The exponential behavior described above is an exact solution for a plane conductor of infinite thickness. For a conductor of finite thickness, the exponential behavior is a good approximation if the conductor is more than a skin depth thick. If the thickness is less than δ, then the DC resistance is usually substituted for the skin depth resistance.

The diffusion equation and the classical skin depth will be derived below. However, the purpose of this Part of the book is to derive time domain solutions of the diffusion equation so that the physics of the skin effect can be simulated accurately for the nonsinusoidal signals present in digital electronic devices. Because practical conductors always have a finite thickness, the time domain methods are developed for plane conductors of finite thickness and cylindrical conductors.

The term "surface current" or "surface current density" (denoted j_s [A/m]) will be used to denote the current per meter width of surface (integrated with respect to depth into the material) flowing near the surface of a conductor. In diffusion problems, it's a useful description of the current flow before the current density penetrates the entire depth of the conductor.

Referring to Figure 20.1, the surface current density in the x direction is defined as $j_{sx} = \int_0^{z_d} j_x(z)dz$ [A/m]. It is the current flowing in the x direction per meter width (y direction).

20.1 DIFFUSION IN A SLAB

The starting point is Maxwell's equations; they can be written (in rationalized MKS units) as

$$\nabla \times \mathbf{H} = \varepsilon \frac{\partial \mathbf{E}}{\partial t} + \mathbf{j} \tag{20-1}$$

and

$$\nabla \times \mathbf{E} = -\mu \frac{\partial \mathbf{H}}{\partial t} \tag{20-2}$$

where
  $\mathbf{H}$ is the magnetic field vector [A/m]
  $\mathbf{E}$ is the electric field vector [V/m]
  $\mathbf{j}$ is the current density vector [A/m^2]
  ε is the permittivity of the material [F/m]
  μ is the permeability of the material [H/m] and
  $\nabla \times$ is the curl operator [m^{-1}]

The application of these equations will be in common conducting metals such as copper, silver, or gold. Hence, it will be assumed that $\mu = \mu_o = 4\pi \times 10^{-7}$ [H/m] and $\varepsilon = \varepsilon_o \cong 8.854 \times 10^{-12}$ [F/m] in the conductor. The current density in the conductor is given by

$$\mathbf{j} = \sigma \mathbf{E} \tag{20-3}$$

where σ [℧/m] is the conductivity of the metal ($\sigma_{Cu} \cong 5.8 \times 10^7$ [℧/m]).

We note that in common metals, the current density term on the right-hand side of Equation (20-1) is dominant, that is

$$\sigma \mathbf{E} >> \varepsilon \frac{\partial \mathbf{E}}{\partial t} \tag{20-4}$$

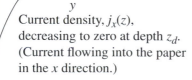

Surface of conductor
y
Current density, $j_x(z)$, decreasing to zero at depth z_d. (Current flowing into the paper in the x direction.)

Figure 20.1 Schematic for "Surface Current Density" definition.

This can be illustrated by considering a (fast) rise time of, say, 1 [ns]. Then, for an electric field rising from 0 to 1 [V/m], we have approximately (in copper), $\sigma E \cong 5.8 \times 10^7$ [A/m^2] and $\varepsilon \frac{\partial E}{\partial t} \cong \frac{8.854 \times 10^{-12} \text{ [F/m]} \times 1 \text{ [V/m]}}{1 \times 10^{-9} \text{ [s]}} = 8.854 \times 10^{-3}$ [A/m^2]. Hence, in this example, the left side of inequality (20-4) is more than 9 orders of magnitude larger than the right side. The result is that the term $\varepsilon \frac{\partial E}{\partial t}$ can be neglected in Equation (20-1) when the solution is in a metal. With this approximation, Equation (20-1) becomes

$$\nabla \times \mathbf{H} = \sigma \mathbf{E} \tag{20-5}$$

For a plane geometry, assume the surface of the metal is in the x-y plane at $z = 0$. Assume that the fields are uniform in the x and y directions; the only spatial dependence is in the z direction. Furthermore, assume that the electric field vector is in the x direction and the magnetic field is in the y direction. With these assumptions, Equations (20-2) and (20-5) become

$$\frac{\partial E_x}{\partial z} = -\mu \frac{\partial H_y}{\partial t} \tag{20-6}$$

and

$$-\frac{\partial H_y}{\partial z} = \sigma E_x \tag{20-7}$$

Substituting Equation (20-7) into (20-6), one obtains

$$\frac{\partial^2 H_y}{\partial z^2} = \mu\sigma \frac{\partial H_y}{\partial t} \tag{20-8}$$

This is the one-dimensional diffusion equation for H_y.

PROBLEM 20.1 Assume that $H_y(z,t)$ has the form $H_y(z,t) = Ae^{-\alpha t} \sin\omega z$, where A is a constant. What is the relationship between α and ω for $H_y(z,t)$ to satisfy the diffusion equation? *Hint*: Substitute $H_y(z,t)$ into Equation (20-8).

20.2 CLASSICAL SKIN EFFECT

To solve Equation (20-8) in the frequency domain, let H_y have the form

$$H_y(z,t) = H_o e^{j\omega t} e^{-\gamma z} \tag{20-9}$$

where H_o is a constant and the propagation constant γ is to be determined (we will find only the solution that attenuates in the z direction; the general solution has two terms). Substitute Equation (20-9) into (20-8) to obtain

$$\gamma^2 = j\omega\mu\sigma \qquad (20\text{-}10)$$

Choose the root of γ^2 whose real part is positive so that the solution attenuates in the positive z direction. This root yields

$$\gamma = (1+j)\sqrt{\frac{\omega\mu\sigma}{2}} = \frac{(1+j)}{\delta} \qquad (20\text{-}11)$$

where

$$\delta = \sqrt{\frac{2}{\omega\mu\sigma}}.$$

The current density flowing in the conductor is given by (using Equation (20-7) and (20-9))

$$j_x(z,t) = \gamma H_y(z,t) = \gamma H_o e^{j\omega t} e^{-\gamma z} \quad [\text{A/m}^2] \qquad (20\text{-}12)$$

The surface current [A/m] flowing in the conductor (i.e., the current density integrated over its depth) is

$$j_{xs} = \int_0^\infty j_x(z,t)dz = H_o e^{j\omega t} \quad [\text{A/m}] \qquad (20\text{-}13)$$

At the surface of the conductor ($z = 0$), the tangential electric field is (using (20-7) and (20-9))

$$E_x = \frac{\gamma}{\sigma}H_y(0,t) = \frac{(1+j)}{\delta\sigma}H_o e^{j\omega t} \quad [\text{V/m}] \qquad (20\text{-}14)$$

The impedance of the metal surface is the ratio of electric field to the surface current density, yielding

$$Z_s = \frac{E_x}{j_{sx}} = \frac{(1+j)}{\delta\sigma} \quad [\Omega] \qquad (20\text{-}15)$$

The impedance per length of a conductor is then approximately

$$Z' = \frac{Z_s}{D_c} \quad [\Omega/\text{m}] \qquad (20\text{-}16)$$

where D_c is the distance around the circumference of the conductor, assuming the current is distributed uniformly around the conductor and that the conductor is thicker than two skin depths.

Equation (20-15) shows that the real part of the impedance corresponds to a conducting slab a skin depth thick. The reactive part of the impedance is equal to the real part so that the phase angle is 45 degrees.

The result given in Equation (20-15) is the classical result; it was derived here to introduce the diffusion equation whose solution yields "skin depth" theory. In the treatment that follows, solutions of the diffusion equation, representing the physics of the skin depth will be derived, but the result will be in the time domain. That is, time domain solutions of Equation (20-8) will be found that lead to time-stepping algorithms to calculate the tangential electric field at the surface of a conductor as a function of the conduction current. The tangential electric field, the resistive voltage drop along a conductor (including diffusion effects), can replace the product Ri in the transmission line equations.

PROBLEM 20.2 A copper trace $\sigma = 5.8 \times 10^7$ is 10^{-4} [m] (about 4 mils) thick. At what frequency is its thickness equal to one skin depth? Note that at higher frequencies, the resistance will be determined by the diffusion process.

Chapter 21
PLANE GEOMETRY SKIN EFFECT SIMULATION

In this chapter the skin effect in a slab of finite thickness will be evaluated. The finite thickness of the conductor leads to the following question: Do the fields diffuse into the slab from one side or both? Either can occur. There's no simple way to determine exactly how the current distributes and determine where the diffusion occurs, except for some special cases that have a high degree of symmetry. An interesting qualitative statement is that, at high frequencies, the current follows the "path of least inductance" (see Johnson and Graham, 1993). At low frequencies, the current tends to follow the path of least resistance. If a current distribution is chosen *a priori* for a given geometry, it will only be an approximation to reality at some simulation times, or frequencies. To calculate accurately the distribution of current as a function of time or frequency requires, in general, a three-dimensional solution and would be very computationally intensive.

21.1 DC CURRENT DENSITY AND MAGNETIC FIELD

The skin effect models derived in this section are exact for an infinite slab; they approximate the skin effect for a rectangular cross-section trace. The wider the trace the more accurate they will be since the edge effect has not been included. The cases of diffusion from one side and from both sides will be derived; they are closely related. Figure 21.1 illustrates the field geometry for diffusion from both sides of a conducting slab of thickness $2d$. A surface current density I_x [A/m] is flowing in the x direction. The magnetic field distribution shown (linearly varying through the slab thickness) indicates that the current density $j_x = \sigma E_x$ is uniform throughout the slab; see Equation (20-7). That is, the current has been flowing for a long time so that the fields have diffused inwardly and attained their DC configuration.

The field configuration illustrated above is the starting point for the derivation of the time domain model. That is, in the metal slab ($z \in [0,2d]$), let

$$E_x(z) = E_o \tag{21-1}$$

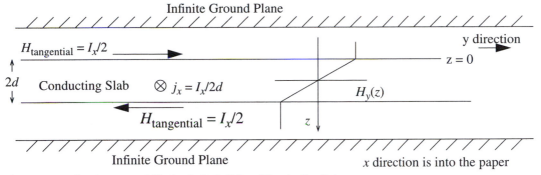

Figure 21.1   Idea Geometry, Diffusion in Both Sides of Conducting Slab.

The surface current flowing in the plane in the x direction (integrated through the slab) is

$$I_x = 2d\sigma E_O \text{ [A/m]}$$ (21-2)

The magnetic field in the metal slab is then

$$H_y(z) = \frac{I_x}{2}\left[1 - \frac{z}{d}\right]$$ (21-3)

EXERCISE 21-1.   Show that the magnetic field given by Equation (21-3) has odd symmetry about $z = d$ and that it satisfies Equation (20-7).

We require the Fourier series of the function $H_y(z)$. The function is sketched below in Figure 21.2 to illustrate its odd symmetry about the point $z = d$. The figure also shows the periodic behavior of the Fourier series of $H_y(z)$ and the resulting discontinuities.

We note that the extended periodic function has odd symmetry about both $z = 0$ and $z = d$. The cosine functions in the Fourier series have *even* symmetry about both of these points; therefore, the coefficients of the cosine in the Fourier series of $H_y(z)$ are zero and the Fourier series has only sine terms.

The Fourier series for $H_y(z)$ is

$$H_y(z) = \sum_{n=1}^{\infty} A_n \sin\frac{n\pi z}{d}$$ (21-4)

where

$$A_n = \frac{2}{d}\int_0^d H_y(z)\sin\frac{n\pi z}{d}dz = \frac{I_x}{d}\int_0^d\left[1 - \frac{z}{d}\right]\sin\frac{n\pi z}{d}dz$$

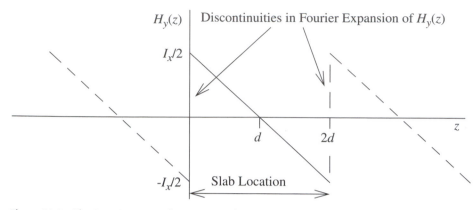

Figure 21.2   The Function $H_y(z)$ Showing Periodic Extrapolation of Fourier Series.

EXERCISE 21-2.   Perform the integration to show that

$$A_n = \frac{I_x}{n\pi} \tag{21-5}$$

There is a subtle, but important, difference between $H_y(z)$ as defined in Equation (21-3) and as it is expressed in terms of a Fourier series in Equation (21-4). The reader may have noticed that the function $H_y(z)$ as given in Equation (21-3) has the values $H_y(0) = I_x/2$ and $H_y(2d) = -I_x/2$ while the Fourier expansion of that function given in Equation (21-4) has the value zero at both $z = 0$ and $z = 2d$ (because the value of the sine function is zero when its argument is zero or a multiple of 2π). This occurred because of the nature of the Fourier series. That is, a Fourier expansion converges to the value of the function at points where the function is continuous. If the function is discontinuous, a Fourier expansion converges to the midpoint of the discontinuity. The Fourier expansion is periodic with period equal to the length of its normal expansion domain (0 to $2d$ in this case). Therefore, if the value of the function at $z = 2d$ is not equal to the value at $z = 0$, the Fourier series will converge to the midpoint of the values at these two points, zero in this case. See Figure 21.2.

Often the value of a function at a single point is of little interest. In this case, this behavior is the crux of the method of applying the boundary condition to the diffusion problem. We note that by integrating Equation (20-7), the total current (per meter width) in the slab can be obtained; that is,

$$I_x = \sigma \int_0^{2d} E_x(z)dz = -\int_0^{2d} \frac{\partial H_y}{\partial z}dz = H_y(0,t) - H_y(2d,t) \quad \text{[A/m]} \tag{21-6}$$

Hence, the total current is the assumed value I_x according to Equation (21-3) but is zero according the Fourier series of Equation (21-4). The Fourier series has created delta functions of current density at each boundary that contain total current equal to, but of opposite sign to, the current in rest of the slab. The two current densities are illustrated below in Figure 21.3.

21.2 DIFFUSION EQUATION SOLUTIONS

Let $\hat{H}_y(z,t)$ be the function

$$\hat{H}_y(z,t) = \begin{cases} \dfrac{I_x}{2}\left[1 - \dfrac{z}{d}\right], & t < 0 \\[2em] \displaystyle\sum_{n=1}^{\infty} A_n e^{-a_n t} \sin\dfrac{n\pi z}{d}, & t \geq 0 \end{cases} \tag{21-7}$$

Substitute $\hat{H}_y(z,t)$ into Equation (20-8) to find that $\hat{H}_y(z,t)$ satisfies the diffusion equation if

$$a_n = \frac{n^2\pi^2}{d^2\mu\sigma} \tag{21-8}$$

The function $\hat{H}_y(z,t)$ represents the magnetic field in the slab where a uniform current density has been established ($t < 0$) and then the total current is suddenly turned off at $t = 0$. We know that the total current is zero for $t > 0$ because $\hat{H}_y(0,t) = 0$ and $\hat{H}_y(2d,t) = 0$. Thus, by superposition, we can find the magnetic field as a function of time when the initial condition is zero magnetic field and zero current density, and the total current is suddenly turned on at $t = 0$, i.e., the step function of current response of the slab. The step function current response is

$$H_{y\text{step}}(z,t) = \frac{I_x}{2}\left[1 - \frac{z}{d}\right] - \hat{H}_y(z,t) \tag{21-9}$$

$$H_{y\text{step}}(z,t) = \begin{cases} 0, & t < 0 \\[1em] \dfrac{I_x}{2}\left[1 - \dfrac{z}{d}\right] - \displaystyle\sum_{n=1}^{\infty} A_n e^{-a_n t} \sin\dfrac{n\pi z}{d}, & t \geq 0 \end{cases} \tag{21-10}$$

Substituting Equation (21-10) into Equation (20-7) the electric field due to a step function of current is obtained. For $t > 0$,

Figure 21.3 Sketch of Current Density as a Function of z.

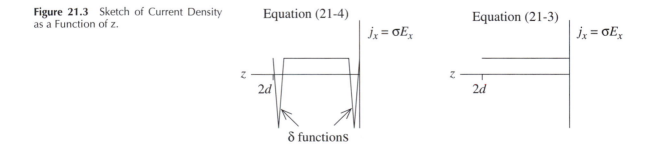

Equation (21-4)   Equation (21-3)

$j_x = \sigma E_x$ $j_x = \sigma E_x$

δ functions

$$E_{x\text{step}}(z,t) = \frac{I_x}{2d\sigma} + \frac{\pi}{d\sigma} \sum_{n=1}^{\infty} A_n n e^{-a_n t} \cos\frac{n\pi z}{d} \qquad (21\text{-}11)$$

Substituting A_n using Equation (21-5) and setting $z = 0$ or $z = 2d$, the electric field at the surface of the slab is found. Setting z to $2d$ yields the same electric field due to the symmetry of the geometry.

$$E_{x\text{step}}(0,t) = E_{x\text{step}}(2d,t) = \frac{I_x}{2d\sigma} + \frac{I_x}{d\sigma} \sum_{n=1}^{\infty} e^{-a_n t} \qquad (21\text{-}12)$$

21.3 EQUIVALENT CIRCUIT FOR TWO-SIDED DIFFUSION

The first term of Equation (21-12) corresponds to the DC resistance of the slab

$$R_{dc} = \frac{1}{2d\sigma} \quad [\Omega] \qquad (21\text{-}13)$$

This must occur since all the other terms approach zero as $t \to \infty$.

Each term under the summation on the right-hand side of Equation (21-12) corresponds to the voltage across an R-L parallel circuit, where

$$R = \frac{1}{d\sigma} = 2R_{dc} \quad [\Omega] \qquad (21\text{-}14)$$

and

$$L_n = \frac{d\mu}{\pi^2 n^2} \quad [\text{H}] \qquad (21\text{-}15)$$

The total internal inductance of the slab can be calculated as[1]

$$L_{int} = \sum_{n=1}^{\infty} L_n = \frac{d\mu}{6} \quad [\text{H}] \qquad (21\text{-}16)$$

An equivalent circuit can be drawn for the impedance of a slab with diffusion; it is given below in Figure 21.4. The voltage given by Equation (21-11) results when a step function of current of amplitude I_x flows through the circuit below.

To obtain the impedance per length or inductance per length for a finite width rectangular strip, divide the impedance or inductance of the slab by the strip width (the decay times of the R-L circuit fragments are unaffected). An interesting observation, obvious from the equiv-

---

1. Note that $\zeta(z) = \sum_{n=1}^{\infty} \frac{1}{n^z}$, Riemann's zeta function. And $\zeta(2) = \sum_{n=1}^{\infty} \frac{1}{n^2} = \frac{\pi^2}{6}$, see Gradshteyn and Ryzhik, 1980, 9.522 and 9.542.

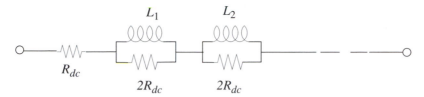

Figure 21.4   Equivalent Circuit of Slab with Diffusion.

alent circuit, is that once a current is established, diffusion can cause an "inductive kick" if the current is suddenly turned off. This is caused by the release of the energy stored in the internal inductance.

21.4 DIFFUSION ON ONE SIDE OF A SLAB

The idealized geometry where diffusion occurs on one side of the slab is shown in Figure 21.5. The slab is defined with thickness d (rather than $2d$ for the two-sided diffusion model). This is to illustrate that the diffusion process takes place over distance d in both cases (in the two-sided model distance d was to the center of the slab). It behaves simply as a symmetrical half of the two-sided case. However, a given surface current will flow through a slab of thickness d, rather than $2d$, so that the electric field will be twice that given in Equation (21-11), therefore, the electric field at the surface of the slab is

$$E_{x\text{step}}(0,t) = \frac{I_x}{d\sigma} + \frac{2I_x}{d\sigma}\sum_{n=1}^{\infty}e^{-a_n t} \qquad (21\text{-}17)$$

If the DC resistance is defined for a slab of thickness d as (twice the value it had for two-sided diffusion)

$$R_{dc} = \frac{1}{d\sigma} \quad [\Omega] \qquad (21\text{-}18)$$

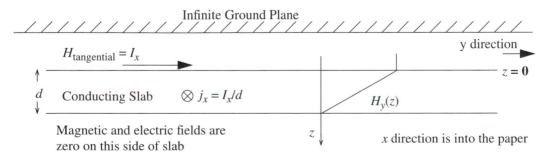

Figure 21.5   Idea Geometry, Diffusion From One Side of Conducting Slab.

Then the equivalent circuit of Figure 21.4 is correct for one-sided diffusion, if the inductors L_n are also doubled to make all the $L_n/2R_{dc}$ $(= 1/a_n)$ time constants the same as they were for the two-sided case. That is,

$$L_n = \frac{2d\mu}{\pi^2 n^2} \quad [\text{H}] \tag{21-19}$$

21.5 ALGORITHM FOR DIFFUSIVE VOLTAGE DROP

The diffusive voltage drop can be determined by using the equivalent circuit of Figure 21.4 and simply solving for the transient circuit response. If that method is used, one must be sure to use time steps that resolve the time constant of the highest order (largest n) circuit fragment that is included. The convergence of the series is discussed in another section.

A superior method for most purposes is the convolution integral. The way the convolution integral is evaluated here, the time-step size can be determined by signal resolution needs without regard for the time constants in the terms being evaluated. The reason is that over each time step the integral will be evaluated analytically (exactly — to the calculational precision used). The only approximation is a straight-line representation of the current over each of the time steps.

To obtain the electric field as a convolution over the current, the delta-function response is needed. The step-function response has already been determined and is given in Equation (21-12) for two-sided diffusion and in (21-17) for one-sided diffusion. Using R_{dc} as a parameter, both of them can be written as

$$E_{x\text{step}}(0,t) = R_{dc}u(t) + 2R_{dc}u(t) \sum_{n=1}^{\infty} e^{-a_n t} \tag{21-20}$$

where the step-function current amplitude I_x has been replaced with the unit step function $u(t)$. Thus, Equation (21-20) is the value of $E_{x\text{step}}(0,t)$ for the case where the current is zero for $t < 0$ and then jumps to one at $t = 0$. To obtain the delta-function response, $E_{x\text{step}}$ is differentiated with respect to time; when the step function $u(t)$ is differentiated (in the sense of distributions), the Dirac delta function $\delta(t)$ is the result. Therefore, the delta-function response of a diffusive slab is

$$E_{x\delta}(0,t) = R_{dc}\delta(t) + 2R_{dc} \sum_{n=1}^{\infty} [\delta(t) - u(t)a_n]e^{-a_n t} \tag{21-21}$$

Applying the convolution theorem, $E_x(0,t)$ can be found for any time function of surface current as

$$E_x(0,t) = \int_0^t E_{x\delta}(0,t-t')I_x(t')dt' \tag{21-22}$$

where $I_x(t)$ is the surface current (zero, for $t < 0$). Substituting Equation (21-21) into Equation (21-22) and performing the (trivial) integration over the delta functions yields

$$E_x(0,t) = R_{dc}I_x(t) + 2R_{dc}\sum_{n=1}^{\infty}[I_x(t) - u(t)I_n(t)] \qquad (21\text{-}23)$$

where

$$I_n(t) = a_n\int_0^t e^{-a_n(t-t')}I_x(t')dt' \qquad (21\text{-}24)$$

EXERCISE 21-3. Show that $I_n(t)$ is the current through the inductor in the nth parallel circuit of Figure 21.4. *Hint:* I_x = (current through the resistor) + (current through the inductor).

To evaluate $I_n(t)$ in a time-stepping procedure, assume $I_n(t)$ has been found at time t and it is desired to find it at time $t + \delta t$. The value at time $t + \delta t$ is

$$I_n(t + \delta t) = a_n\int_0^{t+\delta t} e^{-a_n(t+\delta t - t')}I_x(t')dt' \qquad (21\text{-}25)$$

$$I_n(t + \delta t) = a_n\int_0^t e^{-a_n(t+\delta t - t')}I_x(t')dt' + a_n\int_t^{t+\delta t} e^{-a_n(t+\delta t - t')}I_x(t')dt' \qquad (21\text{-}26)$$

$$I_n(t + \delta t) = e^{-a_n\delta t}I_n(t) + a_n e^{-a_n\delta t}\int_0^{\delta t} e^{a_n t'}I_x(t+t')dt' \qquad (21\text{-}27)$$

For $t \in [t, t + \delta t]$, the current $I_x(t)$ is approximated by the straight line

$$I_x(t + t') = I_x(t) + t'\frac{I_x(t+\delta t) - I_x(t)}{\delta t} \qquad (21\text{-}28)$$

EXERCISE 21-4. Show that if Equation (21-28) is substituted into Equation (21-27) and the integration is performed, one obtains

$$I_n(t + \delta t) = e^{-a_n\delta t}I_n(t) + I_x(t)\left[1 - e^{-a_n\delta t}\right]$$

$$+ [I_x(t + \delta t) - I_x(t)]\left[1 - \frac{1}{a_n\delta t} + \frac{e^{-a_n\delta t}}{a_n\delta t}\right] \qquad (21\text{-}29)$$

Suppose a simulation is performed assuming a conductor is perfectly conducting; the current is obtained as a function of time. Then Equations (21-29) and (21-23) can be used to approximate the voltage drop due to the resistance of the conductor including the effect of diffusion. However, to perform a network simulation including the effect of diffusion, an equivalent circuit is needed that can be included in the network equations (to include the diffusion effect simultaneously with the network solution). As in the case of other circuit elements, such equivalent circuits can be defined for diffusion that are valid over a time step of the calculation. Such Thevenin and Norton equivalents are derived in Chapter 6.

21.6 DIFFUSIVE RESPONSE TO A CURRENT RAMP

A linearly increasing current (ramp) function is an important special case because it approximates the leading edge of a signal pulse. Let a surface current be defined by

$$I_{xramp}(t) = at \tag{21-30}$$

where a [A/ms] is the slope of the ramp function. To obtain the tangential electric field due to this ramp function of current, substitute Equation (21-30) into Equation (21-24) to obtain the inductor currents, $I_n(t)$; that is,

$$I_{nramp}(t) = aa_n\int_0^t t'e^{-a_n(t-t')}\,dt' \tag{21-31}$$

$$I_{nramp}(t) = a\left[t - \frac{1}{a_n} + \frac{e^{-a_n t}}{a_n}\right] \tag{21-32}$$

Substituting Equation (21-32) into (21-23), one obtans

$$E_{xramp}(0,t) = R_{dc}at + 2R_{dc}a\sum_{n=1}^{\infty}\frac{1}{a_n}\left[1 - e^{-a_n t}\right] \tag{21-33}$$

An illustrative graph of E_{xramp} is shown in Figure 21.6.

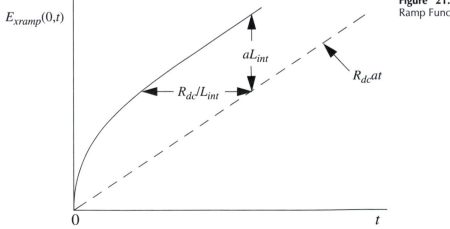

Figure 21.6 Electric Field Due to a Ramp Function of Current.

At very early times, as the diffusion begins, the impedance is very large and the time derivative of the electric field is large. After the magnetic field and current penetrate through the entire thickness of the slab, the electric field increases at the rate implied by the DC resistance. However, the electric field approaches a value larger than that implied by the DC resistance by an amount given by the time derivative of the current times the internal inductance of the slab (see Equation (21-16)). That is, Equation (21-33) becomes

$$E_{xramp}(0,t) \approx a(R_{dc}t + L_{int}), \quad t \gg 1/a_1 \tag{21-34}$$

PROBLEM 21.1 Given a trace with the following description:

Material: copper, $\sigma_{Cu} = 5.8 \times 10^7$
Thickness: 10^{-4} [m] (about 4 mils)
Width: 2×10^{-4} [m] (about 8 mils)
Length: 0.2 [m] (about 8 inches)
Current in the trace: $i = \bar{a}t$ [A], where $\bar{a} = 2 \times 10^7$ [A/s] (note that the dimensions of a are [A/ms].)
Assume two-sided diffusion into the trace.
Find the following:

  1. The slab internal inductance [H] using Equation (21-16).
  2. The internal inductance per length [H/m] of the trace.
  3. The total internal inductance [H] of the trace.
  4. The slab DC resistance R_{dc} [Ω] (Equation (21-13)).
  5. The total DC resistance of the trace.
  6. The decay constant a_1 using Equation (21-8). (Note that $a_n = n^2 a_1$.)
  7. The inductive voltage drop of the trace [V] after the field has diffused into the trace.
  8. For $t = 1, 5, 20, 50$ [ns], find:
     a. The diffusive voltage drop of the trace using Equation (21-33) (include only the first four terms in the summation).
     b. The contribution to the diffusive voltage drop of the first term omitted in a., i.e., the fifth term.
     c. The instantaneous impedance of the trace, i.e., v/i.

21.7 NORTON AND THEVENIN EQUIVALENTS FOR DIFFUSION

A Thevenin equivalent circuit will be derived from Equation (21-23) that is valid over a time step from time t to $t + \delta t$. That is, it is assumed that the diffusive conductor is in a network that has been solved to time t (through order N); therefore, $I_x(t)$ and $I_n(t)$ ($n = 1, 2, \ldots N$) are known. It is desirable to calculate these currents at time $t + \delta t$. A Norton or Thevenin equivalent circuit valid from time t to $t + \delta t$ can be included in the network and the solution can be updated to the new time. The Thevenin equivalent circuit is shown in Figure 21.7.

Algebraically, the equivalent circuit of Figure 21.7 implies that

$$E_x(t + \delta t) = V_{oc} + R_{Te}I_x(t + \delta t) \tag{21-35}$$

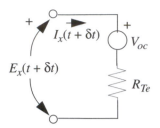

Figure 21.7 Thevenin Circuit for Slab Diffusion Valid over Short Time Interval.

EXERCISE 21-5. Show that Equation (21-23) (with the summation truncated at $n = N$) corresponds to Equation (21-35) if

$$V_{oc} = 2R_{dc}\left\{I_x(t) \sum_{n=1}^{N}\left[e^{-a_n\delta t} - \frac{\left[1 - e^{-a_n\delta t}\right]}{a_n\delta t}\right] - \sum_{n=1}^{N} I_n(t)e^{-a_n\delta t}\right\} \quad (21\text{-}36)$$

$$R_{Te} = R_{dc}\left\{1 + 2\sum_{n=1}^{N}\frac{\left[1 - e^{-a_n\delta t}\right]}{a_n\delta t}\right\} \quad (21\text{-}37)$$

Following a network solution using Equation (21-35) to obtain $I_x(t + \delta t)$, Equation (21-29) is calculated to obtain $I_n(t + \delta t)$ for $n = 1, \ldots, N$, to be used in the next time step.

The Norton equivalent circuit is shown in Figure 21.8. It is related to the Thevenin equivalent of Figure 21.7 by

$$I_{sc} = \frac{V_{oc}}{R_{Te}} \quad (21\text{-}38)$$

$$G_{Ne} = \frac{1}{R_{Te}} \quad (21\text{-}39)$$

$$I_x(t + \delta t) = G_{Ne}E_x(t + \delta t) - I_{sc} \quad (21\text{-}40)$$

This is the convenient form for node-equation solutions of general networks.

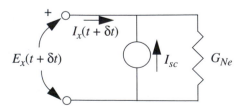

Figure 21.8 Norton Circuit for Slab Diffusion Valid over Short Time Interval.

21.8 CONVERGENCE OF THE SLAB-DIFFUSION SERIES

At $t = 0$, the series of Equations (21-12) or (21-17) does not converge; that is, the instantaneous impedance for a suddenly applied signal is infinite. For $t > 0$, either series converges rapidly, the terms decreasing as $e^{-(a_1 t)n^2}$. However, the decision of how many terms are to be included should be made on the basis of the decay constant of the first term omitted *vis-à-vis* the required resolution of the calculation. Terms that decay much faster than the signal changes significantly, simply have negligible effect because the inductance in the equivalent circuit approximates a short circuit (i.e., an impedance much less than R_{dc}).

Consider a trace 10^{-4} [m] thick (about 4 mils). Assume two-sided diffusion so that $d = 5.0 \times 10^{-5}$ [m]. If the material is copper, $\sigma_{Cu} = 5.8 \times 10^{7}$ [℧/m], then $R_{dc} = 1.724 \times 10^{-4}$ [Ω]. Table 21-1 contains the slab inductances and decay constants for $n = 1$ through 6 for this case. If the fastest signal to be resolved (e.g., rise time) is 5 [ns], then two of the exponential terms would be adequate for most purposes for simulation; the other terms all decay more than two *e*-folds in 5 [ns] and would contribute only a small amount to the voltage drop. Let the trace be twice as wide as it is thick, 2×10^{-4} [m] (about 8 mils).

Then the resistance per length of trace is $R = R_{dc}/2 \times 10^{-4} = 0.862$ [Ω/m].

PROBLEM 21.2 Assume a geometry such that one-sided diffusion is the best approximation. Let the thickness of a trace be 0.75×10^{-4} [m] and the material be copper. What maximum value of n should be included in the diffusion model to include all the terms with relaxation time greater than 1 [ns]? That is, find the largest value of n such that $1/a_n > 1$ [ns].

Computer Code 21-1. Diffusive Conductivity into a Slab.

The computer code listed below contains subroutines that evaluate I_n, V_{oc}, and R_{Te}. The formula that defines each of these quantities has terms in which cancellation exists when $x = a_n \delta t$ is very small. If code is written that straightforwardly evaluates these expressions, a serious problem exists when x is very small. A loss of accuracy exists due to round

TABLE 21-1
Inductance and Decay Time for Two-Sided Slab Diffusion.

n [-]	Slab Inductance L_n [pH]	Time Constants $\frac{L_n}{2R_{dc}} = \frac{1}{a_n}$ [ns]	Inductance per length L_n' [nH/m] Width = 2×10^{-4} [m]
1	6.366	18.46	31.83
2	1.592	4.62	7.96
3	0.707	2.05	3.54
4	0.398	1.15	1.99
5	0.255	0.74	1.27
6	0.177	0.51	0.88

off error; the precision of the result is reduced by about one decimal place for each order of magnitude that x is less than one. When x is about 10^{-7}, all accuracy is gone for single-precision calculations. This is due to the subtraction of quantities that approach one as x approaches zero. This problem is circumvented in the subroutines by evaluating these terms using a series whenever $x < 10^{-2}$. The terms that are evaluated this way are listed as follows.

Equation (21-29), I_n contains the term:

$$1 - \frac{1 - e^{-x}}{x} = x\left[\frac{1}{2} - \frac{x}{6} + \frac{x^2}{24} - \frac{x^3}{120} + \ldots + \frac{(-1)^n}{(n+2)!}x^n + \ldots\right] \qquad (21\text{-}41)$$

Equation (21-36), V_{oc} contains the term:

$$e^{-x} - \frac{1 - e^{-x}}{x} = -x\left[\frac{1}{2} - \frac{x}{3} + \frac{x^2}{8} - \frac{x^3}{30} + \ldots + \frac{(-1)^n(n+1)}{(n+2)!}x^n + \ldots\right] \qquad (21\text{-}42)$$

Equation (21-37), R_{Te} contains the term

$$\frac{1 - e^{-x}}{x} = 1 - \frac{x}{2} + \frac{x^2}{6} - \frac{x^3}{24} + \ldots + \frac{(-1)^n}{(n+1)!}x^n + \ldots \qquad (21\text{-}43)$$

To illustrate the use of these subroutines, the effect of diffusion in the trace defined in Problem 21-1 is simulated. The definition of the trace of Problem 21-1 is used with the driver and load shown in Figure 21.9. The driver voltage $V_{oc}(t)$ is a square pulse.

The circuit of Figure 21.9 is redrawn using three Norton equivalent circuits; it is shown in Figure 21.10. The main program of Computer Code 3-1, originally written for a two-terminal DC network, is revised to perform the time-stepping solution of this circuit.

The code listed below solves the circuit shown in Figures 21.9 and 21.10. The driver voltage is a square pulse with 1 volt amplitude and 5 [ns] width. The structures, function prototypes, and subroutines `SolveNodeEqs`, `SimpleGauss`, and `IndexMatrix` have been omitted from the listing that follows to save space; these can be found in the listing of Computer Code 3-1. The main program is listed, followed by the diffusion subroutines: `SlabIn` (Equation 21-29), `SlabRTe` (Equation 21-37), and `SlabVoc` (Equation 21-36).

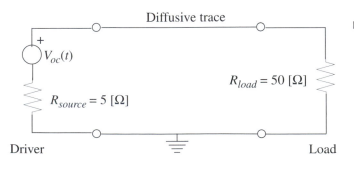

Diffusive trace

$V_{oc}(t)$

$R_{source} = 5\ [\Omega]$

$R_{load} = 50\ [\Omega]$

Driver   Load

Figure 21.9   Physical Model Solved in Test Program.

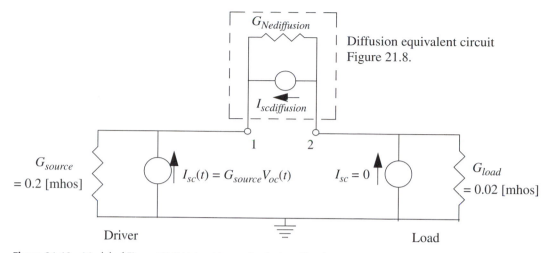

Figure 21.10   Model of Figure 21.9 Using Norton Equivalent Circuits.

```
/* -----Function declaration main ------------------------------- */
void main ()
{
/* ----- Local Variables ------------------------------------- */
// Model parameters
float     Vdrive = 1.0F,          // Driver voltage [V]
          Rsource = 5.0F,         // Source impedance    [ohms]
          Rload = 50.0F,          // Load impedance [ohms]
          sigCu = 5.8e7F,         // Conductivity of copper [mhos/m]
          width = 2.0e-4F,        // Width of trace [m]
          thick = 1.0e-4F,        // Thickness of trace [m]
          d,                      // half the thickness (for 2-sided d)
          length = 0.2F;          // Length of trace [m]
// Diffusion variables
float     a1,                     // First diffusion constant    [1/s]
          mu,                     // Permeability of space
          dt = 0.5e-9F,           // Time step [s]
          time,                   // Calculational time [s]
          pulse_length = 5.0e-9F, // Length of pulse [s]
          tmax = 10.0e-9F,        // Max. time of calculation [s]
          Ixt = 0.0F,             // Ix(t) [A]
          Ixtplus = 0.0F,         // Ix(t+dt) [A]
          *In_ptr = NULL,         // Inductor currents [A]
          Rdc,                    // d.c. resistance of trace [ohms]
          RTe,                    // Thevenin resistance (diffusion)
          Voc;                    // Thevenin voltage (diffusion)
int       Nmax = 10;              // Number of n terms in model
// Solution variables
int       Nnodes;                 // number of nodes
```

```
ISOURCE_PTR  ISlist_ptr = NULL,    // pointer to current sources list
             DriveIs_ptr = NULL,   // pointer to driver Norton current
             DiffuseIs_ptr = NULL,// pointer to diffusion Norton current
              LastIS_ptr = NULL;   // this NULL used to terminate list
GELEMENT_PTR   Glist_ptr = NULL,    // pointer to G element list
               LastGptr = NULL;     // this NULL used to terminate list
float           *Vnodes = NULL,     // node voltages
                V12,                // voltage (v1 - v2)
                pi = 3.141592654F;
int            rcode;               // return code from subroutine calls
int            idx;                 // for-loop index
/* -------------- Executable Code ------------------------------- */
    // Test case for slab diffusion -- see Fig. 21.9
    // Define case with two nodes, 3 g's and 2 source currents
    // similar to Problem 3-2; modification of Computer Code 3-3.
    Nnodes = 2;
    // allocate memory for Vnodes and In_ptr (inductor currents)
    Vnodes = (float*)malloc(sizeof(float) * Nnodes);
    In_ptr = (float*)malloc(sizeof(float) * Nmax);
    for (idx = 0; idx < Nmax; idx++)      // initialize to zero
        *(In_ptr + idx) = 0.0F;
    // Initial calculations
    mu = pi * 4.0e-7F;
    d = 0.5F * thick;                      // 2-sided diffusion
    a1 = pi * pi /(d * d * mu * sigCu);    // Eq. (21-8)
    Rdc = length/(width * thick * sigCu);  // [ohms]
    // Make linked list of current sources
    ISlist_ptr = (ISOURCE_PTR)malloc(sizeof(ISOURCE) * 2);//allocate 2
    //Source 1 -- Driver
    DriveIs_ptr = ISlist_ptr;       // Driver current set each time step
    ISlist_ptr->NodeInto =  1;      //into node 1
    ISlist_ptr->NodeOutof = 0;      //out of ground
    ISlist_ptr->next = LastIS_ptr;  //first time = NULL -- IMPORTANT!
    //Source 2   -- Diffusion model
    LastIS_ptr = ISlist_ptr;
    ISlist_ptr++;                          // go to next link
    DiffuseIs_ptr = ISlist_ptr;     // Diffusion current set each time
    ISlist_ptr->NodeInto =  1;      //into node 1
    ISlist_ptr->NodeOutof = 2;      //out of node 2
    ISlist_ptr->next = LastIS_ptr;
    // Make linked list of G's
    Glist_ptr = (GELEMENT_PTR)malloc(sizeof(GELEMENT) * 3); // allocate 3
    //G 1 -- Driver conductance
    Glist_ptr->g = 1.0F/Rsource;    // Source conductance [mhos]
    Glist_ptr->Node1 = 1;           // node 1
    Glist_ptr->Node2 = 0;           // ground
```

```
Glist_ptr->next = LastGptr;        // first time = NULL
//G 2 -- Diffusion model
LastGptr = Glist_ptr;
Glist_ptr++;                        // go to next link
// calculate diffusion model conductance (doesn't change unless
// dt or other parameters change)
rcode = SlabRTe (a1, dt, Nmax, Rdc, &RTe);
Glist_ptr->g = 1.0F/RTe;           // Diffusion conductance
Glist_ptr->Node1 = 1;              //node 1
Glist_ptr->Node2 = 2;              //node 2
Glist_ptr->next = LastGptr;
//G 3 -- Load impedance
LastGptr = Glist_ptr;
Glist_ptr++;                        // go to next link
Glist_ptr->g = 1.0F/Rload;         //[mhos]
Glist_ptr->Node1 = 2;              //node 2
Glist_ptr->Node2 = 0;              //ground
Glist_ptr->next = LastGptr;
// Output the list of G's to the screen
LastGptr = Glist_ptr;
while (LastGptr != NULL)
{
   printf ("Node1 is %i   Node2 is %i   g = %f\n",
           LastGptr->Node1, LastGptr->Node2, LastGptr->g);
   LastGptr = LastGptr->next;
}
printf("\n");
//*****
//** Top of time-stepping loop
//*****
for (time = dt; time <= tmax+dt; time +=dt)
{
   // Set driver Norton equivalent current
   if (time <= pulse_length)
      DriveIs_ptr->Is = Vdrive/Rsource;
   else
      DriveIs_ptr->Is = 0.0F;
   // Set diffusion Norton equivalent current
   rcode = SlabVoc (a1, dt, Nmax, Rdc, In_ptr, Ixtplus, &Voc);
   if (rcode)
   {
      printf ("Error returned from SlabVoc\n");
      goto DoneMain;
   }
   DiffuseIs_ptr->Is = Voc/RTe;  // Eq. (21-38)
   rcode = SolveNodeEqs ( Nnodes,      // number of nodes
```

```
                             ISlist_ptr,    // pointer to current sources
                              Glist_ptr,     // pointer to g element list
                              Vnodes );      // node voltages address
         if (rcode)
         {
            printf ("Error returned from SolveNodeEqs\n");
            goto DoneMain;
         }
         // Update current through the diffusive trace
         Ixt = Ixtplus;
         V12 = (*Vnodes - *(Vnodes + 1));
         Ixtplus = V12/RTe - DiffuseIs_ptr->Is;   // Eq. (21-40)
         // Update the inductor currents In
         rcode = SlabIn (a1, dt, Nmax, Ixt, Ixtplus, In_ptr);
         //Output data -- Node voltages
         printf ("time = %f [ns] node 1: v = %f, node 2: v = %f\n",
                 time*1.0e9F,Vnodes[0],Vnodes[1]);
      }
   DoneMain:   ;
   // If this were a subroutine, memory should be freed here to avoid
   // memory leak.
}
/* --------- End of main ----------------------------------------- */
/* ------- Function Declaration ---------------------------------- */
// Update inductor currents in slab diffusion model, Equation (21-29).
int SlabIn ( /*   Input variables      */
            float a1,       // First diffusion constant [1/s]
            float dt,       // Time step length [s]
            int Nmax,       // Maximum value of n    [-]
            float Ixt,      // Ix(t) [A/m] or [A]
            float Ixtplus, // Ix(t + dt) [A/m] or [A]
            /*   Input/Output vector    */
            float *In)      // Vector of In(t) input, In(t+dt) output
                            //  [A/m] or [A]
{
/* ----- Local Variables --------------------------------------- */
   float   x,        // = an * dt, temp variable
           an,       // diffusion parameter, nth order
           expf,     // exp(-x)
           f1,f2;    // temp. variables
   int     n;        // for-loop variable, n
/* ------------- Executable Code --------------------------------- */
    // Check input variables
    if ((a1 <= 0.0) || (dt <= 0.0) || (Nmax < 1))
       return (-1);  // input error
    // Begin calculation
```

```
   for (n = 0; n < Nmax; n++)
   {
      an = a1 * (float)((n + 1) * (n + 1));
      x = an * dt;
      expf = (float)exp(-x);
      if (x > 0.01)
      {
        f1 = 1.0F - expf;
        f2 = 1.0F - f1/x;
      }
      else
      {
        f1 =x * (24.0F - x * (12.0F - x * (4.0F -x)))/24.0F;
      f2 =x * (60.0F - x * (20.0F - x * (5.0F - x)))/120.0F;
      }
    *(In + n) = *(In + n) * expf + Ixt * f1 + (Ixtplus - Ixt) * f2;
   }
     return (0);
}
/* -----End of Procedure SlabIn -------------------------------- */
/* ------- Function Declaration -------------------------------- */
  // Calculate Thevenin resistance for diffusion into slab
  // (Valid for a given time step)
int SlabRTe ( /*  Input variables    */
              float a1,    // First diffusion constant [1/s]
              float dt,    // Time step length [s]
              int   Nmax,  // Maximum value of n   [-]
              float Rdc,   // d.c. resistance [ohms]
              /*   Output  variable    */
              float *RTe)  // Thevenin resistance
{
/* ----- Local Variables -------------------------------------- */
    int   n;        // for-loop index over order
    float x,        // = an * dt, temp variable
          an,       // diffusion parameter, nth order
          expf;     // exp(-x)
/* ------------- Executable Code ------------------------------- */
    // Check input variables
    if ((a1 <= 0.0) ||   (dt <= 0.0) || (Nmax < 1))
       return (-1);  // input error
    // Begin calculation
    *RTe = 0.5F;
    for (n = 0; n < Nmax; n++)
    {
       an = a1 * (float)((n + 1) * (n + 1));
       x = an * dt;
```

```c
      if (x > 0.01)
      {
         expf = (float)exp(-x);
         *RTe += (1.0F - expf)/x;
      }
      else
      {
        *RTe += (24.0F - x * (12.0F - x * (4.0F -x)))/24.0F;
      }
    }
    *RTe = Rdc * (*RTe + *RTe);
    return (0);
}
/* -----End of Procedure SlabRTe ------------------------------- */
/* ------- Function Declaration ------------------------------- */
  int SlabVoc (    /*  Input variables    */
               float a1,    // First diffusion constant [1/s]
               float dt,    // Time-step length [s]
               int   Nmax,  // Maximum value of n    [-]
               float Rdc,   // d.c. resistance [ohms]
               float *In,   // Vector of In(t) input [A/m] or [A]
               float Ixt,   // Ix(t) [A/m] or [A]
               /*  Output  variable   */
               float *Voc)  // Thevenin o.c. voltage
  {
/*  ----- Local Variables ------------------------------------ */
   float   x,         // = an * dt, temp variable
           an,        // diffusion parameter, nth order
           expf,      // exp(-x)
           f1;        //  temp. variable
   int     n;         // for-loop variable, n
/* ------------- Executable Code ----------------------------- */
    // Check input variables
    if ((a1 <= 0.0) ||  (dt <= 0.0) || (Nmax < 1))
      return (-1);  // input error
    // Begin calculation
    *Voc = 0.0F;
    for (n = 0; n < Nmax; n++)
    {
      an = a1 * (float)((n + 1) * (n + 1));
      x = an * dt;
      expf = (float)exp(-x);
      if (x > 0.01)
      {
        f1 = expf - (1.0F - expf)/x;
      }
```

```
        else
        {
            f1 = -x * (60.0F - x * (40.0F - x * (15.0F - 4.0F * x)))/
120.0F;
        }
        *Voc += Ixt * f1 - *(In + n) * expf;
    }
    *Voc *= (Rdc + Rdc);
    return (0);
  }
/* -----End of Procedure SlabVoc --------------------------------- */
```

The output from the code with parameters as given in the above listing is given below:

```
Node1 is 2    Node2 is 0  g = 0.020000   Load conductance
Node1 is 1    Node2 is 2  g = 0.398522   Diffusion Norton Conductance
Node1 is 1    Node2 is 0  g = 0.200000   Driver conductance
time = 0.500000 [ns] node 1: v = 0.913058, node 2: v = 0.869425
time = 1.000000 [ns] node 1: v = 0.911579, node 2: v = 0.884207
time = 1.500000 [ns] node 1: v = 0.911023, node 2: v = 0.889769
time = 2.000000 [ns] node 1: v = 0.910725, node 2: v = 0.892752
time = 2.500000 [ns] node 1: v = 0.910533, node 2: v = 0.894674
time = 3.000000 [ns] node 1: v = 0.910396, node 2: v = 0.896045
time = 3.500000 [ns] node 1: v = 0.910291, node 2: v = 0.897086
time = 4.000000 [ns] node 1: v = 0.910209, node 2: v = 0.897912
time = 4.500000 [ns] node 1: v = 0.910141, node 2: v = 0.898588
time = 5.000000 [ns] node 1: v = 0.910085, node 2: v = 0.899154
time = 5.500000 [ns] node 1: v = -0.003021, node 2: v = 0.030212
time = 6.000000 [ns] node 1: v = -0.001585, node 2: v = 0.015849
time = 6.500000 [ns] node 1: v = -0.001066, node 2: v = 0.010655
time = 7.000000 [ns] node 1: v = -0.000800, node 2: v = 0.007998
time = 7.500000 [ns] node 1: v = -0.000637, node 2: v = 0.006368
time = 8.000000 [ns] node 1: v = -0.000526, node 2: v = 0.005261
time = 8.500000 [ns] node 1: v = -0.000446, node 2: v = 0.004458
time = 9.000000 [ns] node 1: v = -0.000385, node 2: v = 0.003851
time = 9.500000 [ns] node 1: v = -0.000338, node 2: v = 0.003376
time = 10.000000 [ns] node 1: v = -0.000299, node 2: v = 0.002994
```

Note that the total DC resistance of the circuit is $R = 5 + 0.1724 + 50 = 55.1724$ [Ω]. So that the DC voltage across the load resistor (node 2) with 1 volt driving the circuit is $v = 50/55.1724 = 0.90625$. This is the voltage being approached when the source voltage is suddenly turned off at 5.0 [ns]. The nonzero voltage following that time is due to the "inductive kick" of the energy stored in the internal inductance of the trace. It decays to zero as the magnetic field diffuses out of the trace.

The lumped circuit model of diffusion presented here neglects propagation delays; it assumes that, at a given time, the same current flows at every point along the trace. For a signal with rise time smaller than the propagation time along the trace, this model yields a larger than realistic effective resistance (and resulting attenuation) for the initial part of the signal. A more accurate simulation is obtained by distributing the diffusion effect along the trace in a propagation model. This is the approach taken in Chapter 23.

Chapter 22

CYLINDRICAL GEOMETRY SKIN EFFECT SIMULATION

A method to simulate the skin effect for a conducting cylinder (e.g., round wire) will be derived in this Chapter. The approach is similar to that used for the slab. The analysis will yield an equivalent circuit and, applying the convolution principle, a time-stepping algorithm will be derived that is identical in form to that of the slab. The values of the diffusion parameters and the equivalent circuit inductances and resistances will be different.

22.1 FIELD PARTIAL DIFFERENTIAL EQUATIONS

We begin with the field Equations (20-2) and (20-5). Those equations are applied using circular cylindrical coordinates. We assume that the fields have symmetry such that the only non-zero field components in this coordinate system are E_z and H_θ and that the fields depend only the r coordinate (i.e., all partial derivatives with respect to z and θ are zero). With these assumptions, Equation (20-2) becomes

$$\frac{\partial E_z}{\partial r} = \mu \frac{\partial H_\theta}{\partial t} \tag{22-1}$$

and Equation (20-5) becomes

$$\frac{1}{r}\frac{\partial}{\partial r}(rH_\theta) = \sigma E_z \tag{22-2}$$

Substituting Equation (22-2) into Equation (22-1), the diffusion equation for H_θ is obtained.

$$\frac{\partial}{\partial r}\left[\frac{1}{r}\frac{\partial}{\partial r}(rH_\theta)\right] = \mu\sigma\frac{\partial H_\theta}{\partial t} \tag{22-3}$$

275

22.2 DC CURRENT DENSITY AND MAGNETIC FIELD

Assume a constant current I_{cyl} has been flowing for a long time so that the field has completely diffused into the cylinder. The current density σE_z is uniform throughout the radius r. Therefore, the current is given by

$$I_{cyl} = \pi d^2 \sigma E_z \tag{22-4}$$

where d is the radius of the cylinder.

The magnetic field that satisfies Equation (22-2) under these conditions is

$$H_\theta = \frac{I_{cyl} r}{2\pi d^2} \tag{22-5}$$

EXERCISE 22-1.   Show that the magnetic field given by Equation (22-5) satisfies both Equations (22-2) and (22-3).

The magnetic field will now be expanded in Bessel functions of order 1. That is, a Fourier-Bessel series of order 1 will be used (see Watson, 1966, p. 594). Using this method, an arbitrary function $f(x)$, for x in the interval $(0,1)$ can be expressed as

$$f(x) = \sum_{n=1}^{\infty} A_n J_1(j_n x) \tag{22-6}$$

where J_1 is the Bessel function of the first kind of order 1 and j_n are the zeros of J_1, that is $J_1(j_n) = 0$. The coefficients A_n are given by

$$A_n = \frac{2}{J_2^2(j_n)} \int_0^1 x f(x) J_1(j_n x) dx \tag{22-7}$$

Applying this expansion to the magnetic field H_θ, over the interval r in $(0,d)$, one obtains

$$H_\theta(r) = \sum_{n=1}^{\infty} A_n J_1(j_n r/d) \tag{22-8}$$

where

$$A_n = \frac{2}{J_2^2(j_n)} \int_0^1 x H_\theta(xd) J_1(j_n x) dx \tag{22-9}$$

Substituting H_θ using Equation (22-5), A_n becomes

$$A_n = \frac{I_{cyl}}{J_2^2(j_n)\pi d} \int_0^1 x^2 J_1(j_n x) dx \qquad (22\text{-}10)$$

Using Gradshteyn, 1980 (5.52) or Dwight, 1947 (801.60), we can evaluate this integral to obtain

$$A_n = \frac{I_{cyl}}{J_2(j_n)\pi d j_n} \qquad (22\text{-}11)$$

22.3 DIFFUSION EQUATION SOLUTIONS

As in the case of diffusion in a slab where the Fourier series expansion of the magnetic field resulted in a zero value on the boundary, so it is with the Fourier-Bessel series in this case. The expansion (22-8) converges to (22-5) at every point r in [0,d] except at the point $r = d$, the boundary, where the expansion yields zero magnetic field and, hence, zero total current. Again, we define the field for the situation where the current has been flowing for a long time and is suddenly set to zero as

$$\hat{H}_\theta(r,t) = \begin{cases} \dfrac{I_{cyl}r}{2\pi d^2}, & t < 0 \\[4mm] \displaystyle\sum_{n=1}^{\infty} A_n e^{-a_n t} J_1\!\left(\dfrac{j_n r}{d}\right), & t \geq 0 \end{cases} \qquad (22\text{-}12)$$

The Bessel function $J_1(x)$ satisfies the differential equation (Bessel's equation of order 1)

$$x^2 \frac{d^2 J_1(x)}{dx^2} + x\frac{dJ_1(x)}{dx} + (x^2 - 1) = 0 \qquad (22\text{-}13)$$

EXERCISE 22-2. Show that $\hat{H}_\theta(r,t)$ given by Equation (22-12) satisfies the diffusion Equation (22-3) if a_n is given by

$$a_n = \frac{j_n^2}{\mu\sigma d^2} \qquad (22\text{-}14)$$

The magnetic field due to a step function of current is now obtained by superposition

$$H_{\theta step}(r,t) = \frac{I_{cyl}r}{2\pi d^2} - \hat{H}_\theta(r,t) \qquad (22\text{-}15)$$

$$H_{\theta step}(r,t) = \begin{cases} 0, & t < 0 \\ \dfrac{I_{cyl}r}{2\pi d^2} - \displaystyle\sum_{n=1}^{\infty} A_n e^{-a_n t} J_1\left(\dfrac{j_n r}{d}\right), & t \geq 0 \end{cases} \tag{22-16}$$

and the electric field due to a step function of current is found by substituting Equation (22-16) into Equation (22-2) to obtain for $t > 0$[1]

$$E_{zstep}(r,t) = \frac{I_{cyl}}{\sigma\pi d^2} - \frac{I_{cyl}}{\sigma\pi d^2} \sum_{n=1}^{\infty} e^{-a_n t} \frac{J_0\left(\dfrac{j_n r}{d}\right)}{J_2\left(\dfrac{j_n r}{d}\right)} \tag{22-17}$$

At the surface of the cylinder, $r = d$, we have[2]

$$E_{zstep}(d,t) = \frac{I_{cyl}}{\sigma\pi d^2} + \frac{I_{cyl}}{\sigma\pi d^2} \sum_{n=1}^{\infty} e^{-a_n t} \tag{22-18}$$

22.4 EQUIVALENT CIRCUIT FOR DIFFUSIVE CYLINDER

Equation (22-18) can be written

$$E_{zstep}(d,t) = R'_{dc} I_{cyl}\left[1 + \sum_{n=1}^{\infty} e^{-a_n t}\right] \tag{22-19}$$

where $R'_{dc} = 1/\sigma\pi d^2$ [Ω/m]. We note that a_n (Equation (22-14)) can be written

$$a_n = R'_{dc}/L'_n \quad [\text{s}^{-1}] \tag{22-20}$$

where

$$L'_n = \mu/j_n^2\pi \quad [\text{H/m}] \tag{22-21}$$

The electric field of Equation (22-19) is that due to a step function of current flowing in the circuit shown below in Figure 22.1

---

1. Note that $\dfrac{d}{dx}[xJ_1(x)] = xJ_0(x)$, Dwight, 1947, (801.6).

2. Note that $J_0(j_n)/J_2(j_n) = -1$, Abramowitz and Stegun, 1970, (9.5.4).

22.5 INTERNAL INDUCTANCE OF A CYLINDRICAL CONDUCTOR

The internal inductance (per length) of a conducting cylinder is given by Smythe, 1950, p. 317 as

$$L'_{int} = \frac{\mu}{8\pi} \quad \text{[H/m]} \tag{22-22}$$

EXERCISE 22-3.   Show that Smythe's value of internal inductance is correct. *Hint*: The magnetic energy stored (per length) is

$$\frac{1}{2}L'_{int}I^2_{cyl} = \int_{Cyl} \frac{1}{2}\mu H_\theta^2 \, dA = \int_0^d \int_0^{2\pi} \frac{1}{2}\mu H_\theta^2 r d\theta dr \quad \text{[J/m]} \tag{22-23}$$

Substitute H_θ, Equation (22-5), and perform the integration.

The equivalent circuit of Figure 22-1 implies that

$$L'_{int} = \sum_{n=1}^{\infty} L'_n = \frac{\mu}{8\pi} \tag{22-24}$$

22.6 NORTON EQUIVALENT CIRCUIT FOR A DIFFUSIVE CYLINDER

The Norton equivalent circuit for the cylinder model is identical to that of the slab (see Figure 21.7). The derivation of the delta-function response and the application of the convolution integral is essentially identical with that for the slab except for the change in some of the coefficients from $2R_{dc}$ to R_{dc} and the different values of the diffusion parameters a_n given in Equation (22-14). In summary, the formula for calculating the inductor currents, Equation (21-29), is unchanged. The equations for the Thevenin V_{oc} and R_{Te} (Equations (21-36) and (21-37)) are slightly different because of the change in the R_{dc} terms. The new equations for the cylinder are

$$V_{oc} = R_{dc}\left\{ I_x(t) \sum_{n=1}^{N} \left[e^{-a_n\delta t} - \frac{\left[1 - e^{-a_n\delta t}\right]}{a_n\delta t} \right] - \sum_{n=1}^{N} I_n(t)e^{-a_n\delta t} \right\} \tag{22-25}$$

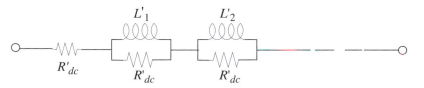

Figure 22.1   Equivalent Circuit for Cylinder Diffusion.

$$R_{Te} = R_{dc}\left\{1 + \sum_{n=1}^{N}\frac{\left[1 - e^{-a_n \delta t}\right]}{a_n \delta t}\right\}$$ (22-26)

where $R_{dc} = R'_{dc} \times (\text{Length of the wire})$ [Ω].

PROBLEM 22.1 Compare the diffusion rates into a slab and a cylinder. That is, let $d = 0.5 \times 10^{-4}$ [m], for both a slab and a cylinder. Assume they are both copper. Calculate the first five values of a_n and $1/a_n$ for the slab and the cylinder; i.e., fill out Table 22-1.

Abramowitz and Stegun (1970, p. 409) list the first 20 zeros of J_1. The first five are given in Table 22-1. The next 15, as given by Abramowitz and Stegun, are:

19.61586	35.33231	51.04354
22.76008	38.47477	54.18555
25.90367	41.61709	57.32753
29.04683	44.75932	60.46946
32.48968	47.90146	63.61136

These zeros of J_1 should be enough for practical simulations. If larger ones are needed, they may be calculated using McMahon's expansion given by Abramowitz and Stegun (1970, p. 371, 9.5.12).

Computer Code 22-1.   Diffusive Conduction of a Cylinder.

Conversion of Computer Code 21-1 from a slab calculation to one for diffusion into a cylinder requires only minor changes. No revised code has been listed here in the text; however, a revised code has been included on the accompanying disk. In essence, the a_n for the cylinder

TABLE 22.1
Cylinder and Slab Compared.

n	$j_n{}^a$	Cylinder diffusion model a_n [s^{-1}]	Cylinder diffusion model $1/a_n$ [ns]	Slab diffusion model a_n [s^{-1}]	Slab diffusion model $1/a_n$ [ns]
1	3.83171				
2	7.01559				
3	10.17347				
4	13.32369				
5	16.47063				

a. Zeros of J_1 from Abramowitz and Stegun, 1970, p. 409.

must be calculated and used instead of those for the slab, and the minor differences between Equations (21-36) and (21-37) and Equations (22-25) and (22-26) must be accounted for. The new subroutines for diffusion in a cylinder are named: `CylIn` (update inductor currents), `CylRTe` (update Thevenin resistance) and `CylVoc` (update Thevenin open circuit voltage). The output of the code is very similar to that for the slab listed at the end of Chapter 21.

Chapter 23
PROPAGATION WITH SKIN EFFECT

The skin effect model can be fully integrated into the propagation algorithm. The basic equations will be written here and a code provided for the single-wire case. Methods will be suggested for including it separately from the propagation algorithm to obtain the signal degradation effects without obtaining its full crosstalk and distributed reflection effects.

23.1 DISTRIBUTED VOLTAGE SOURCE IN THE TRANSMISSION LINE EQUATIONS

The transmission line Equation (1-1) can be modified to include a distributed series voltage source along it as shown schematically in Figure 23.1 (the capacitance and inductance are not shown). The voltage source $V(x,t)$ is in volts per unit length [V/m] distributed along the line.

With such a voltage source included, Equation (1-1) becomes

$$v_x(x,t) = -Li_t(x,t) - Ri(x,t) - V(x,t) \qquad (23\text{-}1)$$

And the voltage transport Equations (2-29) and (2-30) become

$$v_{F1}(x,t-x/c) = -\alpha v_F(x,t-x/c) + \beta v_B(x,t+x/c) - \frac{1}{2}V(x,t) \qquad (23\text{-}2)$$

and

$$v_{B1}(x,t+x/c) = \alpha v_B(x,t+x/c) - \beta v_F(x,t-x/c) - \frac{1}{2}V(x,t) \qquad (23\text{-}3)$$

EXERCISE 23-1. Show that Equations (23-2) and (23-3) are correct by repeating the derivation of Equations (2-29) and (2-30) given in Chapter 2, but including the new distributed voltage source term $V(x,t)$.

The current transport Equations (2-31) and (2-32) become

$$i_{F1}(x,t-x/c) = -\alpha i_F(x,t-x/c) - \beta i_B(x,t+x/c) - \frac{1}{2z_o}V(x,t) \qquad (23\text{-}4)$$

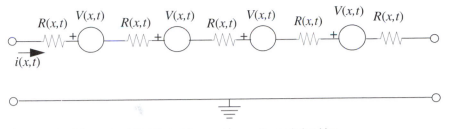

Figure 23.1   Schematic of Distributed Source Along a Transmission Line.

and

$$i_{B1}(x,t + x/c) = \alpha i_B(x,t + x/c) + \beta i_F(x,t - x/c) + \frac{1}{2z_o}V(x,t) \qquad (23\text{-}5)$$

23.2 LOSSY PROPAGATION WITH DIFFUSION

The current propagation equations will be the basis of the algorithm for propagation with diffusion. This method fits nicely with both the diffusion algorithm and the node equation approach to solving the termination network equations. Let the propagation occur from time t_1 to time t_2. For forward propagating signals, the propagation is from $x = x_1$ to $x = x_2$ and τ_F is a constant, where,

$$\tau_F = t_1 - \frac{x_1}{c} = t_2 - \frac{x_2}{c} \qquad (23\text{-}6)$$

For backward propagating signals, the propagation is from $x = x_2$ to $x = x_1$ and τ_B is a constant, where

$$\tau_B = t_1 + \frac{x_2}{c} = t_2 + \frac{x_1}{c} \qquad (23\text{-}7)$$

The difference approximation of Equation (23-4) applicable to an explicit difference solution is

$$i_F(x_2,\tau_F) \qquad (23\text{-}8)$$
$$= i_F(x_1,\tau_F) + (x_2 - x_1)\left[-\alpha\frac{i_F(x_2,\tau_F) + i_F(x_1,\tau_F)}{2} - \beta\frac{i_B(x_2,\tau_B) + i_B(x_1,\tau_B)}{2} - \frac{1}{2z_o}V_{oc}\right]$$

Similarly, the difference approximation of Equation (23-5) is

$$i_B(x_1,\tau_B) \qquad (23\text{-}9)$$
$$= i_B(x_2,\tau_B) + (x_2 - x_1)\left[-\alpha\frac{i_B(x_2,\tau_B) + i_B(x_1,\tau_B)}{2} - \beta\frac{i_F(x_2,\tau_F) + i_F(x_1,\tau_F)}{2} - \frac{1}{2z_o}V_{oc}\right]$$

where V_{oc} is the Thevenin equivalent voltage for the time interval t_1 to t_2 on the segment of the line between x_1 and x_2 (see Figure 21.7 and Equation (21-36)). It will be assumed that the conductance G is zero. Then, α and β are given by

$$\alpha = \beta = \frac{1}{2}\frac{R_{Te}}{z_o} \tag{23-10}$$

where R_{Te} is the Thevenin equivalent resistance (see Figure 21-7 and Equation (21-37)). Unless $\delta t = t_2 - t_1$ or a physical property of the wire changes during the calculation, R_{Te} remains constant and need only be calculated once.

23.3 MODIFIED CIRCULAR ARRAY FOR PROPAGATION WITH DIFFUSION

The array of currents in transit along the transmission line will be circular as described in Part I, but the circle will contain two more storage locations. Instead of the diagram shown in Figure 2.5, that shown in Figure 23.2 will be used. Note that the input of data at points 0 or 4 does not destroy the data at the (fictitious) points (1 and 5) beyond the end of line that are used to calculate the interpolated attenuation and reflection in the last segment of line at each end. This simplifies the algorithm for the diffusion calculation by making the currents at all of the points available at the same time in the calculation. The variable `NhalfMod` will be used for half the number of storage locations in the circle. For the schematic shown in Figure 23.2, its value is 4.

The inductor currents I_n (Equation (21-29)) will be calculated at the midpoint of each line segment (variable nseg; see the segment numbers in Figure 23.2).

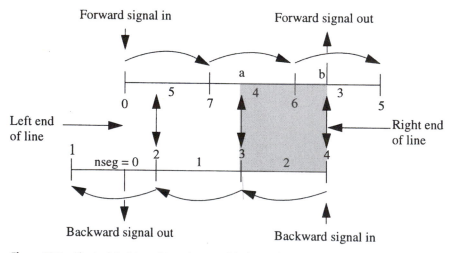

Figure 23.2   Physical Position of Signals in Modified Circular Storage Array.

Computer Code 23-1. Propagation with Diffusion

This computer code demonstrates propagation with diffusion by solving the circuit shown in Figure 21.9, but treating the diffusive trace as a transmission line rather than as a lumped element. The applicable circuit, in terms of Norton equivalent circuits, is shown below in Figure 23.3.

The dimensions of the trace used in the illustration are the same as those used in Computer Code 21-1 (the lumped element diffusive response of a trace). That is, its thickness is 10^{-4} [m] (about 4 mils) and its width is 2×10^{-4} [m] (about 8 mils). The propagation speed of the trace was assumed to be 2×10^{8} [m/s] corresponding to a dielectric constant of 2.25. The characteristic impedance was assumed to be 50 [Ω]. For a trace length of 0.2 [m], the result is very close to that shown following Computer Code 21-1 (for the lumped element diffusive trace response) with the exception of two effects: the time delay due to the propagation and the attenuation at early pulse times. At early times in a pulse, i.e., times less than the propagation time over the line, the lumped element model yields an attenuation larger than is physical. That's because the lumped element model assumes that the signal is everywhere on the trace simultaneously and the conduction depth is shallow (high resistance) at early times everywhere on the trace. The correct model has the corresponding shallow conduction depth only over the physical segment of line that the pulse occupies — less than the entire length of the line at early times — resulting in lower resistance and attenuation.

For short lines, such as the 20 [cm] line used for illustration, the attenuation and distortion due to the diffusive resistance is small. It's little more than 1% of the departing signal. However, this is much more than that calculated using the DC resistance without diffusion. Performing the same simulation without diffusion, the change in the signal is less than 0.5 percent. For much longer lines, because the attenuation and distortion is much larger, simulation of propagation of short pulses without including the effect of diffusion yields attenuation much too small. (Note that the diffusive effect can be removed and a simulation run as a resistive line using the code given below simply by setting the variable Nmax to zero, i.e., using zero terms in the diffusion calculation.)

A simulation was performed for a line 10 [m] long (other parameters the same as those for the 0.2 [m] line). The results are shown in Figure 23.4. The 50 [ns] propagation delay of the line has been removed and the load voltage overlaid with the driver and driven end voltage (the driver has a 5 [Ω] source impedance).

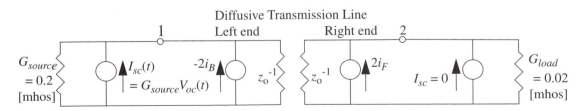

Figure 23.3   Model Including Diffusive Transmission Line.

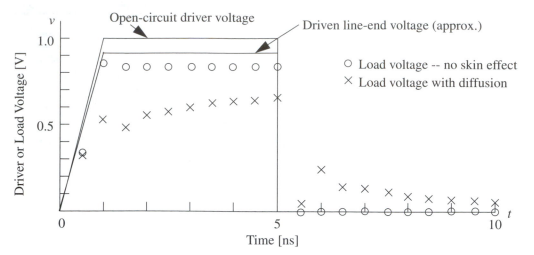

Figure 23.4   Simulation of 10 [m] Line with and without Diffusion.

Besides the sluggishness of the pulse rise at the load end, one should note the diffusion-caused voltage following the pulse for $t = 6$ [ns] and following. This voltage is driven by the internal inductance of the trace; it's not there for the no skin effect calculation.

The main program is shown below; it illustrates the use of the new subroutine `CurrTstepDiff`. This subroutine propagates forward and backward current signals one time step. The time-stepping diffusion algorithm for a slab geometry is applied at each time step. The subroutines `SlabIn`, `SlabTRe`, `SlabVoc` and `InitPropStep` are called but are not listed below because they have previously been listed and described. The new subroutine `IndicesModCirc` is listed below; it calculates the indices needed to carry out the algorithm using the modified circular storage array shown in Figure 23.2.

The main program below follows the general pattern for the time-stepping routines given in this book. The basic problem (constant) parameters are set in the variable declarations. Initialization is then performed including calling subroutine `InitPropTstep` to calculate `Nhalf` and the interpolation weights and the calling of `SlabRTe` to calculate the Thevenin equivalent resistance of a segment (the value of R_{Te} is constant during the numerical solution because the time step and other parameters on which it depends do not change). Arrays are allocated (using `malloc`) to allocate storage for the propagating current ring of storage `ring_ptr`, inductor currents $I_n(t)$, and the segment currents $I_x(t)$; these are initialized to zero. The time-stepping for-loop is then entered in which the following procedures are performed:

1. The driver voltage (and Norton short-circuit current) is calculated.

2. The node voltages v_1 and v_2 are calculated (because the line-end networks are independent, single-node networks, they are solved without use of the network solver used elsewhere).

3. The departing current signals are calculated.

4. The node voltages are output to the screen.

5. The new propagation subroutine `CurrTstepDiff` is called.

```
/* -----Function declaration main --------------------------------- */
void main ()
{
/* ----- Local Variables ------------------------------------------ */
// Model parameters
float      Vdrive = 1.0F,          // Driver voltage [V]
           Rsource = 5.0F,         // Source impedance   [ohms]
           Gsource,                // 1/Rsource
           Rload = 50.0F,          // Load impedance [ohms]
           Gload,                  // 1/Rload
           sigCu = 5.8e7F,         // Conductivity of copper [mhos/m]
           width = 2.0e-4F,        // Width of trace [m]
           thick = 1.0e-4F,        // Thickness of trace [m]
           d,                      // half the thickness (for 2-sided d)
           length = 0.2F,          // Length of trace [m]
           cspeed = 2.0e8F,        // propagation speed [m/s]
           rzo,                    // 1/zo
           zo = 50.0F;             // impedance of line [ohms]
// Time parameters
float      risetime = 1.0e-9F,     // leading edge rise time
           Vslope,                 // voltage slope of leading edge
           dt = 0.5e-9F,           // Time step [s]
           time,                   // Calculational time [s]
           pulse_length = 5.0e-9F,// Length of pulse [s]
           tmax = 10.0e-9F;        // Max. time of calculation [s]
// Diffusion variables
float      a1,                     // First diffusion constant   [1/s]
           mu,                     // Permeability of space
           *Ixt = NULL,            // vector of Ix(t) [A]
           Ixtplus = 0.0F,         // Ix(t+dt) [A]
           *In_ptr = NULL,         // Matrix of Inductor currents [A]
           RTe,                    // Thevenin resistance of segment [ohms]
           Rdc;                    // trace resistance per segment [ohms]
int        Nmax = 10;              // Number of n terms in model
                                   // (set to zero for no-diffusion model)
// Propagation variables
int        Nhalf;                  // ring half dimension
int        TStepOffset = 0;        // offset to current location in ring
float      wgt1,                   // Interpolation weight
           wgt2;                   // Interpolation weight
float      *Ring_ptr = NULL;       // storage ring pointer
float      CurInForward,           // departing signal left end
```

```
                CurInBackward,          // departing signal right end
                CurOutBackward,         // arriving signal left end
                CurOutForward;          // arriving signal right end
// Solution variables
float      Is;                  // Source Norton current
float      v1, v2;              // node voltages
float      pi = 3.141592654F;
int        rcode;               // return code from subroutine calls
int        idx;                 // for-loop index
float      f1000 = 1000.0F;     // 1000
/* -------------- Executable Code -------------------------------- */
    // Test case for propagaton with diffusive trace (skin effect)
    // Adds propagation to diffusive conductivity code V.2-1.
    // See Figure 23.3 for model definition
    //*** Initializations ***
    // Propagation
    rcode = InitPropTstep (  /*   Input variables      */
                      length, dt, cspeed,
                            /*   Output variables     */
                      &Nhalf, &wgt1, &wgt2  );
    if (rcode)
    {
       printf ("Error returned from subroutine InitPropTstep\n");
       goto DoneMain;
    }
    Ring_ptr = (float*)malloc(sizeof(float)*2*(Nhalf + 1));
    for (idx = 0; idx < 2*(Nhalf + 1); idx++)  // initialize to zero
       *(Ring_ptr + idx) = 0.0F;
    // Diffusion
    mu = pi * 4.0e-7F;
    d = 0.5F * thick;                          // 2-sided diffusion
    a1 = pi * pi /(d * d * mu * sigCu);        // Eq. (21-8)
    Rdc = cspeed * dt /(width * thick * sigCu); // [ohms] (segment)
    // Calc. Thevenin resistance of a segment
    rcode = SlabRTe ( /*   Input variables     */
                  a1, dt, Nmax, Rdc,
                  /*   Output   variable     */
                  &RTe);    // Thevenin resistance
    In_ptr = (float*)malloc(sizeof(float) * 2 * Nmax * Nhalf);
    for (idx = 0; idx < 2 * Nmax * Nhalf; idx++)  // initialize to zero
       *(In_ptr + idx) = 0.0F;
    Ixt = (float*)malloc(sizeof(float) * 2 * Nhalf);
    for (idx = 0; idx < 2 * Nhalf; idx++)         // initialize to zero
       *(Ixt + idx) = 0.0F;
    // Circuit
    Gsource = 1.0F/Rsource;
```

```
Gload = 1.0F/Rload;
rzo = 1.0F/zo;
CurOutBackward = 0.0F;
CurOutForward = CurOutBackward;
Vslope = Vdrive/risetime;
// write header to screen
printf
("Thickness = %5.3f [mm] Width = %5.3f [mm] Length = %5.3f [mm]\n",
    f1000 * thick,f1000 * width,f1000 * length);
printf ("No. of diffusion terms, Nmax = %i\n",Nmax);
//*****
//** Top of time stepping loop
//*****
for (time = 0.0F; time <= tmax+dt; time +=dt)
{
   //**** Solve network at left end -- node 1 *****
   // Set driver Norton equivalent current   (square pulse)
   if (time <= risetime)
   {
      Is = time * Vslope * Gsource;
      if (Is > Vdrive * Gsource) Is = Vdrive * Gsource;
   }
   else if (time <= pulse_length)
      Is = Vdrive * Gsource;
   else
      Is = 0.0F;
   // voltage at node 1
   v1 = (Is - 2.0F * CurOutBackward)/(rzo + Gsource);
   // departing current at left end
   CurInForward = Is - v1 * Gsource - CurOutBackward;
   //**** Solve network at right end -- node 2 *****
   // voltage at node 2
   v2 = 2.0F * CurOutForward / (rzo + Gload);
   // departing current at right end
   CurInBackward = v2 * Gload - CurOutForward;
   //Output data -- Node voltages
   printf ("time = %6.2f [ns] node 1: v = %f, node 2: v = %f\n",
           time*1.0e9F,v1,v2);
   // Propagation with diffusion
   rcode = CurrTstepDiff(   /* Input variables  */
        Ring_ptr, Nhalf, CurInForward, CurInBackward, wgt1, wgt2,
          Nmax, a1, dt, zo, Rdc, RTe,
             /* Input/Output variables (t -> t + dt)  */
        &TStepOffset, In_ptr, Ixt,
             /* Output variables  */
          &CurOutBackward, &CurOutForward);
```

```
        if (rcode)
        {
            printf ("Error returned from subroutine CurrTstepDiff\n");
            goto DoneMain;
        }
    }
    DoneMain:   ;
    // If this were a subroutine, memory should be freed here to avoid
    // memory leak (Ring_ptr, In_ptr, Ixt).
}
/* --------- End of main --------------------------------------------- */
```

The subroutine `CurrTstepDiff` is listed below. It propagates a current signal one time step including the effect of diffusion. That is, it implements Equations (23-8) and (23-9). It begins by storing the input (departing) current signals in the ring storage array. It then allocates temporary memory for a vector of V_{oc} at each forward and backward line segment. It then enters a for-loop to call `SlabVoc` to calculate V_{oc} at the center of each segment. It then allocates memory for a temporary vector `rhtemp` into which the right-hand sides of Equations (23-8) and (23-9) are accumulated (for each segment). The values in the vector `rhtemp` are then used to augment the values in the circular store array. Incrementing the time step offset in the circular storage array completes the calculation of Equations (23-8) and (23-9). The vectors of segment currents $I_x(t)$ and the inductor currents $I_n(t)$ are then updated in preparation for the calculation of V_{oc} during the next call. The departing current signal (output) variables are then set. Temporary memory for the arrays `Voc` and `rhtemp` are freed by calling `free`. The subroutine then returns to the calling program. One should note that to propagate without diffusion, that is, using a simple resistive trace model, one need merely set `Nmax`, the number of diffusion terms to zero.

```
/*------- Function Declaration ------------------------------------ */
  /*****   Current Propagation with Diffusion (1 wire line)   *****/
  /*****          Slab model used for diffusion model            *****/
int CurrTstepDiff(   /* Input variables   */
            float *Ring_ptr,      // ring pointer
            int Nhalf,            // number of current segments
                                  // NhalfMod = Nhalf + 1 = no.
                                  // of locations in st. circle
            float CurInForward,   // forward from left (t) [A]
            float CurInBackward,  // backward from right (t)[A]
            float Wgt1,           // interpolation wgt
            float Wgt2,           // interpolation wgt
            int Nmax,             // number of diffusion terms
            float a1,             // first diff. parameter [1/s]
            float Tstep,          // time step [s]
            float zo,             // characteristic impedance
            float Rdc,            // d.c. resistance of segment[ohms]
```

```
            float RTe,                 // Thevenin resistance of a
                                       // segment [ohms]
                    /* Input/Output variables (t -> t + dt)   */
            int *TStepOffset,          // Time step offset
            float *In,                 // Matrix of inductor currents
            float *Ixt,                // Vector of segment currents
                /* Output variable   */
            float *CurOutBackward,  // backward at left (t + dt)
        float *CurOutForward)  // forward at right
{
/* --- Local Variable Declarations ------------------------------------
  Type            Name                    Description & range
  ----            ----         ------------------------------------ */
  int             Offset,                 // offset to point in ring
                  th,                     // 2*NhalfMod
                  rcode,                  // return code, =0, normal exit
                  NhalfMod,               // locations in half the circle
                  nseg;                   // segment index
  float           *Voc = NULL,            // vector of o.c. voltages
                  NewIx;                  // new segment current
  float           MHalfAlpha_dx;          // -0.25*RTe/zo = -0.5*alpha
  float           *rhtemp = NULL;         // temporary vector
  int             opst1,opst2,opst3,      // opposite direction offsets
                  same1,same2,            // ends of segment offsets
                  oseg1,oseg2;            // opposite segments
/*------------------------- Error Conditions Trapped ------------------
  Fatal/Nonfatal    Description
  --------------    -----------
  Fatal             CurrTstepDiff             =-1, Nhalf < 2
                                              or error returned from
                                              SlabVoc or SlabIn
-------------------- Executable Code ------------------------------- */
    if (Nhalf < 2)
    {
      rcode = -1;
      goto alldone;
    }
    NhalfMod = Nhalf + 1;
    th = 2 * NhalfMod;
    //*** Input signals at t = 0
    *(Ring_ptr + *TStepOffset) = CurInForward; // Forward wave input
    Offset = (*TStepOffset + NhalfMod) % th;   // Backward wave input
    *(Ring_ptr + Offset) = CurInBackward;
    //*** Calc. Voc (diffusion model, Eq. 21 -36)
    // Allocate storage for Voc
    Voc = (float*)malloc(sizeof(float) * 2 * Nhalf);
```

```
for (nseg = 0; nseg < 2 * Nhalf; nseg++)
{
    rcode = SlabVoc (/*   Input variables     */
                a1, Tstep, Nmax, Rdc, &In[nseg*Nmax], Ixt[nseg],
                    /*   Output variable    */
            &Voc[nseg]);       // Thevenin o.c. voltage
    if (rcode) goto alldone; // error returned from SlabVoc
}
//*** Calculate attenuation, reflection, and voltage source terms
// Allocate temp right-hand-side vector
rhtemp = (float*)malloc(sizeof(float)*2*Nhalf);
MHalfAlpha_dx = -0.25F * RTe/zo;  // same as HalfBeta_dx (G=0)
for (nseg = 0; nseg < 2 * Nhalf; nseg++)
{
    IndicesModCirc (/*   Input variables     */
            nseg, *TStepOffset, Nhalf,
                    /*   Output variables    */
            &opst1,&opst2,&opst3,       // opposite direction offsets
            &same1,&same2,              // ends of segment offsets
            &oseg1,&oseg2 );            // opposite segments
    if ((nseg == 0) || (nseg == Nhalf)) // end of line
    {
        *(rhtemp+nseg) = MHalfAlpha_dx *
                (*(Ring_ptr + same1) + *(Ring_ptr + same2)); // atten.
        *(rhtemp+nseg) += MHalfAlpha_dx *
            (Wgt1* *(Ring_ptr + opst1) + *(Ring_ptr + opst2)); //reflect
        *(rhtemp+nseg) -= 0.5F/zo * (Voc[nseg]
                                    + Wgt1 * Voc[oseg1]); // voltage src
    }
    else
    {
        *(rhtemp+nseg) = MHalfAlpha_dx *
                (*(Ring_ptr + same1) + *(Ring_ptr + same2)); // atten.
        *(rhtemp+nseg) += MHalfAlpha_dx *
            (Wgt1* *(Ring_ptr + opst1) + *(Ring_ptr + opst2)
                            + Wgt2* *(Ring_ptr + opst3)); //reflect
        *(rhtemp+nseg) -= 0.5F/zo * (Voc[nseg]
                + Wgt1 * Voc[oseg1] + Wgt2 * Voc[oseg2]); // voltage src
    }
}
//*** Modify stored currents
// Backward signals
for (nseg = 0; nseg < Nhalf; nseg++)
{
  Offset = (nseg + 2 + *TStepOffset) % th;
  *(Ring_ptr + Offset) += *(rhtemp+nseg);
```

```
    }
    // Forward signals
    for (nseg = Nhalf; nseg < 2 * Nhalf; nseg++)
    {
      Offset = (nseg + 3 + *TStepOffset) % th;
      *(Ring_ptr + Offset) += *(rhtemp+nseg);
    }
    //*** Increment timestep offset around circle
    //    (I.E. Propagate signals)
    *TStepOffset = (*TStepOffset + 1) % th;
    //*** Update Inductor currents and current array for each segment
    for (nseg = 0; nseg < 2 * Nhalf; nseg++)
    {
      IndicesModCirc (/*   Input variables    */
              nseg, *TStepOffset, Nhalf,
                      /*   Output variables   */
              &opst1,&opst2,&opst3,    // opposite direction offsets
              &same1,&same2,           // ends of segment offsets
              &oseg1,&oseg2 );         // opposite segments
      NewIx = 0.5F * (*(Ring_ptr + same1) + *(Ring_ptr + same2));
      rcode = SlabIn ( /*   Input variables    */
                a1, Tstep, Nmax, Ixt[nseg], NewIx,
                      /*   Input/Output vector */
            &In[nseg*Nmax]);//Vector of In(t) input, In(t+dt) output [A]
      if (rcode) goto alldone; // error returned from SlabIn
      Ixt[nseg] = NewIx;
    }
    //*** Output signals at t = Tstep
    // Forward signal
    Offset = (*TStepOffset + NhalfMod + 1) % th;
    *CurOutForward = Wgt1 * *(Ring_ptr + Offset);
    Offset = (*TStepOffset + NhalfMod + 2) % th;
    *CurOutForward += Wgt2 * *(Ring_ptr + Offset);
    // Backward signal
    Offset = (*TStepOffset + 1) % th;
    *CurOutBackward = Wgt1 * *(Ring_ptr + Offset);
    Offset = (*TStepOffset + 2) % th;
    *CurOutBackward += Wgt2 * *(Ring_ptr + Offset);
alldone:
    free(Voc);
    free(rhtemp);
    return(rcode);
}
/* -----End of Procedure CurrTstepDiff ---------------------------- */
/*------- Function Declaration ---------------------------------------
  *****  Calculate indices for Modified Circular Storage   *****/
```

```
        // see Figure 23.2.
     void IndicesModCirc (  /*   Input variables    */
              int segin,              // line segment index
              int TStepOffset,      // time offset in storage ring
              int Nhalf,            // segments each direction
                      /*   Output variables    */
              int *opst1, int *opst2, int *opst3, // opposite
                                                  // direction offsets
              int *same1, int *same2,      // ends of segment
                                           // offsets
              int *oseg1, int *oseg2 )     // opposite segments
{
/*   --- Local Variable Declarations -------------------------------
  Type        Name                 Description & range
  ----        ----        -------------------------------- */
  int         NhalfMod,    // half-length of storage ring
              th;          // 2 * NhalfMod
/*--------------------- Error Conditions Trapped ------------------
   none
----------------- Executable Code ------------------------------ */
     NhalfMod = Nhalf + 1;
     th = 2 * NhalfMod;
     // segments indices -- opposite direction propagation
     *oseg1 = 2 * Nhalf - 1 - segin;
     *oseg2 = *oseg1 + 1;
     if (segin < Nhalf)
     { /*** backward signal segments ***/
        // storage offsets at ends of segment
       *same1 = (segin + 1 + TStepOffset) % th;
       *same2 = (*same1 + 1) % th;
       // storage offsets opposite direction
       *opst1 = (2 * Nhalf + 1 - segin + TStepOffset) % th;
       *opst2 =  (*opst1 + 1) % th;
       *opst3 =  (*opst2 + 1) % th;
     }
     else
     {   /*** forward signal segments ***/
        // storage offsets at ends of segment
       *same1 = (segin + 2 + TStepOffset) % th;
       *same2 = (*same1 + 1) % th;
       // storage offsets opposite direction
       *opst1 = (2 * Nhalf - segin + TStepOffset) % th;
       *opst2 =  (*opst1 + 1) % th;
       *opst3 =  (*opst2 + 1) % th;
     }
}
/* -----End of Procedure IndicesModCirc ------------------------- */
```

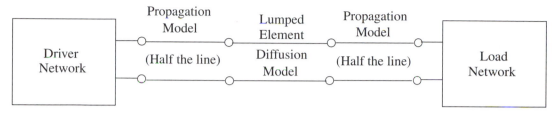

Figure 23.5   Use of Lumped Element Diffusion Model.

23.4 APPROXIMATIONS USING LUMPED ELEMENT DIFFUSION MODEL

The lumped element diffusion model can be spliced into propagation models as shown in Figure 23.5. This method is especially applicable for short lines where the signal rise time is longer than the time of propagation on the line. If the rise time is shorter than the propagation time, the attenuation due to diffusion will be overemphasized for the early part of the pulse (as explained in the previous Section). This method can be used to include diffusion effects in multi-wire lines. It will approximate the propagation crosstalk without the complexity of a full multi-wire diffusive propagation algorithm.

This approach can be used in two ways. The propagation model can be lossless; then the diffusion model would contain the entire resistance of the line (DC resistance and the diffusion terms). A more accurate approach would use a lossy line with the DC resistance included. The lumped element model would then have the DC resistance removed. (R_{Te} would simply be replaced with $(R_{Te} - R_{dc})$). The rest of the diffusion algorithm remains unchanged.

Appendix A

EQUIVALENCE OF TIME DOMAIN AND FREQUENCY DOMAIN METHODS

INTRODUCTION

The reader familiar with classical frequency domain transmission line theory may feel uncomfortable when studying Part I of this book. Can it be that termination reflections can be correctly calculated using a real characteristic impedance? The familiar frequency domain expression for the characteristic impedance is

$$z_o = \sqrt{\frac{R + j\omega L}{G + j\omega C}} \tag{A-1}$$

Often, time domain calculations use some approximation of it to calculate reflections at terminations. If this is done consistently with frequency domain theory, all is well (or, at lease approximately well). The impedance given above represents the ratio of voltage to current at the terminals of a semi-infinite line at the frequency ω. For this ratio to become established, the current must have been flowing for a long time (theoretically an infinite time). At times shortly after a signal is turned on (shortly must be defined in various ways for different lines and frequencies) the ratio of current to voltage will be different until "the transients die out" and a steady state is attained. That is, during the time of early important transients, either the time domain method should be used or a Fourier transform should be performed over all the important frequencies in the spectrum of the signal if accurate results are to be obtained.

In this Appendix, some of the relationships between the time domain approach and the frequency domain approach are given. The impedance given in Equation (A-1) is derived using the time domain approach, but assuming a sinusoidal signal. The relationship between classical forward and backward waves and the instantaneous forward and backward signals defined in Part I is given.

SINUSOIDAL SIGNALS ON A SINGLE-CONDUCTOR LINE

For a lossy line treated using the time domain methods but carrying a sinusoidal signal, the forward and backward voltages have the general form

$$v_F(x, t - x/c) = f(x)e^{j\omega(t - x/c)} \tag{A-2}$$

and

$$v_B(x, t + x/c) = b(x)e^{j\omega(t + x/c)} \tag{A-3}$$

where $f(x)$ and $b(x)$ are functions to be determined. The physical time domain forward and backward signals are the real parts of Equations (A-2) and (A-3). Following the convention used in frequency domain analysis, the complex form (of expressions for v_F and v_B) is carried through the algebraic manipulations for convenience, and to demonstrate at the end of this Appendix that the complex impedance for a semi-infinite line is the classical value though the time domain, transport equations of Part I have been used to define signal propagation.

Substituting Equations (A-2) and (A-3) into the transport Equations (2-29) and (2-30), one finds (by combining and solving the resulting pair of ordinary differential equations) that the functions $f(x)$ and $b(x)$ have the forms

$$f(x) = e^{j\omega x/c}\left[A_f e^{\gamma x} + B_f e^{-\gamma x}\right] \tag{A-4}$$

$$b(x) = e^{-j\omega x/c}\left[A_b e^{\gamma x} + B_b e^{-\gamma x}\right] \tag{A-5}$$

where γ is the classical propagation constant given by

$$\gamma = \sqrt{(R + j\omega L)(G + j\omega C)} \tag{A-6}$$

Assuming that γ is chosen to be the root with positive real part, the terms with $e^{-\gamma x}$ attenuate in the positive x direction and are *defined in the classical frequency domain theory* as forward moving waves. Conversely, the terms with $e^{\gamma x}$ attenuate in the negative x direction and are *defined in the classical frequency domain theory* as backward moving waves. However, the instantaneous forward signal contains contributions from classical waves of both forward and backward type (and the classical forward wave contains instantaneous signals of both forward and backward type). Substitution of Equations (A-4) and (A-5) into (A-2) and (A-3) yields precisely the classical form of the transmission line solution except that the time dependence $e^{j\omega t}$ is usually suppressed in the classical theory because it appears as a multiplier in every equation.

The constant coefficients, $A_f, B_f, A_b,$ and $B_b,$ are arbitrary as obtained (independently) in the solution for $f(x)$ and $b(x)$. However, a general solution for a single frequency can have only two independent arbitrary constants. Substituting Equation (A-2) and (A-3) into Equation (2-29) gives an equation that will relate the constants.

$$f'(x) = -\alpha f(x) + \beta b(x)e^{\frac{2j\omega}{c}x} \tag{A-7}$$

Substituting Equation (A-4) into Equation (A-7) and equating coefficients of $e^{-\gamma x}$ and $e^{\gamma x}$ yields the equations

$$A_b = \frac{A_f}{\beta}\left[\frac{j\omega}{c} + \alpha + \gamma\right] \tag{A-8}$$

$$B_b = \frac{B_f}{\beta}\left[\frac{j\omega}{c} + \alpha - \gamma\right] \tag{A-9}$$

Note that for the sinusoidal signal case, the following useful identity obtains (and can be easily verified by the reader)

$$\beta^2 = \left[\frac{j\omega}{c} + \alpha + \gamma\right]\left[\frac{j\omega}{c} + \alpha - \gamma\right] \tag{A-10}$$

Hence, Equation (A-9) can also be written

$$B_b = \frac{B_f\beta}{\frac{j\omega}{c} + \alpha + \gamma} \tag{A-11}$$

THE SEMI-INFINITE LINE

Consider a line with its left end at $x = 0$ and extending indefinitely in the positive x direction. It is assumed that there are no drivers except at $x = 0$. The constants A_f and A_b must be zero, else the signal becomes infinite as $x \rightarrow \infty$. Hence, using Equations (A-4), (A-5), and (A-11),

$$f(x) = B_f e^{\left[\frac{j\omega}{c} - \gamma\right]x} \tag{A-12}$$

$$b(x) = \frac{B_f\beta}{\frac{j\omega}{c} + \alpha + \gamma} e^{-\left[\frac{j\omega}{c} + \gamma\right]x} \tag{A-13}$$

Substituting Equations (A-12) and (A-13) into Equations (A-2) and (A-3), the time domain form of the forward and backward signals is obtained

$$v_F(x, t - x/c) = B_f e^{j\omega t} e^{-\gamma x} \tag{A-14}$$

and

$$v_B(x, t + x/c) = \frac{B_f\beta}{\frac{j\omega}{c} + \alpha + \gamma} e^{j\omega t} e^{-\gamma x} \tag{A-15}$$

Therefore, the instantaneous forward signal is accompanied by an instantaneous backward signal given by Equation (A-15) on the semi-infinite line (with a sinusoidal signal). Note that if the line is distortionless ($\beta = 0$), no backward signal exists.

The voltage is then

$$v(x,t) = v_o e^{j\omega t} e^{-\gamma x} \tag{A-16}$$

where

$$v_o = B_f \left[1 + \frac{\beta}{\dfrac{j\omega}{c} + \alpha + \gamma} \right]$$

The current is then

$$i(x,t) = \frac{1}{z_o}(v_F(x,t - x/c) - v_B(x,t + x/c)) \tag{A-17}$$

$$i(x,t) = i_o e^{j\omega t} e^{-\gamma x} \tag{A-18}$$

where

$$i_o = \frac{B_f}{z_o} \left[1 - \frac{\beta}{\dfrac{j\omega}{c} + \alpha + \gamma} \right]$$

and $z_o = \sqrt{\dfrac{L}{C}}$, (not Equation (A-1)).

The impedance looking into the semi-infinite line is

$$z_{in} = \frac{v_o}{i_o} = z_o \frac{\dfrac{j\omega}{c} + \alpha + \gamma + \beta}{\dfrac{j\omega}{c} + \alpha + \gamma - \beta} \tag{A-19}$$

To reduce this to the standard form (i.e., Equation (A-1)), first square both sides to obtain

$$z_{in}^2 = z_o^2 \frac{\left[\dfrac{j\omega}{c} + \alpha + \gamma\right]^2 + 2\left[\dfrac{j\omega}{c} + \alpha + \gamma\right]\beta + \beta^2}{\left[\dfrac{j\omega}{c} + \alpha + \gamma\right]^2 - 2\left[\dfrac{j\omega}{c} + \alpha + \gamma\right]\beta + \beta^2} \tag{A-20}$$

Then, substitute β^2 using the identity Equation (A-10). The numerator and denominator then have the factor $\left[\dfrac{j\omega}{c} + \alpha + \gamma\right]$ in every term, which can be divided out. Noting that

$\alpha + \beta = R/z_o$ and $\alpha - \beta = Gz_o$, one obtains

$$z_{in}^2 = z_o^2 \frac{\dfrac{j\omega}{c} + \dfrac{R}{z_o}}{\dfrac{j\omega}{c} + Gz_o} \tag{A-21}$$

Equation (A-21) is easily reduced to

$$z_{in}^2 = \frac{R + j\omega L}{G + j\omega C} \tag{A-22}$$

Hence, the impedance looking into a semi-infinite line using the time domain theory agrees with the classical frequency domain theory for a sinusoidal signal. Any signal can be expressed as a sum over all of its frequencies (the inverse Fourier transform); hence, the time domain theory is in exact agreement with classical frequency domain theory.

Appendix B
EFFECT OF RESISTANCE IN REFERENCE CONDUCTOR

A two-conductor transmission line with a resistive reference conductor is illustrated in Figure B.1. The voltage drop due to the resistance is derived to show the effect of the resistance R_s (the resistance per unit length of the reference conductor). The algebraic effect is to add R_s to each of the elements in the resistance matrix (see Equations (B-3) and (B-4)).

$$dv_1 = v_1(x+dx) - v_1(x) = [-R_1 i_1 - R_s(i_1 + i_2)]dx \tag{B-1}$$

$$dv_2 = v_2(x+dx) - v_2(x) = [-R_2 i_2 - R_s(i_1 + i_2)]dx \tag{B-2}$$

$$\begin{bmatrix} dv_1 \\ dv_2 \end{bmatrix} = \begin{bmatrix} -R_1 - R_s & -R_s \\ -R_s & -R_2 - R_s \end{bmatrix} \begin{bmatrix} i_1 \\ i_2 \end{bmatrix} dx \tag{B-3}$$

$$\begin{bmatrix} \dfrac{dv_1}{dx} \\ \dfrac{dv_2}{dx} \end{bmatrix} = -\left(\begin{bmatrix} R_1 & 0 \\ 0 & R_2 \end{bmatrix} + \begin{bmatrix} R_s & R_s \\ R_s & R_s \end{bmatrix} \right) \begin{bmatrix} i_1 \\ i_2 \end{bmatrix} \tag{B-4}$$

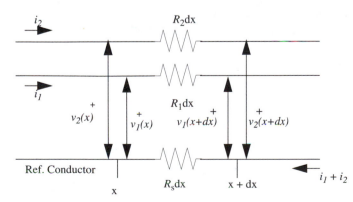

Figure B.1 Infinitesimal Length of Line with Resistive Reference Conductor.

301

SOLUTIONS OF PROBLEMS

Problem 2-1. Given a lossless transmission line 0.1 [m] long, $z_o = 60$ [Ω] with propagation speed $c = 3 \times 10^8$ [m/s]. Let the delay time be $\tau = 0.1/c = 1/3$ [ns]. Let the left end of the line be driven by a unit step function voltage ($v = 0$, for $t < 0$, $v = 1$, $t > 0$).

    a. What is the inductance per unit length and the total inductance of the line?
        $L = z_o/c = 60 / 3 \times 10^8 = 0.20$ [μH/m]
        Total $L = 0.1$ [m] x 0.20 [μH/m] = 0.020 [μH]

    b. What is the capacitance per unit length and the total capacitance of the line?
        $C = 1/cz_o = 1/ (3 \times 10^8 \times 60) = 55.5$ [pF/m]
        Total $C = 0.1$ x 55.5 = 5.55 [pF]

    c. What is the current at the left end of the line and the voltage at the right end of the line for the first 2 [ns] (table or graph) if:

        i. The line is terminated at the right end with $R = 60$ [Ω]?
            Termination in z_o, no reflection. $i_{left} = 1/60$ [A], t > 0. $v_{right} = 1$ volt, t > 1/3 [ns]

        ii. The line is terminated at the right end with a short ($R = 0$) (see Table S-1)?

        iii. The line is terminated at the right end with $R = 180$ [Ω] (−2) (see Table S-2)?

$$\text{Reflection coefficient} = \frac{\dfrac{180}{60} - 1}{\dfrac{180}{60} + 1} = \frac{1}{2} \qquad (2\text{-}11)$$

TABLE S-1
Problem 2-1c, $R = 0$.

		$0<t<\tau$	$\tau<t<2\tau$	$2\tau<t<3\tau$	$3\tau<t<4\tau$	$4\tau<t<5\tau$	$5\tau<t<6\tau$
Left end	v_F	1	1	1/2	1/2	3/4	3/4
	v_B	0	0	1/2	1/2	1/4	1/4
	I	1/60	1/60	0	0	1/120	1/120
Right end	v_F	0	1	1	1/2	1/2	3/4
	v_B	0	1/2	1/2	1/4	1/4	3/8
	v	0	3/2	3/2	3/4	3/4	9/8

d. If the line were replaced by an inductor equal to the total inductance of the line, what is the current through the inductor for the first 2 ns (see Table S-3)? Compare this to the results of c.ii.

$$v = 1 = L\frac{dI}{dt}$$

$$I(t) = \frac{t}{L} = \frac{t}{0.020 \times 10^{-6}}$$

Problem 2-2   Given a finite length transmission line with 1 volt applied at its left end for a very long time. Its right end is open circuit. Assume time enough has elapsed for all the transients to die out (the line is not quite lossless). The current through the line is zero and the voltage everywhere on it is 1 volt. What are the forward and backward currents and voltages? Does this result agree with Equation (2-11) (applied at the right end)?

TABLE S-2
Problem 2-1c, $R = 180$ [Ω].

		$0<t<\tau$	$\tau<t<2\tau$	$2\tau<t<3\tau$	$3\tau<t<4\tau$	$4\tau<t<5\tau$	$5\tau<t<6\tau$
Left end	v_F	1	1	1/2	1/2	3/4	3/4
	v_B	0	0	1/2	1/2	1/4	1/4
	I	1/60	1/60	0	0	1/120	1/120
Right end	v_F	0	1	1	1/2	1/2	3/4
	v_B	0	1/2	1/2	1/4	1/4	3/8
	v	0	3/2	3/2	3/4	3/4	9/8

Table S-3
Problem 2-1d, Current Through Inductor 0.020 [μH].

		$t = \tau$ $= 1/3$ ns	$t = 2\tau$ $= 2/3$ ns	$t = 3\tau$ $= 1$ ns	$t = 4\tau$ $= 4/3$ ns	$t = 5\tau$ $= 5/3$ ns	$t = 6\tau$ $= 2$ ns
Current in L	I	1/60	2/60	3/60	4/60	5/60	6/60

Using Equations (2-7) and (2-8):

$$i_F + i_B = 0 \Rightarrow v_F - v_B = 0 \Rightarrow v_F = v_B$$

$$v_F + v_B = 1 \Rightarrow v_F = \frac{1}{2}, v_B = \frac{1}{2}$$

hence,

$$i_F = \frac{1}{2z_o}, i_B = \frac{-1}{2z_o} \quad , \text{Yes, this agrees with Equation (2-11).}$$

Problem 2-3   Assume the small backward signal approximation:

a. If $v_o(t)$ is a unit step function, that is, $v_o(t) = u(t)$, where $u(t)$ is defined by

$$u(t) = \begin{cases} 0, & t < 0 \\ 1, & t \geq 0 \end{cases} \tag{2-44}$$

show that the arriving voltage at the left end is

$$v_B(0,t) = \begin{cases} \dfrac{\beta}{2\alpha}(1 - e^{-\alpha ct}), & t < \dfrac{2l}{c} \\ \dfrac{\beta}{2\alpha}(1 - e^{-2\alpha l}), & t \geq \dfrac{2l}{c} \end{cases} \tag{2-45}$$

Show that if α is small (as it usually is) and if the arguments of the exponential functions are small compared to 1, then Equation (2-45) can be written

$$v_B(0,t) \cong \begin{cases} \dfrac{\beta ct}{2}, & t < \dfrac{2l}{c} \\ \beta l, & t \geq \dfrac{2l}{c} \end{cases} \tag{2-46}$$

Solution of a.:

Note that $v_o(t - 2x'/c) = 0$ if $x' > ct/2$ hence, for $t < 2l/c$ one has (for Equation 2-42)

$$v_B(0,t) = \beta \int_0^{ct/2} e^{-2\alpha x'} dx'$$

For $t \geq 2l/c$, $v_o(t - 2x'/c) = 1$ for $x \in (0,l)$, hence,

$$v_B(0,t) = \beta \int_0^{l} e^{-2\alpha x'} dx'$$

b. If $v_o(t)$ is a unit amplitude square pulse with width t_w, show that the arriving voltage at the left end is

$$v_a(t) = v_B(0,t) - u(t - t_w)v_B(0,t - t_w) \qquad (2\text{-}47)$$

where $v_B(0,t)$ is the result for the unit step function driver $u(t)$.

Simply recognize that the unit amplitude square pulse $= u(t) - u(t - t_w)$.

c. Let $R = 10$ [Ω/m], $G = 0$, $z_o = 50$ [Ω], $l = 0.1$ [m]. For $v_o(t)$ equal to a five-volt step function, what is the maximum of the arriving voltage at the right end? At the left end?

$$\alpha = \beta = \frac{R}{2z_o} = 0.1[\text{m}^{-1}]$$

At the right end, $x = l$ (use Equation 2-39)

$$v_F(l,t) = 5 \times e^{-0.01} u(t - l/c) = 4.95u(t - l/c)$$

At the left end, the maximum voltage is (use Equation 2-46)

$$v_B = 5\beta l = 0.05 \text{ [V]}$$

Problem 3-1   Assume that the circuit of Figure 3-8 has only 2 nodes (as shown); let $i = 1$ and $j = 2$. Write out explicitly the matrix G and the column vector i_s.

$$G = \begin{bmatrix} g_{11} + g_{12} & -g_{12} \\ -g_{12} & g_{22} + g_{12} \end{bmatrix}$$

$$i_s = \begin{bmatrix} I_{11} - I_{12} \\ I_{22} + I_{12} \end{bmatrix}$$

Problem 4-1   Refer to the transmission line shown above in Figure 4.2. The transmission line has the following parameters:

$l = 0.1$ [m]
$c = 10^8$ [m/s]
(hence, $t_{delay} = l/c = 1$ [ns] for the line)

$z_o = 50$ [Ω]
$R = 10$ [Ω/m]
$G = 0$ [Ω^{-1}/m]

At the left end,
$V_1(t) = u(t) =$ unit step function
$R_1 = 5$ [Ω]

At the right end,
$V_2(t) = 0$
$R_2 = 50$ [Ω]

   a. Calculate the voltage propagation every 0.5 [ns] from 0.0 to 2.5 [ns] using Equations (2-49) and (2-50) and at each time step use the line end equivalent circuit to obtain the departing signals. Divide the line into two segments so that time steps are 0.5 [ns] and the x grid points are 0, 0.05, and 0.10 [m]. Fill in the missing voltages in Table 4-1 (see Table S-4).

   Before starting, review the general pattern of transmission line solutions as given in Section (2.5). The equations mentioned in step 3 are replaced in this problem with Equations (4-9) and (4-11) (that were derived using the ones mentioned in step 3).

   Solution:

   a.
   $\alpha = 0.5 * 10/50 = 0.1$ [m^{-1}]
   $(x_2 - x_1)\alpha = 0.005$ [-]

TABLE S-4
Problem 4-1.

Time [ns]	$x_1 = 0.00$ [m] Left Termination			$x_2 = 0.05$ [m] Middle			$x_3 = l = 0.10$ [m] Right Termination		
	$v_F(x_1)$	$v_B(x_1)$	$v(x_1)$	$v_F(x_2)$	$v_B(x_2)$	$v(x_2)$	$v_F(x_3)$	$v_B(x_3)$	$v(x_3)$
0.0	0.909	0	0.909	0	0	0	0	0	0
0.5	0.907	.0023	0.910	0.907	0	0.907	0	0	0
1.0	0.905	.0045	0.910	0.902	0.0023	0.904	0.905	0	0.905
1.5	0.904	.0068	0.910	0.900	0.0045	0.905	0.898	0	0.897
2.0	0.902	.0090	0.911	0.899	0.0045	0.903	0.896	0	0.896
2.5	0.902	.0090	0.911	0.897	0.0045	0.901	0.895	0	0.895
$t \to \infty$	0.902	0.009	0.911	0897	0.005	0.902	0.893	0	0.893

   b. Calculate a row in the above table at $t \to \infty$. That is, assume all the transients have
   decayed and the voltages and currents are d.c. *Hint*: Refer to *Exercise (2-8)*.

   Voltages Left, Middle, Right: 51/56, 50.5/56, 50/56.
   Current everywhere I = 1/(5 + 1 + 50) = 1/56.
   See the last row of Table S-4.

Problem 4-2   For the transmission line defined for *Problem 4-1*, calculate the current propagation for $t = 0$ to $t = 2.5$ [ns]. Use the Norton equivalent circuits for the line and its terminations. Use Equations (2-49) and (2-50) with the sign of β reversed to propagate the forward
and backward current. The Norton equivalent for the driver will have $I_1(t) = 0.2u(t)$ [A]. and
$G_1 = 0.2$ [℧]. Fill in Table S-5 with the missing currents.

Do you know of any relationships between the various currents calculated in this problem and
the voltages calculated (for the same model) in the preceding problem? If so, use them to see
if your results for the current are consistent with the voltages.

Problem 4-3   Perform the calculations to generate Table 4-3. Create an additional row in the
table for $t \to \infty$, i.e., for the case when the DC solution is applicable (see Table S-6).

At $t \to \infty$, on both lines, V = 90/95 = 0.947 [V], I = 1/95 = 0.0105 [A].

Apply Equations (2-35) to (2-38).

TABLE S-5
Problem 4-2.

Time [ns]	$x_1 = 0.00$ [m] Left Termination			$x_2 = 0.05$ [m] Middle			$x_3 = l = 0.10$ [m] Right Termination		
	$i_F(x_1)$	$i_B(x_1)$ (10^{-6})	$i(x_1)$	$i_F(x_2)$	$i_B(x_2)$ (10^{-6})	$i(x_2)$	$i_F(x_3)$	$i_B(x_3)$ (10^{-6})	$i(x_3)$
0.0	.0182	0	.0182	0	0	0	0	0	0
0.5	.0182	-45.5	.0182	.0182	0	.0182	0	0	0
1.0	.0181	-91.0	.0180	.0181	-45.5	.0181	.0181	0	.0181
1.5	.0181	-136.3	.0180	.0180	-90.5	.0179	.0180	0	.0180
2.0	.0180	-180.3	.0178	.0180	-90.0	.0179	.0179	0	.0179
2.5	.0180	-179.8	.0178	.0179	-90.25	.0179	.0179	0	.0179

Problem 5-1 Assume a 5 [V] signal departs the right end of a three-wire transmission line. The signal appears on wires 1 and 3; $v_{B2} = 0$. What are the three departing currents assuming the characteristic impedance of the line is given by Equation (5-16)?

$$\mathbf{i}_B = -\mathbf{Z}_o^{-1}\mathbf{v}_B = -\begin{bmatrix} 0.0200 & -0.00674 & -0.00110 \\ -0.00674 & 0.0224 & -0.00674 \\ -0.00110 & -0.00674 & 0.0200 \end{bmatrix}\begin{bmatrix} 5 \\ 0 \\ 5 \end{bmatrix} = \begin{bmatrix} -0.0945 \\ +0.0674 \\ -0.0945 \end{bmatrix}$$

TABLE S-6
Problem 4-3, $t \to \infty$.

Time [ns]	$x_1 = 0.00$ [m] Left Termination			$x_2 = l_1$ Middle			$x_3 = l_1 + l_2$ Right Termination		
t	v_{left}	i_{F1} dep.	i_{B1} arr.	$i_1 = i_2$	i_{F1} arr.	i_{B1} dep.	i_{F2} dep.	i_{F2} arr.	v_{right}
inf.	0.947	.0211	-.0105	.0105	.0211	-.0105	.0105	.0105	0.947

Problem 5-2   A 2-wire line exists with:

Characteristic impedance, $\mathbf{Z}_o = \begin{bmatrix} 50 & 30 \\ 30 & 50 \end{bmatrix}$ [Ω]

Wire resistances: $R_1 = 8$ [Ω/m], $R_2 = 10$ [Ω/m]
Conductance $\mathbf{G}$ = zero matrix.
Calculate $\mathbf{A}$ and $\mathbf{B}$ (since $\mathbf{G}$ is zero, they are equal).

$$\mathbf{Z}_o^{-1} = \begin{bmatrix} 0.03125 & -0.01875 \\ -0.01875 & 0.03125 \end{bmatrix}$$

$$\mathbf{A} = \mathbf{B} = \frac{1}{2}\mathbf{R}\mathbf{Z}_o^{-1}$$

$$\mathbf{A} = \mathbf{B} = \begin{bmatrix} 0.125 & -0.075 \\ -0.09375 & 0.15625 \end{bmatrix}$$

Problem 5-3   Assume a 3-wire line with $\mathbf{A}$ and $\mathbf{B}$ given by Equation (5-27). A forward signal is launched from the left end with (forward voltages): $v_{F1} = 5$, $v_{F2} = 0$, $v_{F3} = 0$ [V]. The backward voltage signal is zero. Evaluate the right-hand sides of Equations (5-9) and (5-10). Explain the physical meaning of each element of the resulting two-column vectors.

Equation (5-9):

$$-\mathbf{A}\mathbf{v}_F = -\begin{bmatrix} 0.08 & -0.02696 & -0.0044 \\ -0.0337 & 0.112 & -0.0337 \\ -0.0066 & -0.04044 & 0.12 \end{bmatrix}\begin{bmatrix} 5 \\ 0 \\ 0 \end{bmatrix} = \begin{bmatrix} -0.4 \\ 0.1685 \\ 0.033 \end{bmatrix} \begin{matrix} \text{Attenuation per meter wire 1} \\ \text{Coupled forward on wire 2} \\ \text{Coupled forward on wire 3} \end{matrix}$$

Equation (5-10):

$$-\mathbf{B}\mathbf{v}_F = -\begin{bmatrix} 0.08 & -0.02696 & -0.0044 \\ -0.0337 & 0.112 & -0.0337 \\ -0.0066 & -0.04044 & 0.12 \end{bmatrix}\begin{bmatrix} 5 \\ 0 \\ 0 \end{bmatrix} = \begin{bmatrix} -0.4 \\ 0.1685 \\ 0.033 \end{bmatrix} \begin{matrix} \text{Reflected per meter wire 1} \\ \text{Coupled backward on wire 2} \\ \text{Coupled backward on wire 3} \end{matrix}$$

Problem 5-4   Consider a three-wire transmission line of length $l = 0.05$ [m], $c = 1.0\times10^8$ [m/s] and $\mathbf{A}$ and $\mathbf{B}$ given by Equation (5-27). At $t = 0$ a step function of amplitude 5 [V] is launched from the left end on wire 3; wires 1 and 2 are held at ground potential. Calculate the forward voltage on wire 2 at the right end at t = 0.5, 1.0 and 1.5 [ns] using the small coupling approximation, i.e., Equation (5-39) (see Table S-7).

$i = 2, k = 3$
$\alpha_{ii} = 0.112, \alpha_{kk} = 0.12, (\alpha_{ii} + \alpha_{kk}) = 0.132, \alpha_{ik} = \beta_{ik} = -0.0337$
$l/c = 0.5$ [ns]

Problem 5-5 Consider a two-wire line with characteristic impedance matrix $\mathbf{Z}_o = \begin{bmatrix} 50 & 30 \\ 30 & 50 \end{bmatrix}$. Let the line be terminated with 50 [Ω] resistors to ground from both terminal 1 and terminal 2. Write out the matrix $(\mathbf{Z}_o + \mathbf{R}_L)$. Calculate the inverse of this matrix; i.e., write out the matrix $(\mathbf{Z}_o + \mathbf{R}_L)^{-1}$. If $\mathbf{v}_{arr.} = \begin{bmatrix} 5 \\ 0 \end{bmatrix}$, calculate the terminal voltage vector, v. Calculate the departing voltage vector $\mathbf{v}_{dep.} = \mathbf{v} - \mathbf{v}_{arr.}$.

Solution:

$$\mathbf{Z}_o + \mathbf{R}_L = \begin{bmatrix} 100 & 30 \\ 30 & 100 \end{bmatrix}$$

det = 9100

$$(\mathbf{Z}_o + \mathbf{R}_L)^{-1} = \frac{1}{9100}\begin{bmatrix} 100 & -30 \\ -30 & 100 \end{bmatrix} = \begin{bmatrix} 1.099\times10^{-2} & -3.297\times10^{-3} \\ -3.297\times10^{-3} & 1.099\times10^{-2} \end{bmatrix}$$

TABLE S-7
Problem 5-4.

Quantity	t = 0.5 [ns]	t = 1.0 [ns]	t = 1.5 [ns]
τ (right end)	0	0.5 [ns]	1.0 [ns]
$I_{ik}(\tau)$	0	0.0250	0.0498
$-\beta_{ik}e^{-\alpha_{ii}x}I_{ik}(\tau)$, term 1	0	0.00084	0.00167
term 2 of (5-39)	0.00168	0.00168	0.00168
5 x sum of terms 1 and 2	0.0084	0.0126	0.01675
$\mathbf{v}_{Fi}(l,\tau) =$			

$$\mathbf{v} = 2\mathbf{R}_L(\mathbf{Z}_o + \mathbf{R}_L)^{-1}\mathbf{v}_{arr.}$$

$$\mathbf{v} = 2\begin{bmatrix} 50 & 0 \\ 0 & 50 \end{bmatrix}\begin{bmatrix} 1.099\times10^{-2} & -3.297\times10^{-3} \\ -3.297\times10^{-3} & 1.099\times10^{-2} \end{bmatrix}\begin{bmatrix} 5 \\ 0 \end{bmatrix} = 2\begin{bmatrix} 0.5495 & -0.16485 \\ -0.16485 & 0.5495 \end{bmatrix}\begin{bmatrix} 5 \\ 0 \end{bmatrix} = \begin{bmatrix} 5.495 \\ -1.6485 \end{bmatrix}$$

$$\mathbf{v}_{dep.} = \mathbf{v} - \mathbf{v}_{arr.}$$

$$\mathbf{v}_{dep.} = \begin{bmatrix} 5.495 \\ -1.6485 \end{bmatrix} - \begin{bmatrix} 5 \\ 0 \end{bmatrix} = \begin{bmatrix} 0.495 \\ -1.6485 \end{bmatrix}$$

Problem 5-6 For a two-wire line with characteristic impedance matrix $\mathbf{Z}_o = \begin{bmatrix} 50 & 30 \\ 30 & 50 \end{bmatrix}$ find the characteristic conductance matrix, i.e., the inverse of $\mathbf{Z}_O$. If the load is 50 [Ω] to ground on each of the two conductors, write the load conductance matrix $\mathbf{G}_L$. Find the matrix $(\mathbf{G}_o + \mathbf{G}_L)^{-1}$ and the matrix $\mathbf{G}_L(\mathbf{G}_o + \mathbf{G}_L)^{-1}$. For an arriving current vector $\mathbf{i}_{arr.} = \begin{bmatrix} 0 \\ 0.1 \end{bmatrix}$ [A], find the current out of the terminals into the load.

Solution:

$\det(\mathbf{Z}_O) = 1600$

$$\mathbf{G}_o = \frac{1}{1600}\begin{bmatrix} 50 & -30 \\ -30 & 50 \end{bmatrix} = \begin{bmatrix} 3.125\times10^{-2} & -1.875\times10^{-2} \\ -1.875\times10^{-2} & 3.125\times10^{-2} \end{bmatrix}$$

$$\mathbf{G}_L = \begin{bmatrix} 0.02 & 0 \\ 0 & 0.02 \end{bmatrix}$$

$$\mathbf{G}_o + \mathbf{G}_L = \begin{bmatrix} 5.125\times10^{-2} & -1.875\times10^{-2} \\ -1.875\times10^{-2} & 5.125\times10^{-2} \end{bmatrix}$$

$$\det(\mathbf{G}_o + \mathbf{G}_L) = 2.275 \times 10^{-3}.$$

$$(\mathbf{G}_o + \mathbf{G}_L)^{-1} = \frac{1}{2.275\times10^{-3}}\begin{bmatrix} 5.125\times10^{-2} & 1.875\times10^{-2} \\ 1.875\times10^{-2} & 5.125\times10^{-2} \end{bmatrix} = \begin{bmatrix} 22.53 & 8.242 \\ 8.242 & 22.53 \end{bmatrix}$$

$$\mathbf{G}_L(\mathbf{G}_o+\mathbf{G}_L)^{-1} = \begin{bmatrix} 0.02 & 0 \\ 0 & 0.02 \end{bmatrix}\begin{bmatrix} 22.53 & 8.242 \\ 8.242 & 22.53 \end{bmatrix} = \begin{bmatrix} 0.4506 & 0.16484 \\ 0.16484 & 0.4506 \end{bmatrix}$$

$$2\mathbf{G}_L(\mathbf{G}_o+\mathbf{G}_L)^{-1}i_{arr.} = 2\begin{bmatrix} 0.4506 & 0.16484 \\ 0.16484 & 0.4506 \end{bmatrix}\begin{bmatrix} 0 \\ 0.1 \end{bmatrix} = \begin{bmatrix} 0.03297 \\ 0.09012 \end{bmatrix}$$

Problem 6-1   Given a two-terminal network with the following resistors: $R_{11} = 60$ [Ω], $R_{22} = 40$ [Ω], $R_{12} = 100$ [Ω]. Find the matrix G and the matrix $\mathbf{R} = \mathbf{G}^{-1}$.

Solution:

$$\mathbf{G}_{11} = \frac{1}{60} + \frac{1}{100} = 0.02667$$

$$\mathbf{G}_{22} = \frac{1}{40} + \frac{1}{100} = 0.035$$

$$\mathbf{G}_{12} = \mathbf{G}_{21} = \frac{-1}{100} = -0.01$$

$det = 8.333\times10^{-4}$

$$\mathbf{R} = \frac{1}{det}\begin{bmatrix} 0.035 & 0.01 \\ 0.01 & 0.02667 \end{bmatrix} = \begin{bmatrix} 42.0 & 12.0 \\ 12.0 & 32.0 \end{bmatrix}$$

Problem 6-2   The characteristic impedance matrix of a two-wire line is $\mathbf{Z}_o = \begin{bmatrix} 50 & 30 \\ 30 & 50 \end{bmatrix}$.
What are the values of R_{11}, R_{22}, and R_{12}, the resistors that form a matching network for the line?

Solution:

$det = 1600$

$$\mathbf{G} = \mathbf{Z}_o^{-1} = \frac{1}{det}\begin{bmatrix} 50 & -30 \\ -30 & 50 \end{bmatrix} = \begin{bmatrix} 0.03125 & -0.01875 \\ -0.01875 & 0.03125 \end{bmatrix}$$

$$R_{12} = \frac{-1}{\mathbf{G}_{12}} = \frac{det}{30} = 53.33 \text{ [}\Omega\text{]}$$

$$R_{11} = R_{22} = \frac{1}{0.03125 - 0.01875} = 80 \; [\Omega]$$

Problem 6-3   A (nonsymmetric) two-wire line with conductors loosely coupled has a characteristic impedance matrix given by $\mathbf{Z}_o = \begin{bmatrix} 50 & 5 \\ 5 & 40 \end{bmatrix}$. What are the values of R_{11}, R_{22}, and R_{12}, the resistors that form a matching network for this line?

Solution:

det = 1975

$$\mathbf{G} = \mathbf{Z}_o^{-1} = \frac{1}{det} \begin{bmatrix} 40 & -5 \\ -5 & 50 \end{bmatrix} = \begin{bmatrix} 0.02025 & -2.532 \times 10^{-3} \\ -2.532 \times 10^{-3} & 0.02532 \end{bmatrix}$$

$$R_{12} = \frac{-1}{\mathbf{G}_{12}} = \frac{det}{5} = 395 \; [\Omega]$$

$$R_{11} = \frac{1}{0.02025 - 2.532 \times 10^{-3}} = 56.4 \; [\Omega]$$

$$R_{22} = \frac{1}{0.02532 - 2.532 \times 10^{-3}} = 43.9 \; [\Omega]$$

Problem 6-4   Suppose that wire 2 of the two-wire line of the preceding problem is left open circuit at both ends (i.e., the current in wire 2 is very close to zero), what value of resistance should be used to terminate wire 1 to minimize reflection?

Solution:

The propagating voltage on wire 1 is

$$v_1 = \mathbf{Z}_{11} i_1 + \mathbf{Z}_{12} i_2$$

but $i_2 = 0$. Hence, the characteristic impedance of wire 1 is $\mathbf{Z}_{11}$. The load resistance should be $\mathbf{Z}_{11} = 50 \; [\Omega]$.

Problem 6-5   Suppose that wire 2 of the two-wire line of the preceding problem is shorted to ground at both ends (i.e., the voltage of wire 2 is very close to zero), what value of resistance should be used to terminate wire 1 to minimize reflection? *Hint:* What is the solution of

$$\begin{bmatrix} v_1 \\ 0 \end{bmatrix} = \begin{bmatrix} 50 & 5 \\ 5 & 40 \end{bmatrix} \begin{bmatrix} i_1 \\ i_2 \end{bmatrix}$$

Solution:

The propagating current on wire 1 is

$$i_1 = \mathbf{G}_{11}v_1 + \mathbf{G}_{12}v_2$$

but $v_2 = 0$. Hence, the characteristic impedance of wire 1 in this case is $1/\mathbf{G}_{11}$. The load resistance should be $1/0.02025 = 49.4$.

Problem 6-6   A capacitor connected directly to a resistor has voltage history

$$v(t) = v_o e^{-t/RC} \tag{6-49}$$

where, v_o is the voltage at $t = 0$. The current through the capacitor is

$$i(t) = C\frac{dv}{dt} = \frac{-v_o}{R}e^{-t/RC} \tag{6-50}$$

What is g and I_{sc} for the chord approximation (Equation (6-10))? Draw a sketch of the i-v curve (see Figure S.1). Does the negative value of g have a physical interpretation?

Solution:

$$g = \frac{i(t + \delta t) - i(t)}{v(t + \delta t) - v(t)} = \frac{-1}{R}$$

$$I_{sc} = v(t)g - i(t) = 0$$

The negative value of g reflects the fact that C is the power source; it delivers power to the resistor.

Problem 6-7   Let a capacitor be discharging through a shunted resistor. The problem parameters are $C = 10,000$ [pF] $= 10^{-8}$ [F], $R = 100$ [Ω] and the initial voltage $v(0)$ is 5 [V]. For the time step $t = 0$ to $\delta t = 0.25 \times 10^{-6}$ [s], calculate the following: the RC decay time constant, g_C, I_{scC}, the value of $v(\delta t)$ using a. the Norton approximation for the capacitor, and, b. the exact exponential decay for the R-C circuit. What can be done to improve the agreement between the approximate model (a.) and the exact model (b.)?

Figure S.1   Sketch of Voltage-Current Curve.

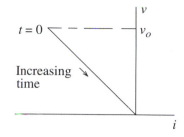

Solution:

$$RC = 10\text{-}6 \ [\text{s}]$$

$$g_C = \frac{10^{-8}}{0.25 \times 10^{-6}} = 0.04 \ \ [\text{mhos}]$$

$$I_{scC} = 0.04 \times 5 = 0.2 \ \ [\text{A}]$$

a. Norton approximation

$$v(\delta t) = \frac{I_{scC}}{\dfrac{1}{R} + g_C} = \frac{0.2}{0.01 + 0.04} = 4 \ \ [\text{V}]$$

b. Exact solution

$$v(\delta t) = v_o e^{-t/RC} = 5e^{-0.25 \times 10^{-6} / 10^{-6}} = 3.894 \ \ [\text{V}]$$

To improve the approximation, one can use a smaller δt and multiple steps to the same time. For instance, set $\delta t = 0.125 \times 10^{-6}$ [s] and take two time steps.

Problem 6-8   An inductor its driven by a driver as shown below in Figure 6.4. The driver is defined by $V_{oc}(t) = 5u(t)$ [V], where $u(t)$ is the unit step function and the driver impedance is $R = 5$ [Ω]. The inductor $L = 5 \times 10^{-6}$ [H].

a. What is the initial voltage $v(0^+)$ and current $i(0^+)$?

b. What is the exact solution for $v(t)$?

c. Convert the driver to a Norton equivalent circuit. What is g and I_{sc} for the driver?

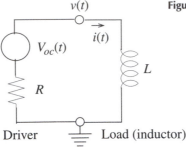

Figure 6.4   Problem 6-8 Driver and Load Inductor.

Driver   Load (inductor)

d. Assume a time step of $\delta t = 0.2 \times 10^{-6}$ [s]. Find the Norton equivalent circuit of the load L for a time step from $t = 0$ to $t = \delta t$.

e. Using the Norton equivalent for the driver (c.) and the Norton equivalent for the inductor (d.), calculate the voltage $v(\delta t)$.

f. Calculate the exact solution at $t = \delta t$ using (b.).

Solution:

a. $v(0^+) = 5$ [V], $i(0^+) = 0$ [A]

b. $\quad v(t) = v(0)e^{-Rt/L} = 5e^{-t/10^{-6}}$ [V]

c. For the driver, $g = 1/R = 0.2$ [mhos], $I_{sc} = V_{oc}(t)/R = u(t)$ [A]

d. For the load (L), $g_L = \delta t/L = 0.04$ [mhos], $I_{scL} = 0$ [A].

e. $\quad v(\delta t) = \dfrac{I_{sc} + I_{scL}}{g + g_L} = 4.167$ [V]

f. $\quad v(\delta t) = 5e^{-\delta t/10^{-6}} = 4.094$ [V]

Problem 6-9   Refer to the AC termination shown in Figure 6.6.

a. What is the time constant (RC) of the termination?

b. What is the Norton conductance g_{AC} for the AC termination?

c. Assume the capacitor voltage is initially zero; what is the Norton current source I_{scAC} at $t = 0$ for the AC termination?

d. Let $\delta t = 1$ [ns], $v_C(0) = 0$ and the forward voltage $v_F = 5$ [V] (constant). Calculate $v(t)$ and $v_C(t)$ for the first four time steps. *Hint*: Convert the circuit of Figure 6.6 to Norton equivalents using the Norton equivalent for the AC termination (Figure 6-2 and Equations (6-25) and (6-26) for the AC termination).

Solution:

a. $\quad RC = 50 \times 2 \times 10^{-9} = 0.1 \times 10^{-6}$ [s]

b.

$$g_{AC} = \frac{1}{R + \dfrac{\delta t}{C}} = \frac{1}{50 + \dfrac{10^{-9}}{2 \times 10^{-9}}} = \frac{1}{50.5} = 0.0198\ldots\,[\mho]$$

c.

$$I_{scAC} = \frac{v_C(t)}{R + \dfrac{\delta t}{C}} = \frac{0}{R + \dfrac{\delta t}{C}} = 0$$

d. Use Norton circuit for the left line end and the AC termination (see Figure S.2):

This circuit implies that

$$v(t + \delta t) = \frac{2I_{sc} + I_{scAC}}{z_o^{-1} + g_{AC}} = \frac{0.2 + I_{scAC}}{0.0398}$$

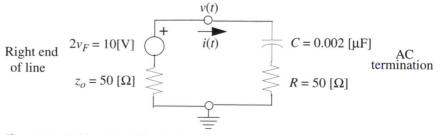

Figure 6.6   Problem 6-9, AC Termination.

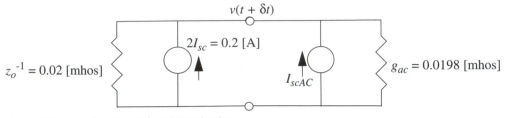

Figure S.2   Equivalent Circuit for AC Termination.

Then, (Equation 6-29):

$$v_C(t+\delta t) = v(t+\delta t)(1-Rg_{AC}) + RI_{scAC} = v(t+\delta t)0.0099 + 50I_{scAC}$$

And, for the next time step (Equation 6-26),

$$I_{scAC} = \frac{v_C(t)}{R+\dfrac{\delta t}{C}} = \frac{v_C(t+\delta t)}{50.5}$$

v(0) = 5 [V].
step 1. $t = \delta t = 1$ [ns]:
($I_{scAC} = 0$)

$$v(\delta t) = \frac{2I_{sc}+I_{scAC}}{z_o^{-1}+g_{ac}} = \frac{0.2}{0.0398} = 5.025$$

$$v_C(\delta t) = v(\delta t)0.0099 = 5.025 \times 0.0099 = 0.04975$$

$$I_{scAC} = \frac{v_C(\delta t)}{50.5} = \frac{0.04975}{50.5} = 9.85\times10^{-4}$$

step 2. $t = 2\delta t = 2$[ns]:

$$v(2\delta t) = \frac{2I_{sc}+I_{scAC}}{z_o^{-1}+g_{AC}} = \frac{0.2+9.85\times10^{-4}}{0.0398} = 5.050$$

$$v_C(2\delta t) = v(2\delta t)0.0099 + 50I_{scAC} = 5.050 \times 0.0099 + 50 \times 9.85\times10^{-4} = 0.0992$$

$$I_{scAC} = \frac{v_C(2\delta t)}{50.5} = \frac{0.0992}{50.5} = 1.964\times10^{-3}$$

step 3. $t = 3\delta t = 3$[ns]:

$$v(3\delta t) = \frac{2I_{sc} + I_{scAC}}{z_o^{-1} + g_{AC}} = \frac{0.2 + 1.964 \times 10^{-3}}{0.0398} = 5.074$$

$$v_C(3\delta t) = v(3\delta t)0.0099 + 50I_{scAC} = 5.074 \times 0.0099 + 50 \times 1.964 \times 10^{-3} = 0.148$$

$$I_{scAC} = \frac{v_C(3\delta t)}{50.5} = \frac{0.148}{50.5} = 2.93 \times 10^{-3}$$

step 4. $t = 4\delta t = 4$[ns]:

$$v(4\delta t) = \frac{2I_{sc} + I_{scAC}}{z_o^{-1} + g_{AC}} = \frac{0.2 + 2.93 \times 10^{-3}}{0.0398} = 5.099$$

$$v_C(4\delta t) = v(4\delta t)0.0099 + 50I_{scAC} = 5.099 \times 0.0099 + 50 \times 2.93 \times 10^{-3} = 0.197$$

$$I_{scAC} = \frac{v_C(4\delta t)}{50.5} = \frac{0.197}{50.5} = 3.90 \times 10^{-3}$$

Problem 6-10   Assume the line parameters and termination of Problem 6-9 (with the exception of v_F). The line is lossless and is assumed driven by a 5 [V] step-function driver with a 5 [Ω] source impedance. Calculate the departing voltage taking into account the driver impedance. (This is the forward voltage signal v_F.) When this pulse arrives at the right end of the line, what is the expected overshoot (assume delay time $l/c = 5$ [ns]).

Solution:

$$v_F = 5\frac{50}{50 + 5} = 4.54$$

$$v_{overshoot} \cong v_o \frac{l}{cz_o C} = 4.54\frac{5 \times 10^{-9}}{50 \times 2 \times 10^{-9}} = 0.227 \ [V]$$

For the circuit shown in Figure 7.2, the characteristic conductance matrices (inverse of their characteristic impedance matrices) are:

$$\mathbf{G}_{o1} = \begin{bmatrix} 0.030 & -0.015 & -0.005 \\ -0.015 & 0.035 & -0.012 \\ -0.005 & -0.012 & 0.040 \end{bmatrix}, \ [\mho], \text{ Line 1}$$

and

$$G_{o2} = \begin{bmatrix} 0.025 & -0.010 \\ -0.010 & 0.025 \end{bmatrix}, \text{ [ʊ], Line 2}$$

Problem 7-1   For the network shown in Figure 7.2, let $r_1 = r_2 = r_3 = 50$ [Ω]. What is the circuit conductance matrix G at the left end of the line?

Solution:

$$G = \begin{bmatrix} 0.030 + \dfrac{1}{50} & -0.015 & -0.005 \\ -0.015 & 0.035 + \dfrac{1}{50} & -0.012 \\ -0.005 & -0.012 & 0.040 + \dfrac{1}{50} \end{bmatrix} = \begin{bmatrix} 0.050 & -0.015 & -0.005 \\ -0.015 & 0.055 & -0.012 \\ -0.005 & -0.012 & 0.060 \end{bmatrix}$$

Problem 7-2   For the network shown in Figure 7.2, let $r_4 = 40$ [Ω], $r_5 = 100$ [Ω] and $r_6 = 50$ [Ω]. What is the circuit conductance matrix G for the center part of the network (the junction of Lines 1 and 2 and resistors r_4 and r_5). *Hint:* The matrix G should be a 4×4 matrix. The auxiliary node is node 4.

Solution:

$$G = \begin{bmatrix} 0.030 + 0.025 & -0.015 - 0.010 & -0.005 & 0 \\ -0.015 - 0.010 & 0.035 + 0.025 + \dfrac{1}{50} & -0.012 & \dfrac{-1}{50} \\ -0.005 & -0.012 & 0.040 & \dfrac{-1}{40} \\ 0 & \dfrac{-1}{50} & \dfrac{-1}{40} & \dfrac{1}{40} + \dfrac{1}{40} + \dfrac{1}{100} + \dfrac{1}{50} \end{bmatrix}$$

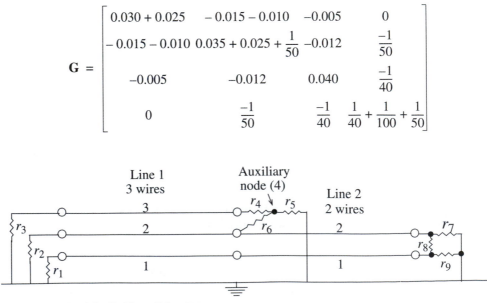

Figure 7.2   Circuit for Problems 7-1 to 7-4.

$$G = \begin{bmatrix} 0.055 & -0.025 & -0.005 & 0 \\ -0.025 & 0.075 & -0.012 & -0.02 \\ -0.005 & -0.012 & 0.040 & -0.025 \\ 0 & -0.020 & -0.025 & 0.055 \end{bmatrix}$$

Problem 7-3 For the network shown in Figure 7.2, let the arriving voltages at the center part of the network be, $v_{arr1} = \begin{bmatrix} 5 \\ 0 \\ 0 \end{bmatrix}$ [V] and $v_{arr2} = \begin{bmatrix} 0.5 \\ 0 \end{bmatrix}$ [V]. What are the arriving current vectors? What is the Norton current vector for the circuit that represents the junction (both lines combined, positive into the nodes)?

Solution:

$$i_{arr1} = G_{o1}v_{arr1} = \begin{bmatrix} 0.030 & -0.015 & -0.005 \\ -0.015 & 0.035 & -0.012 \\ -0.005 & -0.012 & 0.040 \end{bmatrix} \begin{bmatrix} 5 \\ 0 \\ 0 \end{bmatrix} = \begin{bmatrix} 0.15 \\ -0.075 \\ -0.025 \end{bmatrix} \text{ (Forward signal)}$$

$$i_{arr2} = -G_{o2}v_{arr2} = -\begin{bmatrix} 0.025 & -0.010 \\ -0.010 & 0.025 \end{bmatrix} \begin{bmatrix} 0.5 \\ 0 \end{bmatrix} = \begin{bmatrix} -0.0125 \\ 0.005 \end{bmatrix} \text{ (Backward signal)}$$

Norton short circuit current at junction

$$I_{scmid} = i_{arr1} - i_{arr2} = \begin{bmatrix} 0.15 \\ -0.075 \\ -0.025 \end{bmatrix} - \begin{bmatrix} -0.0125 \\ 0.005 \\ 0 \end{bmatrix} = \begin{bmatrix} 0.1625 \\ -0.080 \\ -0.025 \end{bmatrix}$$

Problem 7-4 For the network shown in Figure 7.2, let $r_7 = r_8 = 66.7$ [Ω] and $r_9 = 100$ [Ω]. What is the circuit conductance matrix for the right end termination? What are the resistors required for a matching termination at the right end?

Solution:

Right end termination matrix

$$G = \begin{bmatrix} 0.025 + \dfrac{1}{66.7} + \dfrac{1}{100} & -0.010 - \dfrac{1}{66.7} \\ -0.010 - \dfrac{1}{66.7} & 0.025 + \dfrac{1}{66.7} + \dfrac{1}{66.7} \end{bmatrix} = \begin{bmatrix} 0.0500 & -0.0250 \\ -0.0250 & 0.0550 \end{bmatrix}$$

Termination resistors (Equation 6-6)

$$R_{ij} = \begin{cases} -\dfrac{1}{G_{ij}}, & i \neq j \\[2em] \dfrac{1}{\sum\limits_k G_{ik}}, & i = j \end{cases}$$

$$R_{11} = \frac{1}{G_{11} + G_{12}} = \frac{1}{0.025 - 0.010} = 28.6$$

$$R_{22} = \frac{1}{G_{21} + G_{22}} = \frac{1}{-0.010 + 0.025} = 28.6$$

$$R_{12} = \frac{-1}{G_{12}} = \frac{-1}{-0.010} = 100$$

Problem 10-1   For the two-trace cross section discussed above, let the length of a pair of traces be $l = 0.1$ [m]. Assume a step function signal departs the left end at $t = 0$. What is the difference of the arrival times of the two modal step-function signals at the right end?

Solution:

$$\gamma_1 = 4.56 \times 10^{-9} \text{ and } \gamma_2 = 6.77 \times 10^{-9} \text{ [s/m]}$$

$$t_{difference} = l(\gamma_2 - \gamma_1) = 0.221 \times 10^{-9} \text{ [s]}$$

Problem 10-2   Two closely coupled (i.e., very close to each other) traces as shown below in Figure 10.2 have capacitance and inductance matrices:

$$\mathbf{C} = \begin{bmatrix} 72.5 & -35.1 \\ -35.1 & 72.5 \end{bmatrix} \text{ [pF/m]} \tag{10-25}$$

and

$$\mathbf{L} = \begin{bmatrix} 0.597 & 0.349 \\ 0.349 & 0.597 \end{bmatrix} \text{ [µH/m]} \tag{10-26}$$

Figure 10.2   Two Closely Coupled Traces, Problem 10-2. $\varepsilon_r = 1$ (air)

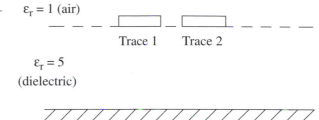

Trace 1   Trace 2

$\varepsilon_r = 5$

(dielectric)

Ground Plane

1. Calculate the **LC** matrix.
2. Find the two propagation speeds (c_1 and c_2) and the associated eigenvectors ($\mathbf{u}_1$ and $\mathbf{u}_2$) normalized so that their absolute values are 1 (write them as the matrix **U**).
3. Find the inverse of **U**.
4. For some purposes, a balanced differential drive is used on two traces (i.e., the traces are driven with equal and opposite voltages). Would such a signal experience time dispersion if launched on the traces shown in Figure 10.2?
5. If a driver launched a signal on one of the traces, could time dispersion be a problem?

1. Calculate the **LC** matrix.

$$\mathbf{LC} = \begin{bmatrix} 0.597\times10^{-6} & 0.349\times10^{-6} \\ 0.349\times10^{-6} & 0.597\times10^{-6} \end{bmatrix} \begin{bmatrix} 72.5\times10^{-12} & -35.1\times10^{-12} \\ -35.1\times10^{-12} & 72.5\times10^{-12} \end{bmatrix}$$

$$\mathbf{LC} = \begin{bmatrix} 0.310326\times10^{-16} & 0.43478\times10^{-17} \\ 0.43478\times10^{-17} & 0.310326\times10^{-16} \end{bmatrix} \; [\text{s}^2/\text{m}^2]$$

2. Find the two propagation speeds (c_1 and c_2) and the associated eigenvectors ($\mathbf{u}_1$ and $\mathbf{u}_2$) normalized so that their absolute values are 1 (write them as the matrix **U**).

The characteristic equation is

$$det\begin{bmatrix} 0.310326\times10^{-16} - \gamma^2 & 0.43478\times10^{-17} \\ 0.43478\times10^{-17} & 0.310326\times10^{-16} - \gamma^2 \end{bmatrix} = 0$$

$$(0.310326\times10^{-16} - \gamma^2)^2 = (0.43478\times10^{-17})^2$$

$$\gamma^2 = 0.310326\times10^{-16} \pm 0.43478\times10^{-17}$$

$$\gamma_1 = 5.95 \times 10^{-9} \quad \text{or} \quad \gamma_2 = 5.17 \times 10^{-9}$$

$$c_1 = 1.68 \times 10^8 \qquad c_2 = 1.93 \times 10^8$$

Let $u_{11} = 1$,

$$0.310326 \times 10^{-16} + 0.43478 \times 10^{-17} u_{21} = 3.54 \times 10^{-17}$$

$$u_{21} = 1.00$$

Normalized, $\mathbf{u}_1 = \begin{bmatrix} 0.707 \\ 0.707 \end{bmatrix}$

Let $u_{12} = 1$,

$$0.310326 \times 10^{-16} + 0.43478 \times 10^{-17} u_{22} = 2.67 \times 10^{-17}$$

$$u_{22} = -1.00$$

Normalized, $\mathbf{u}_2 = \begin{bmatrix} 0.707 \\ -0.707 \end{bmatrix}$

$$\mathbf{U} = \begin{bmatrix} 0.707 & 0.707 \\ 0.707 & -0.707 \end{bmatrix}$$

3. Find the inverse of $\mathbf{U}$.

$$\mathbf{U}^{-1} = \begin{bmatrix} 0.707 & 0.707 \\ 0.707 & -0.707 \end{bmatrix}$$

4. For some purposes, a balanced differential drive is used on two traces (i.e., the traces are driven with equal and opposite voltages). Would such a signal experience time dispersion if launched on the traces shown in Figure 10-2?

No, the differential drive produces a pure mode (2).

5. If a driver launched a signal on one of the traces, could time dispersion be a problem?

Yes.

Problem 11-1   Calculate Z_{o1}, for the line defined by the matrices C and L of Equations (10-3) and (10-4).

Solution:

$$Z_{o1} = c_1 L = 2.19 \times 10^8 \begin{bmatrix} 0.511 \times 10^{-6} & 0.160 \times 10^{-6} \\ 0.160 \times 10^{-6} & 0.370 \times 10^{-6} \end{bmatrix} = \begin{bmatrix} 112 & 35 \\ 35 & 81 \end{bmatrix} \ [\Omega]$$

Problem 11-2   Calculate the characteristic impedance matrix for the line of Problem 10-2.

$$Z_o = U\Gamma^{-1}U^{-1}L$$

$$Z_o = \begin{bmatrix} 0.707 & 0.707 \\ 0.707 & -0.707 \end{bmatrix} \begin{bmatrix} 1.68 \times 10^8 & 0 \\ 0 & 1.93 \times 10^8 \end{bmatrix} \begin{bmatrix} 0.707 & 0.707 \\ 0.707 & -0.707 \end{bmatrix} \begin{bmatrix} 0.597 \times 10^{-6} & 0.349 \times 10^{-6} \\ 0.349 \times 10^{-6} & 0.597 \times 10^{-6} \end{bmatrix}$$

$$Z_o = \begin{bmatrix} 103.4 & 55.5 \\ 55.5 & 103.4 \end{bmatrix}$$

Problem 11-3   Calculate Z'_o for the line of *Problem 10-2*.

$$Z'_o = U^{-1}Z_o(U^{-1})^T$$

$$Z'_o = \begin{bmatrix} 0.707 & 0.707 \\ 0.707 & -0.707 \end{bmatrix} \begin{bmatrix} 103.4 & 55.5 \\ 55.5 & 103.4 \end{bmatrix} \begin{bmatrix} 0.707 & 0.707 \\ 0.707 & -0.707 \end{bmatrix}$$

$$Z'_o = \begin{bmatrix} 158.8 & 0 \\ 0 & 47.9 \end{bmatrix}$$

Problem 11-4   For the transmission line of Figure 10.1 with C and L matrices given by Equations (10-3) and (10-4), the matrix U^{-1} is given by Equation (10-23). The transpose of U^{-1} has column vectors that are eigenfunctions of $\mathbf{CL}$ and, therefore, correspond to propagation modes of current. The $n = 1$ current propagation mode is proportional to the first column of $(\mathbf{U}^{-1})^T$ and is given by $\mathbf{i}_1 = \begin{bmatrix} 1.008 \\ -0.412 \end{bmatrix}$. Calculate a. $\mathbf{v}_a = \mathbf{Z}_o \mathbf{i}_1$, and, b. $\mathbf{v}_b = \mathbf{Z}_{o1} \mathbf{i}_1$. Note that $\mathbf{Z}_{o1}$ was calculated in Problem 11-1 and $\mathbf{Z}_o$ is given by Equation (11-14).

Solution:

a.

$$\mathbf{v}_a = \begin{bmatrix} 107.5 & 24.3 \\ 24.3 & 54.8 \end{bmatrix} \begin{bmatrix} 1.008 \\ -0.412 \end{bmatrix} = \begin{bmatrix} 98.3 \\ 1.92 \end{bmatrix}$$

b.

$$\mathbf{v}_b = \begin{bmatrix} 112 & 35 \\ 35 & 81 \end{bmatrix} \begin{bmatrix} 1.008 \\ -0.412 \end{bmatrix} = \begin{bmatrix} 98.5 \\ 1.91 \end{bmatrix}$$

Not equal because of round-off errors.

Problem 12-1   Assume $\mathbf{Z}'_o = \begin{bmatrix} 158.8 & 0 \\ 0 & 47.9 \end{bmatrix}$, $\mathbf{U}^{-1} = \begin{bmatrix} 0.707 & 0.707 \\ 0.707 & -0.707 \end{bmatrix}$, each conductor has a resistance of 10 [Ω/m] and the matrix $\mathbf{G} = \mathbf{0}$. Calculate the matrix $\mathbf{A}' = \mathbf{B}'$.

Solution

$$\mathbf{R} = \begin{bmatrix} 10 & 0 \\ 0 & 10 \end{bmatrix}$$

$$\mathbf{R}' = \mathbf{U}^{-1}\mathbf{R}(\mathbf{U}^{-1})^T = \begin{bmatrix} 0.707 & 0.707 \\ 0.707 & -0.707 \end{bmatrix} \begin{bmatrix} 10 & 0 \\ 0 & 10 \end{bmatrix} \begin{bmatrix} 0.707 & 0.707 \\ 0.707 & -0.707 \end{bmatrix} = \begin{bmatrix} 10 & 0 \\ 0 & 10 \end{bmatrix}$$

$$\mathbf{A}' = \frac{1}{2}\left(\mathbf{R}'\mathbf{Z}'^{-1}_o + \mathbf{Z}'_o\mathbf{G}' \right)$$

$$= \frac{1}{2}\begin{bmatrix} 10 & 0 \\ 0 & 10 \end{bmatrix} \begin{bmatrix} 6.297\times10^{-3} & 0 \\ 0 & 2.088\times10^{-2} \end{bmatrix} = \begin{bmatrix} 0.031485 & 0 \\ 0 & 0.1044 \end{bmatrix} [\text{m}^{-1}]$$

Problem 16-1   A trace is 10 [cm] long. Its measured capacitance is 10 [pF]. If the trace is held at a potential of 1 [V], what is the charge per length [C/m]? What is its capacitance in MKS units [F/m]?

Solution:

Its capacitance per meter is:

$$C = 10 \times 10^{-12} / 0.1 = 100 \times 10^{-12} \text{ [F/m]}$$

Its charge per meter is:

$$q = CV = 100 \times 10^{-12} \text{ [C/m]}$$

Problem 16-2   A trace is divided into 10 segments (as in Figure 16.4). How many potential functions must be evaluated to calculate its capacitance? (That is, how many elements are in the matrix **V**?) If it were divided into 100 segments, how many potential functions would be required?

Solution:

If the trace has 10 segments, **V** is the 10 x 10 matrix with elements $v_{i,j}$. It has 100 elements. If it had 100 segments, the number of elements would be 10,000.

Problem 16-3   Suppose the capacitance is approximated by using only one segment per side, a total of four segments. Write out explicitly the matrix **V** and the four linear equations that the matrix Equation (16-7) represents.

Solution:

$$\mathbf{V} = \begin{bmatrix} v_{1,1} & v_{1,2} & v_{1,3} & v_{1,4} \\ v_{2,1} & v_{2,2} & v_{2,3} & v_{2,4} \\ v_{3,1} & v_{3,2} & v_{3,3} & v_{3,4} \\ v_{4,1} & v_{4,2} & v_{4,3} & v_{4,4} \end{bmatrix}$$

$$\begin{bmatrix} 1 \\ 1 \\ 1 \\ 1 \end{bmatrix} = \begin{bmatrix} v_{1,1} & v_{1,2} & v_{1,3} & v_{1,4} \\ v_{2,1} & v_{2,2} & v_{2,3} & v_{2,4} \\ v_{3,1} & v_{3,2} & v_{3,3} & v_{3,4} \\ v_{4,1} & v_{4,2} & v_{4,3} & v_{4,4} \end{bmatrix} \begin{bmatrix} q_1 \\ q_2 \\ q_3 \\ q_4 \end{bmatrix}$$

Problem 16-4   Consider a segment parallel to the x axis as shown in Figure 16.7. Let the length of the segment be 1.25×10^{-4} [m] (about 5 mils) and let the center of the segment be located at the point $(0, 2.5 \times 10^{-5})$ [m], about 1 mil above the ground plane. If the charge on the segment is unity (1 [C/m]), what is the potential at the center of the segment due to this charge?

Solution:

Set $x = x_s = 0$, $y = y_s = 2.5 \times 10^{-5}$ [m] and $l = 1.25 \times 10^{-4}$ [m] in Equation (16-19).
From computer code:
Segment parameters:
Length: 1.2500e-004 [m]
Segment center (xs, ys) = (0.0000e+000, 2.5000e-005) [m]
Observer point (xp, yp) = (0.0000e+000, 2.5000e-005) [m]
Potentials:
Direct potential: 1.9198e+011 [V]
Image potential: -1.7465e+011 [V]
Combined potential: 1.7331e+010 [V]

Problem 16-5   If only two points are used in a fit to Equation (16-24), an exact fit can be made (because two points determine a straight line). Using calculations indexed 1 and 2 from Table 16-1, what value of $C(0)$ is obtained? How accurate is that value?

Solution:

Let
$$C(\delta l) = a\delta l + C(0)$$

Index 1: $C(1.0) = 130.3763 = a + C(0)$

Index 2: $C(0.5) = 132.2749 = 0.5a + C(0)$

$$C(0) = 2C(0.5) - C(1.0) = 134.1735$$

Problem 17-1   Referring to Figure 17.2 and Equations (17-4) and (17-5), let $\varepsilon_1 = \varepsilon_o$, the permittivity of free space, $\varepsilon_2 = 5\varepsilon_o$, and $E_{n1} = 5,000$ [V/m]. What is E_{n2}?

Solution:

Equation (17.5) yields

$$E_{n2} = \frac{\varepsilon_1}{\varepsilon_2}E_{n1} = \frac{1}{5}(5,000) = 1,000 \ [\text{V/m}]$$

Problem 17-2   Assume that the normal electric field due to all charges present at a point on a dielectric boundary is $\bar{E}_n = 5,000$ [V/m]. Let $\varepsilon_1 = \varepsilon_o$, the permittivity of free space and $\varepsilon_2 = 5\varepsilon_o$ (as in the preceding problem). Use Equation (17-11) to obtain the equivalent surface charge density in [C/m^2]. Use $\varepsilon_o = 8.854\times10^{-12}$ [F/m].

Solution:

Equation (17-11) is

$$\tilde{\rho}_s = 2\varepsilon_o \bar{E}_n \frac{\varepsilon_1 - \varepsilon_2}{\varepsilon_1 + \varepsilon_2} = 1.7708\times10^{-11} \cdot 5,000\frac{1-5}{1+5} = -5.9027\times10^{-8} \ [C/m^2]$$

Problem 17-3   Assume the normal electric field adjacent to a conductor (in the dielectric) is $E_n = 5,000$ [V/m]. Let the permittivity of the dielectric be $5\varepsilon_o$. What part of the electric field is due to the physical charge and what part is due to the equivalent surface charge on the dielectric. *Hint*: The total electric field (5,000 [V/m]) is given by Equation (17-15); the part due to the physical charge is given by ρ_s/ε_o .

Using Equation (17.15), the physical surface charge density is.

$$\rho_s = \varepsilon_1 E_{n2}$$

In our model, the electric field due directly to the physical surface charge is

$$E_{phys} = \frac{\rho_s}{\varepsilon_o} = \frac{\varepsilon_1}{\varepsilon_o}E_{n2} = 5E_{n2} = 25,000 \ [V/m]$$

The equivalent surface charge density is given by Equation (17-19)

$$\tilde{\rho}_s = \rho_s\frac{\varepsilon_o - \varepsilon_1}{\varepsilon_1} = \varepsilon_1 E_{n2}\frac{\varepsilon_o - \varepsilon_1}{\varepsilon_1}$$

The electric field due to the equivalent surface charge density is

$$E_{equiv} = \frac{\tilde{\rho}_s}{\varepsilon_o} = \frac{\varepsilon_1}{\varepsilon_o}E_{n2}\frac{\varepsilon_o - \varepsilon_1}{\varepsilon_1} = \frac{\varepsilon_o - \varepsilon_1}{\varepsilon_o}E_{n2} = -4E_{n2} = -20,000 \ [V/m]$$

Problem 17-4   If $\varepsilon_{r1} = \varepsilon_{r2}$, what is $\bar{\varepsilon}_r$? Let $l_1 = l_2$, $\varepsilon_{r1} = 2\varepsilon_o$ and $\varepsilon_{r2} = 5\varepsilon_o$, What is $\bar{\varepsilon}_r$?

If $\varepsilon_{r1} = \varepsilon_{r2}$, then using Equation (17-43),

$$\bar{\varepsilon}_\gamma = \varepsilon_{\gamma 1} = \varepsilon_{\gamma 2}$$

$$\bar{\varepsilon}_r = \frac{l_1 \varepsilon_{r1} + l_2 \varepsilon_{r2}}{(l_1 + l_2)} = \frac{\varepsilon_{r1} + \varepsilon_{r2}}{2} = \frac{7}{2}\varepsilon_o$$

Problem 20-1   Assume that $H_y(z,t)$ has the form $H_y(z,t) = Ae^{-\alpha t} \sin \omega z$, where A is a constant. What is the relationship between α and ω for $H_y(z,t)$ to satisfy the diffusion equation? *Hint:* Substitute $H_y(z,t)$ into Equation (20-8).

Answer: $\omega^2 = \alpha\mu\sigma$

Problem 20-2   A copper trace $\sigma = 5.8 \times 10^7$ is 10^{-4} [m] (about 4 mils) thick. At what frequency is its thickness equal to one skin depth? Note that at higher frequencies, the resistance will be determined by the diffusion process.

$$\delta = \sqrt{\frac{2}{\omega\mu\sigma}}$$

$$\omega = \frac{2}{\mu\sigma\delta^2}$$

Solution:

$$\omega = \frac{2}{4\pi \times 10^{-7} 5.8 \times 10^7 (10^{-4})^2} = 2.744 \times 10^6 \quad [\text{radians/s}]$$

$$f = \frac{\omega}{2\pi} = 4.367 \times 10^5 \quad [\text{hz}]$$

Problem 21-1   Given a trace with the following description:

Material: copper, $\sigma_{Cu} = 5.8 \times 10^7$
Thickness: 10^{-4} [m] (about 4 mils)
Width: 2×10^{-4} [m] (about 8 mils)
Length: 0.2 [m] (about 8 inches)
Current in the trace: $i = \bar{a}t$ [A], where $\bar{a} = 2 \times 10^7$ [A/s] (note that the dimensions of a are [A/ms].)

Assume two-sided diffusion into the trace.

Find the following:

1. The slab internal inductance [H] using Equation (21-16).

$$L_{int} = \frac{d\mu}{6} = \frac{0.5\times10^{-4} \times 4\pi\times10^{-7}}{6} = 1.047\times10^{-11} \text{ [H]}$$

2. The internal inductance per length [H/m] of the trace.

$$L' = \frac{L_{int}}{\text{Width}} = 5.2360\times10^{-8} \text{ [H/m]}$$

3. The total internal inductance [H] of the trace.

$$L = L' \times \text{Length} = 1.047\times10^{-8} \text{ [H]}$$

4. The slab d.c. resistance R_{dc} [Ω] (Equation (21-13)).

$$R_{dc} = \frac{1}{2d\sigma} = \frac{1}{2 \times 0.5\times10^{-4} \times 5.8\times10^{7}} = 1.724\times10^{-4} \text{ [Ω]}$$

5. The total d.c. resistance of the trace. $R_{trace} = R_{dc} \times \frac{\text{Length}}{\text{Width}} = 0.1724 \text{ [Ω]}$

6. The decay constant a_1 using Equation (21-8). (Note that $a_n = n^2 a_1$.)

$$a_1 = \frac{\pi^2}{d^2\mu\sigma} = \frac{\pi^2}{(0.5\times10^{-4})^2 \times 4\pi\times10^{-7} \times 5.8\times10^{7}} = 5.417\times10^{7} \text{ [s}^{-1}\text{]}$$

7. The inductive voltage drop of the trace [V] after the field has diffused into the trace.

$$v = \bar{a} \times L = 2\times10^{7} \times 1.047\times10^{-8} = 0.2094 \text{ [V]}$$

8. For $t = 1, 5, 20, 50$ [ns], find:

a. The diffusive voltage drop of the trace using Equation (21-33) (include only the first 4 terms in the summation).
Multiply Equation (21-33) by the Length to get

$$V_{trace}(t) = R_{trace}\bar{a}t + 2R_{trace}\bar{a}\sum_{n=1}^{\infty}\frac{1}{a_n}\left[1 - e^{-a_n t}\right]$$

$$V_{trace}(t) = R_{trace}\bar{a}\left\{t + 2\sum_{n=1}^{\infty}\frac{1}{a_n}\left[1 - e^{-a_n t}\right]\right\}$$

$$R_{trace}\bar{a} = 3.448\times10^6$$

$$t = 1 \text{ [ns]:} \quad \left\{t + 2\sum_{n=1}^{4}\frac{1}{a_n}\left[1 - e^{-a_n t}\right]\right\} = 1 + 1.95 + 1.80 + 1.58 + 1.34 = 7.67 \text{ [ns]}$$

$V_{trace}(t) = 0.026$ [V]

$t = 5$ [ns]: $5 + 8.76 + 6.11 + 3.74 + 2.28 = 25.9$ [ns], $V_{trace}(t) = 0.089$ [V]

$t = 20$ [ns]: $20 + 24.4 + 9.11 + 4.10 + 2.31 = 59.8$ [ns], $V_{trace}(t) = 0.207$ [V]

$t = 50$ [ns]: $50 + 34.5 + 9.23 + 4.10 + 2.31 = 100.1$ [ns], $V_{trace}(t) = 0.345$ [V]

b. The contribution to the diffusive voltage drop of the first term omitted in a., i.e., the fifth term.

$n = 5$

$t = 1$ [ns]: $V_{trace(5th\ term)} = 0.003$ [V]

$t = 5$ [ns]: $V_{trace(5th\ term)} = 0.005$ [V]

$t = 20$ [ns]: $V_{trace(5th\ term)} = 0.005$ [V]

$t = 50$ [ns]: $V_{trace(5th\ term)} = 0.005$ [V]

c. The instantaneous impedance of the trace, i.e., v/i. ($\bar{a} = 0.02$ [A/ns])

$t = 1$ [ns]: $\quad \dfrac{v}{i} = \dfrac{0.026}{0.02 \times 1} = 1.3 \quad [\Omega]$

$t = 5$ [ns]: $\quad \dfrac{v}{i} = \dfrac{0.089}{0.02 \times 5} = 0.89 \quad [\Omega]$

$t = 20$ [ns]: $\quad \dfrac{v}{i} = \dfrac{0.207}{0.02 \times 20} = 0.52 \quad [\Omega]$

$t = 50$ [ns]: $\quad \dfrac{v}{i} = \dfrac{0.345}{0.02 \times 50} = 0.35 \quad [\Omega]$

Problem 21-2   Assume a geometry such that one-sided diffusion is the best approximation. Let the thickness of a trace be 0.75×10^{-4} [m] and the material be copper. What maximum value of n should be included in the diffusion model to include all the terms with relaxation time greater than 1 [ns]? That is, find the largest value of n such that $1/a_n > 1$ [ns] .

Solution:

Use Equation (21-8):

$$a_n = \frac{n^2\pi^2}{d^2\mu\sigma}$$

$$n = \frac{d}{\pi}\sqrt{a_n\mu\sigma} = \frac{0.75\times10^{-4}}{\pi}\sqrt{10^9 \times 4\pi\times10^{-7} \times 5.8\times10^7} = 6.45$$

Max. value of n needed is 6.

Problem 22-1 Compare the diffusion rates into a slab and a cylinder. That is, let $d = 0.5\times10^{-4}$ [m], for both a slab and a cylinder. Assume they are both copper. Calculate the first five values of a_n and $1/a_n$ for the slab and the cylinder; i.e., fill out Table S-8.

TABLE S-8
Problem 22-1 Cylinder and Slab Compared.

n	Cylinder diffusion model			Slab diffusion model	
	j_n^a	a_n [s-1]	$1/a_n$ [ns]	a_n [s-1]	$1/a_n$ [ns]
1	3.83171	8.06×10^7	12.4	5.42×10^7	18.5
2	7.01559	2.90×10^8	3.70	2.17×10^8	4.62
3	10.17347	5.68×10^8	1.76	4.87×108	2.05
4	13.32369	9.74×10^8	1.03	8.67×10^8	1.15
5	16.47063	1.49×10^9	0.672	1.35×10^9	0.738

a. Zeros of J_1 from Abramowitz and Stegun, Ref. V-1, p. 409.

REFERENCES

Abramowitz, Milton, and I.A. Stegun, *Handbook of Mathematical Functions*, National Bureau of Standards AMS 55, U.S. Government Printing Office, 1970.

Anderson, E., Z. Bai, C. Bischof, J. Demmel, J. Dongarra, J. Du Croz, A. Greenbaum, S. Hammarling, A. McKenney, S. Ostrouchov, and D. Sorensen, *LAPACK Users' Guide*, SIAM, 1992.

BoardSim User's Guide, HyperLynx, Inc., P.O. Box 3578, Redmond, WA 98073-3578, 1995.

Dongarra, J.J., C.B. Moler, J.R. Bunch, and G.W. Stewart, *LINPACK Users' Guide*, SIAM, 1979.

Dwight, Herbert Bristol, *Tables of Integrals and Other Mathematical Data*, MacMillan, 1947.

Dworsky, L.N., *Modern Transmission Line Theory and Applications*, Robert E. Krieger Publishing Company, 1988.

Everitt, W.L., and G.E. Anner, *Communication Engineering*, McGraw-Hill, 1956.

Faché, Niels, F. Olyslager, and D. De Zutter, *Electromagnetic and Circuit Modelling of Multiconductor Transmission Lines*, Oxford, 1993.

Finkbeiner, Daniel T., II, *Introduction to Matrices and Linear Transformations*, W. H. Freeman and Company, 1978.

Golub, Gene, and C.F. Van Loan, *Matrix Computations*, John Hopkins University Press, 1989.

Gradshteyn, I.S., and I.M. Ryzhik, *Table of Integrals, Series, and Products*, Academic Press, 1980.

Granzow, K.D., and D.E. Jones, *Close-In Transmission-Line EMP Coupling with User Manual for Program TWODIM*, AFWL TR-74-321, Air Force Weapons Laboratory, KAFB, NM (also DC-FR-2224, The Dikewood Corporation, Albuquerque, NM), Dec. 9, 1974.

Jackson, J.D., *Classical Electrodynamics*, John Wiley & Sons, 1975.

Johnson, Howard W., and M. Graham, *High-Speed Digital Design, A Handbook of Black Magic*, Prentice Hall PTR, 1993.

Morse, Philip M., and H. Feshbach, *Methods of Theoretical Physics*, McGraw-Hill, 1953.

Press, William H., B.P. Flannery, S.A. Teukolsky and W.T. Vetterling, *Numerical Recipes*, Cambridge University Press, 1986.

Richtmyer, R.D., and K.W. Morton, *Difference Methods for Initial-Value Problems*, John Wiley & Sons, 1967.

Scott, E.J., *Transform Calculus*, Harper & Brothers Publishers, 1955.

Smith, G.D., *Numerical Solution of Partial Differential Equations*, Oxford University Press, 1965.

Smith, B. T., J. M. Boyle, J. J. Dongarra, B. S. Garbow, Y. Ikebe, V. C. Klema, and C. B. Moler, *Lecture Notes in Computer Science, Matrix Eigensystem Routines — EISPACK Guide*, Second Edition, Springer-Verlag, 1976.

Smythe, William R., *Static and Dynamic Electricity*, McGraw-Hill, 1950.

Stratton, J.A., *Electromagnetic Theory*, McGraw-Hill, 1941.

Watson, G.N., *A Treatise on the Theory of Bessel Functions*, Cambridge University Press, 1966.

INDEX

Skills in collaborative classroom consultation

As the integration of children with special educational needs into ordinary classrooms progresses, most special needs professionals spend an increasing amount of time in mainstream schools.

Based on materials extensively used on in-service courses with classroom and support teachers, *Skills in Collaborative Classroom Consultation* is a practical guide to the tools and techniques required to work effectively with colleagues in defining goals, allocating responsibility and formulating strategies. Emphasizing the interpersonal factors that contribute to success in working together, Anne Jordan focuses on school problems and hands-on activities to deal with them. She shows how consultative skills can be used to solve particular educational problems and also how the consultant professional can act in a more far-reaching way as an agent of change within an institution.

This book will appeal to special education professionals and mainstream classroom teachers, as well as head teachers and senior management teams involved in implementing a whole school policy on special educational needs.

Anne Jordan is Chairperson and Associate Professor in the Department of Instruction and Special Education at the Ontario Institute for Studies in Education. She is the author of numerous articles on special education.

Skills in collaborative classroom consultation

Anne Jordan

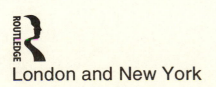

London and New York

First published 1994
by Routledge
11 New Fetter Lane, London EC4P 4EE

Simultaneously published in the USA and Canada
by Routledge
29 West 35th Street, New York, NY 10001

Typeset in Palatino by LaserScript, Mitcham, Surrey
Printed and bound in Great Britain by
TJ Press (Padstow) Ltd, Padstow, Cornwall

British Library Cataloguing in Publication Data
A catalogue record for this book is available from the British Library.

Library of Congress Cataloging in Publication Data
Jordan, Anne, 1944–
  Skills in collaborative classroom consultation/Anne Jordan.
  p. cm.
  Includes bibliographical references (p.) and index.
  1. Educational consultants. 2. Mainstreaming in education.
  3. Interpersonal relations. 4. Special education. I. Title.
  LB2799.J67   1994
  371.9 – dc20   93-5800
                                                                      CIP

ISBN 0–415–03863–4 (hbk)
ISBN 0–415–03864–2 (pbk)

Contents

Illustrations

Preface

Collaboration and consultation are the subjects of a number of recent textbooks and manuals for educators. Many of these texts are designed to equip people with the skills to be effective together in solving the problems of a third party, usually the student. I did not want to write a 'how-to' book, however, which focused on remediating students. Instead, I see classroom-based consultation as one tool in the more interesting endeavour of creating lasting change in educational practice. While the techniques of consulting and inter-personal communication are notably absent from educational texts on resource support and consulting and need to be made explicit, the techniques cannot be acquired and used unless one has a larger purpose. Negotiating with a teacher to provide a specific service for a pupil would be simple and therefore not worthy of documenting, if teachers did not differ from their negotiators in what they want and in what they are prepared to do. At the root of skilled educational consulting is a goal of supplying another person, the client, with the skills to work with children in new and different ways. Sometimes the client is reluctant or even refuses to change. In a broad sense, the consultant is a change agent, seeking to bring about permanent changes in another's actions, and in that person's beliefs about his or her own ability and effectiveness. I try to argue in this book that, as a change agent, the consultant must have a plan of where changes will be effective, and therefore what it is that one is moving towards. My own vision of this goal is that one is working to develop teachers' attitudes and beliefs about their responsibilities towards exceptional children. I see teachers as being at different stages of development; some fully cognizant of the needs of exceptional children and equipped to find ways to surmount the communicational, behavioural or cognitive barriers between them and their students. Others, however, are unable to cope with the diversity of a modern classroom and are therefore resistant to what they see as taking on the added load of an exceptional child's problems. I describe the developmental continuum of teachers towards working with exceptional children in terms of its two extremes or poles. One, the restorative viewpoint, vests the responsibility for learning squarely on the child. The preventive viewpoint sees the responsibility as shared between teacher and child. Our recent work has shown that preventive teachers have a strong sense of their own ability to get through to even the most challenging children, while restorative teachers are much less sure of their own effectiveness, even with non-exceptional students. It seems, then, that part of the role of the consultant, in supplying teachers with the skills to solve students' problems, is to assist teachers to increase their confidence in their own effectiveness as teachers.

Although this book is written from the perspective of the resource or support teacher for special-needs children, the larger goal or plan is to help the support teacher to become a teacher-developer within the context of the teaching staff in a particular school. To some, this goal is presumptuous. What right have we to train support teachers to be change agents on the assumption that classroom teachers need developing? Indeed some schools of thought hold that teachers themselves must generate their own developmental pathways.

I have tried in this book to show that acting as a change agent is not antithetical to these schools of thought. I believe that the professional growth of teachers takes both push and pull: another person can open the possibilities of alternative ways of tackling problems provided it is done with respect and openness and, as Little (1985) says, with recognition that skilful pairs build trust and recognition by deferring to each other's knowledge and skills. Much of the skill of consulting is to demonstrate empathy for the feelings of the other and one's respect for the skills and experiences which that person brings to the consulting event. The approach taken in the book is that of 'process–product' where one aims to create a pre-designed outcome through working together. For that I make no apology, since it does not negate the right of the teacher to seek his or her own fulfilment in the relationship.

The book is offered, then, not as a 'how-to' guide, but as a set of ideas and techniques which consultants and resource teachers might try, adapt and adopt for their own purpose. I have included some case studies and exercises for use in the manner of a workbook, to help people who are reading the book individually or in groups to find their own comfort levels with techniques of interaction.

Acknowledgements

I am grateful to the many people who have offered practical, theoretical and stylistic suggestions in the preparation of this book. Drafts of the book were read and helpful ideas offered in Canada by Helen Osborne, Hastings Board, Bob Kennedy and Barlow Patten, Nipissing Board, Annelli Kerr, Toronto Board, and Patti Shutak now at OISE. Students in several sections of course 4273 at OISE and in professional development courses in the Simcoe County Separate and Wellington County Separate Boards read the book, tried out and wrestled with the suggestions for developing interpersonal skills and improved the sections on contracting, restating client needs, and dealing with resistance. In Britain, a delightful two months at OUDES in Oxford enabled me to discuss my ideas with Neville Jones, and continue an ongoing discussion of consultation with Mel Lloyd-Smith, University of Warwick. My British reviewers offered suggestions for adapting the style of the text to the 'ethos' of British education.

By far my closest associate in preparing the book, and certainly the person who has supported me for the longest time, is Letty Guirnela, who typed, retyped and re-retyped the drafts with careful attention to her high standards and my errors.

I have drawn heavily on the work of Peter Block, whose book *Flawless Consulting* is essential reading for anyone in a consulting role. I have adapted many of Peter's ideas to the educational context, and over the course of the last few years, I and my colleagues have had the opportunity to try out his suggestions in school settings. I am grateful for Peter's permission to adapt his material for educators.

Finally, my thanks are due to my family and friends, and especially Bob, who have supported my writing and indulged my work habits.

Chapter 1

Collaborative consultation in context

A leader is best when people know he exists, not so good when people acclaim him, worse when they despise him, but of a good leader who talks little, when his work is done and his aim fulfilled, they will say 'we did it ourselves'.

(Lao Tse, 565 BC)

The education literature on consultation has burgeoned in the last ten years. The movement to locate all pupils in integrated or mainstreamed[1] classes has given rise to delivery systems in schools in which special-education-trained personnel, formerly deployed in direct remedial work with pupils in segregated or withdrawal settings, are now required to provide support to regular or ordinary classroom teachers. This resource support is increasingly taking the form of collaboration and consultation within the ordinary classroom. Collaboration, as a means for teachers to learn from each other's experience, and consultation, as an in-school, in-service procedure, has recently become prominent in all forms of teacher training and upgrading (e.g., Hunt, 1989; Yarger, 1990).

This book is about the development of teachers' skills during their regular work day. The purpose is to describe some practical approaches to working collaboratively and to equip the consultant to plan for success-ful collaborative projects and to anticipate and deal with possible resist-ance and even sabotage of the projects. The description of consulting skills, however, is couched in a larger plan for creating changes in the way in which classroom teachers have traditionally worked with special-needs and at-risk students. For several decades, special and remedial needs have been the exclusive domain of specially trained personnel who work outside of the ordinary classroom. With the advent of integration or

---

1 Integration and mainstreaming are used in this book to indicate the trend to place special-needs and exceptional pupils in an educational setting which is as close to an ordinary classroom setting as is possible to maximize each child's opportunity to learn. In some cases an optimal learning environment may not be an ordinary classroom, but the term implies that a 'press to integration', or the goal of reducing restrictions due to segregation and categorical groupings, is a component of an optimal learning environment.

mainstreaming, some important changes are taking place in teachers' understanding about their roles and responsibilities, and in their skills for teaching special-needs pupils and for meeting their educational and physical needs. The potential power of collaborative consultation, as represented in this book, is to scale down the traditional way of providing special education, by changing people's expectations about who is responsible for meeting the needs of pupils, and by supporting ordinary teachers to acquire the skills to meet the needs.

A major movement in current teacher training is based on the initial work of Schön (1983, 1987) who claims that the acquisition of skills in the human services professions requires an apprenticeship component in which the novice learner works alongside a master or mentor colleague in the profession. The purpose of the collaboration is to reflect upon and make explicit the personal knowledge and theories which teachers are developing about their art in practice, and to tackle ongoing practical problems as cases-in-point to enrich their reflections. Essentially the collaborative relationship is reciprocal, for each learns to make explicit his or her knowledge and skills in interaction.

As a result, both pre-service and in-service teacher education programmes have begun to incorporate such collaborative endeavours. Research and practices in teaching have involved mentoring and coaching as a series of events in the classroom (Joyce and Showers, 1980, 1988). One-time only or 'one-shot' professional development activities or INSET programmes are unfavourably compared to a series of in-service activities in which participants apply their learning in their daily classroom settings, often in conjunction with a peer with whom they can discuss the form and outcome of their practice (Fullan, 1982).

A large area of research has also evolved in the exploration of teachers' personal professional knowledge (Connelly and Clandinin, 1988; Diamond, 1988). Through encouragement to 'tell their stories' and to provide a spoken narrative of their implicit perspectives about their work, teachers come to know their own competencies and their art. This approach has given rise, among other things, to teacher support groups which meet outside of school to discuss ongoing professional problems in the manner documented by Miller (1990).

This book will not attempt to synthesize these developments, although I do not view them as incompatible with the approach I have taken. Teachers' practice is probably influenced not only by their own experiences, but by the knowledge and practice of others. Therefore, I will attempt to import some ideas and techniques from fields as disparate as psychotherapy and industrial management and to apply them to the field of teacher in-service development. But in so doing, I hope to contribute to Yarger's (1990) requirement that modern teacher education acknowledges that teachers are experts and that their self-determined needs are an essential prerequisite to success. In the chapters to follow, I suggest tools and techniques for consulting teachers to use in order to arrive at a consensus with others about the objectives to be achieved, the division of labour and each person's responsibilities in solving day-to-day educational problems. The book also discusses how to deal with resistant, entrenched colleagues, how to work at changing the attitude of others, and even how to confront people bent on sabotaging your joint endeavours. I do not intend, however, to offer a guaranteed prescription for how people should behave in consulting situations. I am attempting here to offer some suggestions, tips and techniques which you might try in your own work setting, and if useful, incorporate into your own professional repertoire. More than that, I hope that you will derive a framework and rationale for the consulting which you undertake. I anticipate that you will wish to adapt or modify much of what I propose, if only because your style of interacting, your personality, and indeed your language are unique to you. I therefore see one purpose of this book as that of providing some tools of the teaching trade which, upon reflection and with use in your own setting, you may wish to adopt and adapt.

In an attempt to make the book as practical as possible, I have not discussed the theoretical context, no matter how interesting, if I thought it did not have direct practical application. Idol and West (1987) and West and Idol (1987) have already provided much of this context.

In assisting you to generate an overall framework in which to conduct your daily work, and a rationale for your selection, I favour an approach to the school-based delivery of services[2] to students which is here termed 'prevention'. As will be discussed, its focus is on the establishment of a school ethos in which all staff members share responsibility for all pupils, regardless of the differences of each child from the norm, or the causes of such differences. It is an approach in which the consultant or support teacher defines his or her role and then sets out to model that role and create expectations among staff about how to collaborate in a way which reflects a shared responsibility. The prevention approach, its underlying assumptions about pupils and teachers, and the means to implement it are offered, once again, with anticipation that local school structures, staffs and social contexts will result in different forms of implementation and outcome. Prevention is offered as a framework in which you might review your current role, take stock of your strengths and those of your

2 The term service delivery, for which there may be no British equivalent, is used in American and Canadian schools to indicate the organization of staff responsibilities for special-needs pupils, and the steps to be taken to admit new pupils to special-education provisions or to demit pupils who are to be served wholly or in part in the ordinary classroom. 'School-based delivery models' vary from school to school as a function of the philosophy and beliefs of the staff and administration, the availability of additional helpers, the composition of the pupil population, and so forth.

colleagues, and identify areas for change. Much of the book proposes the techniques and skills which might be useful to you in setting out upon a course of change, which ultimately could improve the delivery of services to pupils throughout the school.

I am inviting you then to consider how your tacit and explicit knowledge about teaching, and your beliefs about your purpose as a colleague consulting to fellow adults, might be linked to how to structure your role and how to be effective in implementing it in the broader context of school improvement.

The prevention approach is a way of considering the special needs of pupils. It will be used as the context for the discussion of consultation in this book. One aspect of prevention is often termed 'pre-referral intervention' in the North American literature. The resource teacher assists the ordinary or regular classroom teacher to deal with minor learning and behavioural problems of pupils before they become so severe as to merit referral of the pupil to a special programme. The purpose of this type of intervention is preventive in that it enables the teacher to meet the pupil's learning needs before performance deteriorates to the point where the pupil can no longer remain in the classroom. Prevention implies more than early intervention, however. Later in this chapter, the beliefs and attitudes of teachers towards their responsibilities in prevention will be considered as a part of the issues which the resource consultant may need to address.

The second aspect of the prevention approach that can draw upon collaborative consultation is the complement or flipside of the first. It commences when a pupil is already designated as having special needs, but is being integrated for a portion or for all of the school day into the ordinary classroom. Increasingly, integration is seen to require more than the placement of a child in the classroom. As Thomas (1986, p. 22) states, 'with the best will in the world, it does not seem possible for the class teacher on her(his) own adequately to meet the needs of children who are experiencing difficulty when they are part of a large class. She(he) will need effective assistance.' Such assistance takes the form of support, resource, ancillary and peripatetic staff, who will work within the context of that classroom to enable the teacher to address the needs of special, at-risk and ordinary children. Integration is viewed as part of the prevention approach here, because the purpose of collaboration is to ensure that the child's gains which were previously made with special-needs teachers are sustained in the ordinary class, and built upon in the context of the ordinary classroom teacher's programme.

In both the pre-referral and the integrative cases, the resource teacher will draw upon two types of skills; the technical skills of assessment and designing an intervention which are applied to the pupil's difficulties in the context of the teacher's classroom, and the interpersonal or consulting skills. The consulting skills of contracting, reporting and developing an action plan will form the 'how-to' section of this book, in the context of prevention and integration. These skills will enable the support teacher to collaborate with the ordinary teacher, to ensure that the teacher is successful in meeting the needs of the special-needs and at-risk pupils and the class as a whole.

There are at least two dangers inherent in the application of collaborative consultation to classroom problem-solving. The first danger is that, by promoting collaborative consultation as a method for increasing mainstreaming, teachers will inevitably reject it because other problems of mainstreaming will be left unsolved. Indeed, effective consultation in ordinary classrooms should lead to a decrease in the referral rates for withdrawal or 'pull out' procedures, and segregated, remedial and special-education programmes. The decrease in demand for the removal of pupils from ordinary classrooms will be a slow process, however, as the transfer of skills through collaboration enhances teachers' comfort levels and confidence to work with a wider range of pupils.

Collaborative consultation should not imply that all pupils will eventually be served in the ordinary classroom. It is not a substitute for segregated or other remedial programmes, and should not be equated with mainstreaming as a process. Ideally, a variety of placement options exist which range from segregated to ordinary class placements. It is the criteria used to select these options, not the placements themselves, which may be affected by the consultative process.

The second concern is that consultation skills cannot be learned in isolation. They require that the consultant has a purpose and a direction to which the skills are applied. One purpose may be to introduce or refine the manner in which pupils are mainstreamed or integrated in a school, but not necessarily so. In this book, I will argue that the use for the consultative skills which are outlined is one which enhances the functional objectives set by ordinary classroom teachers for all of the pupils in their care. Consultation, as a tool for enhancing classroom practice, has three goals: to solve an immediate problem about a learning situation as defined by a colleague (ordinary classroom teacher or other professional or parent), to assist the colleague to master the skills and knowledge to deal with similar problems in the future, and ultimately to change the way in which that person works.

Consultation skills can be applied to assist others to solve educational problems. But they can also be applied to make overt the perception of teachers (and others) about their roles in adapting instruction to meet pupil needs, and to achieve a consensus about the nature and purpose of their roles. This requires that consultants view themselves as change agents within their schools. They will need to make an assessment of the current attitudes and beliefs of the staff towards pupil differences and difficulties. They will then be in a position to plan the delivery approach which their school might adopt and to work towards implementing the approach through consultation and collaboration with administrators, parents and other education professionals.

Before describing the specifics of consultation, therefore, the reader is asked to consider the purpose and the plan for which consultative skills are to be used. The essence of consultation is to give away expertise, so that others will be able to assume ownership of their successes in dealing with the problems of learning and behaviour of the children in their charge. This implies a turnaround in many schools in the expectations which both teachers and parents hold about the ways in which they provide instruction and resources to children.

In this book, the subject of our discussion, the consultant, will be viewed from the perspective of the support (British) or resource (North American) teacher role, not because consulting skills are most applicable to this role, but because this role is used to illustrate consultation skills as a case in point. However, consultation skills should not be considered as limited to this person's work. Special- and remedial-education teachers who supervise segregated classes of pupils are increasingly assuming a consulting role as they negotiate with ordinary classroom and subject teachers to align the instruction which they provide with that offered in the regular classroom. System-level support service personnel, that is, psychologists, speech and language specialists, social workers, and others deployed directly from the central office of the school system or LEA, also find themselves increasingly involved in educational consultation. Psychologists no longer merely pull pupils out of classrooms for testing, but focus on pupils' difficulties within the classroom setting (Reschly, 1988). Speech and language pathologists, itinerant resource teachers for hearing-impaired, visually handicapped, and culturally different pupils, system-level curriculum leaders and programme specialists all spend a part of their day consulting with teachers to provide follow-up training for pupils within the regular programme. School librarians and teacher assistants are increasingly aware of the need to provide resource support to fellow teaching staff. The skills to be discussed are applicable to any professional who works collaboratively with teachers. But for simplicity, 'you', the audience, will be addressed as a support teacher, and your colleague or client, with whom you are working, will be addressed as 'teacher' or 'client'.

In this book, the term 'support teacher' is used as a generic term to indicate many different educational roles. In North America, most school boards deploy one or more resource support teachers in each elementary school. This person is designated by one of many acronyms; Special Education Resource Teacher (SERT), Methods and Resource Teacher (M and R), Developmental Teacher, School-based Support Teacher (SBST), etc. In Great Britain, also, the remedial or peripatetic teacher and teachers who formerly taught segregated classes are increasingly becoming support persons for ordinary classroom teachers (Bines, 1986; Garnett, 1988; Thomas, 1985, 1986). Acronyms include Special Needs Assessment and Support Teachers (SNASTs), Teacher Consultant (TC) and Designated Teachers (Hodgson *et al.*, 1984; Heward and Lloyd-Smith, 1990).

There are some differences in special education provisions, however, between Great Britain and North America. In Great Britain, only 2 to 3 per cent of pupils have been statemented or designated as having special needs, although Warnock indicated that 20 per cent of children in ordinary classrooms may have required special help at some stage in their school lives (DES, 1978; Gipps *et al.*, 1987). In Canada, and in North America in general, approximately 12 per cent of pupils are designated as 'exceptional', and are eligible for special education services. In addition, as many as 20 per cent of children might be termed 'at risk' of failing to reach their academic potential. With the popularity of mainstreaming, many segregated special-needs schools and special classes in ordinary schools have been closed, and special-education teachers, trained to teach in segregated settings, are finding themselves redeployed in support roles in regular or ordinary classrooms.

A second major difference between British and North American schools can be seen at the secondary level. American and Canadian schools have placed little emphasis on the secondary school support role, preferring to create learning centers staffed by teachers who tend to work directly with pupils who have either been referred by teachers, or who themselves select some remedial assistance. In Great Britain, the secondary schools, and in particular the comprehensive schools, have chosen to construct a more collaborative approach to meeting pupil needs through the pastoral care system. Staff work together to ensure that pupils' progress is monitored through a cooperative network. Designated support staff may then be drawn upon through the pastoral care network. Postlethwaite *et al.* (1986) describe how the majority of secondary and middle schools in Oxfordshire, for example, combine mixed-ability pastoral groups with support in ordinary classes. Special-needs staff are deployed in Oxfordshire schools on a ratio that varies from 1 per 200 pupils to 1 per 500 pupils. Staffs of North American secondary schools might consider borrowing from the innovations of the British secondary system and, in turn, the resource consulting models in some Canadian and US elementary schools might be informative for British primary school support staff.

The function of the support teacher, then, is evolving in both North American and British schools. There is considerable variation from school to school and school system to school system as to how these teachers operate. The proportion of time which they spend in direct service to pupils, compared to indirect service through consultation to ordinary teachers, depends largely on the philosophy and practices operating in that school, and the assumptions and beliefs of the teachers and administrators about the role (Wilson and Silverman, 1991).

It is assumed that you, the reader, will be working in some capacity as a consultant to teachers. You have an interest in learning more about collaborative consultation, and in honing your skills. These topics are discussed in Chapters 3 to 5. You will, however, be attempting to bring about changes in your school or school system for which collaboration and consultation are tools. You have in mind some objectives, and a philosophy about the ways in which school personnel could be more effective in meeting the needs of pupils, exceptional and otherwise. You are a change agent (Fullan, 1982; Fullan and Stiegelbauer, 1991; Miles *et al.*, 1988) and you face a variety of staff, some of whom are supportive of your direction, and others who are disinterested, or even resistant to the changes you propose. Not only do you wish to increase the effectiveness of your consultations, you wish to do so by influencing teachers' perceptions of their own roles in working with you, and ultimately in working with their pupils. In order to do this, you have a goal in mind for the form which staff collaboration will take, and a plan of how to get there.

The structure of this book, therefore, is to explore each component of your master-plan in order. First the issues involved in educational provisions for special-needs children are considered. This leads to objectives which can be translated into a plan for the educational provisions which you hope to achieve within your school or group of schools. The specific skills for bringing about consultative interactions are then proposed. Finally, we will explore the change process and ask how you set about reaching your objectives from your current vantage point.

RESTORATIVE AND PREVENTIVE BELIEFS ABOUT THE RESOURCE ROLE

A teacher is concerned about a child. After trying to adapt his instructional approach to the child and seeing no improvement over three weeks, the teacher calls in the resource teacher and asks for a psychological and educational assessment. Asked what he would like to get out of the assessment results, the teacher replies that he wants confirmation that the child has a learning problem. The child is tested and has a score of ninety on the performance and seventy-five on the verbal components of an IQ test. She is three years behind norms for her age group in reading and two years behind in mathematics. The child is confirmed as exceptional, and she is recommended for placement in a remedial group in the mornings, and integrated with her agemates in the afternoons for music, art and physical education.

Teachers with these restorative viewpoints may justify the decisions made about this child in several ways:

(a) I suspected she had a learning disability/was a slow learner, and I know that with these problems she'd never be able to keep up in my class. She needs some individual help to (either) modify the programmes to her level and rate (or) help her make up the gap so she can get back on track. In my class, she'd demand so much of my time just to keep her going that it wouldn't be fair to the other children.

(b) I suspected she was a learning disabled/educationally sub-normal child. Children like her deserve to get extra help and specialized attention. She should be allowed to remain as much as possible with her friends, but only for those subjects in which she can compete on their ground. She shouldn't be stigmatized by failing in my classroom, so I referred her on to someone who can help her.

These two explanations reflect different values but result in the same outcome. Both assume that the child's problems reside within her. Both are categorical, and depend on specialized curricula to meet the needs of children like her. However, if one agrees with Dessent (1987) that special needs should be viewed as relative, non-categorical and as created in interactions with the current context, both of these perspectives are flawed. Consider a third teacher, who, when asked what he expects to get out of the assessment replies as follows:

> I'm concerned that she's falling behind. While she's having difficulty with comprehending the words she reads, I'm not sure how much of that is due to her ability level, and how much to the emphasis her previous teacher placed on accurate word identification. By analysing her reading miscues, the resource teacher could give me a handle on how she understands the task of reading. This will give us a start to explore other ways of teaching her, to fill in those conceptual 'gaps' she seems to have.

In this example the request for the help of the resource teacher takes a very different focus. Instead of requesting a confirmation of the deficit, the teacher seeks additional ways to tackle the child's difficulties. The requested assessment is not normative but formative; the yardstick is the current profile of the child's learning ability and 'miscues' (Goodman, 1976), in the ongoing flow of curriculum goals set by the teacher. This teacher takes some responsibility for the child's progress, and he will not conclude that the child has a 'disability' until he has tried every method possible to fulfill his instructional obligations to their partnership.

Belief systems are fluid and intangible. Very few professionals have one fixed perspective about exceptionality. One suspects that most teachers vary in the extent to which they will invest their time and resources to explore the learning potential of their pupils. When their personal and professional resources run low, the 'explanation' for the problem shifts to the pupil. Many teachers will articulate beliefs about the necessity of sharing responsibility for instruction with the pupil. It is in practice, under the heavy task demands of day-to-day teaching, that these beliefs become compromised. It is simplistic to talk about the 'match, batch and dispatch' restorative delivery models in schools as being due to the negative attitudes of staff and parents. Changing the delivery model depends on making it possible for the deeper beliefs of teachers to find expression in their actions. Many of them require a tacit permission to use their skills effectively, and the time, resources and conditions to allow them to do so.

From the perspective of the resource teacher, in the average school, there will be at least two groups of teachers. There are those who feel so out-of-control of their own time and resources that their main requests of you will be to remove the lowest-functioning child from their classes in order to make things more manageable. Usually, removal of this child will leave another at the bottom of their list and he or she will be the next candidate for referral. Their rationale will be 'restorative' and categorically-based. The second, 'preventive' group will request your help with working more effectively with a particular child. Unless the help they receive permits them to take control of both the child's achievement and the progress of the rest of the class, these teachers will gradually cease to ask for help. It is this second group of teachers with preventive beliefs about their role who offer the greatest potential for commencing a collaborative service delivery model in a school. You will respond differently to each group. For restorative teachers, the task of the resource teacher is to negotiate a sharing of the responsibility for the problem. Initially, the sharing may be unevenly weighted upon the resource teacher, as is the case when the resource teacher complies with the expectation to remove the pupil from the class and 'take over' that child's progress. In the case of the 'preventive oriented' classroom teacher, your task will be to negotiate the conditions and resources which you will provide in order to free the classroom teacher to work directly with the pupil. By starting with this latter collaboration, you are able to model your role with as much publicity as possible, in order to start the process of forming staff expectations about the resource possibilities open to them. In Figure 1.1 this idea is presented in a diagram. By modelling the services and skills which you hope to provide, the teacher-clients begin to shape their requests to fit the role. This in turn leads to more opportunities for you to model the role. One of your earliest objectives therefore is to design a clear plan of the type of role which the resource teacher will ultimately provide.

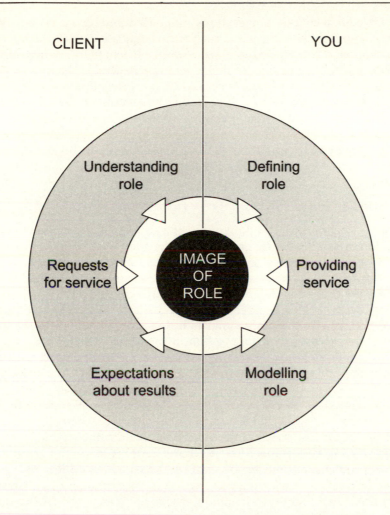

CLIENT   YOU

Understanding   Defining
role   role

Requests   IMAGE   Providing
for service   OF   service
                  ROLE

Expectations   Modelling
about results   role

Figure 1.1 Consultant and client define and shape the consulting role

Any contract with another staff member will be designed to enhance and develop the role. A contract which falls short of that role will be the product of a negotiation in which other goals are being set. For example, the resource role may be defined as working only with other staff and parents, and with no direct service to pupils. In order to achieve this, the resource teacher may compromise by withdrawing a child from the class of a teacher who requests it, but with agreement to continue the withdrawal arrangement only long enough for the resource teacher to report back about the child's learning capabilities and needs, and on condition that the ordinary teacher works one-to-one with the child on a specific objective while the resource teacher conducts the rest of the class.

In subsequent chapters suggestions are made about how to work collaboratively with colleagues, including how to negotiate the initial contract. These chapters are intended to assist resource teachers to put into effect the plans which they design for their role in their school. The first step in designing the plan is to take a census of the prevailing attitudes and beliefs of the staff, and of how these beliefs are expressed in their day-to-day activities. The next step includes designing the roles which the resource teacher, administrative and teaching staff will ideally play. The third stage will be to 'sell' the potential for changing in this direction, a lengthy task in which role modelling will be a major tool.

THE GOALS OF CONSULTATION

One major goal of the consultative role of a support teacher, therefore, is to influence the attitudes of a considerable portion of colleagues towards a more pragmatic perspective about pupil difficulties. Discussions in the research literature about the attitudes of professionals towards the handicapped have tended to over-simplify the task. Most people, including teachers, do not hold negative or prejudicial attitudes towards handicapped people, and therefore it makes little sense to try to train them to feel more positive by exposure to handicapped people, or by exercises designed to tackle their hostility. Teachers may feel reticent to have exceptional pupils in their classrooms, and justifiably so. The feedback which they have received about their competence to deal with differences has been limited by the prevailing expectations of the dual-track delivery system in most schools. Furthermore, if classroom teachers have never had their categorical, medical perspective of handicap challenged, then they too are likely to focus on the handicapping condition, and the practical problems of time, energy and resources which it presents. Only by rethinking their expectations about this child, and their own competencies as teachers will they be able to focus on the many other characteristics of that child which lie behind the handicap. From a different perspective there are new ways to tackle the very real problems of working with children with learning difficulties, and of meeting the needs of the other children in the class. This will be the topic of Chapter 7.

Most teachers are willing to acknowledge that the exceptional child has a right to all the classroom resources and services available to the average child. They just don't know how to go about providing them. They may settle therefore for a physical presence of the child in their classroom, in the hope that some social learning will brush off in the course of the modelling behaviour of peers. Or they may attempt to modify parts of their programme or to lower their expectations about this child's success relative to those of the other pupils. Usually, someone else deals with the educational needs of the child in a setting removed from the classroom.

Few classroom teachers express satisfaction with this arrangement. They state that exceptional children integrated for social reasons are 'missing out' on full classroom participation. They express concern that, in the withdrawal setting, exceptional children experience a totally different curriculum from their own students in the ordinary classroom. They have little or no say in the objectives of the individual programme and little information about how it relates to their own curriculum, and as a result, the exceptional students receive a fragmented education. There is little continuity in the teaching of concepts and ideas and, consequently, no sequence to understanding and communication. Teachers may also complain of token collaboration in which support teachers and health professionals involved with the child keep the others informed about what they are doing, but there are no commonly agreed-upon objectives for the child, or criteria whereby all can agree that the child is progressing. Progress is a benchmark which each keeps in mind in working with the child, but there may be little or no reporting of progress to others working with the child, and therefore the gains are not generalized. This phenomenon is common in our schools in Ontario; each professional works as intensively as possible with their special-needs pupils, often in separate locations, but frequently there is no overall shared understanding about how this effort should impact upon the child's other learning needs. Collaboration and consultation are conducted only if there is enough time left at the end of the lessons, a rare occurrence. Unable to provide this carry-over alone, the child does not sustain the gains he or she has made, and often slips back into needing more remedial help shortly after a period of apparent success. It is frustrating to teachers to try so hard to assist the child, then see the fruits of their labour lost so quickly. Part of the potential for working collaboratively is that such fragmentation of service delivery will not occur, and each professional will be working in a team to create, transfer and sustain the pupil's gains in progress.

Yet working together implies exposing one's own efforts to the scrutiny of others. It also implies the loss of autonomy in deciding what is best for the child. Smith (1982) interviewed classroom teachers about working collegially with a resource teacher in order to integrate an exceptional pupil into their ordinary classes. The list of fears which they expressed illustrates the bind in which the teachers find themselves. They fear that asking for extra help implies their incompetence. They fear that this incompetence will be relayed to their superiors or to the wider profession. Their understanding of the 'specialness' of exceptional children gives rise to the fear that they will have to work harder to learn new skills and remedies. They fear that extra work will detract from their own priorities; that the 'experts' will not pay attention to their own needs and perspectives. It is this fear of embarking upon the unknown, of being vulnerable to others and of losing control of the consequences which makes teachers reticent to work with others in order to teach exceptional children. One simple way to resolve the dilemma is to view the difficulties of the child as a characteristic resident within the child. The child's inability to sustain gain can then be 'explained' in terms of the deficit assumed to be within him or her. Failure to improve is due to a condition which is beyond the scope of being restored by external interventions. This notion that the difficulty resides within the child is termed by Sarason and Doris (1979) as 'the search pathology' within the child, and gave rise to Warnock's criticism (DES, 1978) that schools treated special-needs pupils along the lines of medical management.

It is the thesis of this book that the major task of the resource or support teacher is to tackle and change these 'restorative' attitudes of teachers which focus on student 'deficits' and the underlying fears of the teachers, not by confrontation, but by a gradual process of change, through solving day-to-day difficulties. As Thomas (1985) states, the process is accomplished by enabling teachers to allow them to experience success. The process is slow, however, and in the average school staff will possibly take three to five years before they are completely committed to a changed perspective (Goodlad, 1975).

Collaborative consultation, then, can be a process for teacher development which takes place within a larger plan. The plan itself defines the delivery of services to teachers and pupils. It deemphasizes the 'gate-keeping' or restorative models of special-education delivery in which children do or do not qualify for special provisions according to the degree of their 'inherent' deficit. The term 'restorative model' is used for such delivery systems since it emphasizes the child's deficits and the provision of services by specially trained professionals to restore the child to the mainstream. The plan of services in the school is based on a 'preventive model' of service delivery. The understanding in this model is that the delivery of services is applied to ordinary classroom teachers to assist them to prevent exceptional and at-risk pupils from sliding further behind their peers by coordinating their efforts with those others at school and at home who will be able to support the child's progress. It acknowledges individual differences and developmental growth in pupils, and therefore the diversity of needs of all pupils within the ordinary classroom. The purpose of the plan is to assess the current attitudinal climate and delivery system in a school, to establish goals for a preventive model, and to work through various means, including collaborative consultation, to achieve those goals.

The goals are slowly achieved, and the resource teacher, acting without direct services to children, also needs resources and support for him or herself – networks of colleagues, mental health breaks, his or her own support systems. The change process is lonely and often there are setbacks. But, as will be discussed in the final chapters, by monitoring changes, the staff and the resource teacher can be helped to see the impact of their work evolve.

Chapter 2

Elements of alternative school delivery models

But I was thinking of a plan
to dye one's whiskers green.

(Lewis Carrol, *Through the Looking Glass*)

A school staff may go about choosing its own approach for the delivery of special education and remedial services in various ways. The school board may mandate the approach. An individual teacher or administrator or an entire school system may subscribe to one approach and undertake a 'pilot' project with a few colleagues, a school or group of schools. Staff members may decide that priority must be given to revising their current approach, and may seek resources to help them. Whatever the impetus for re-examining the delivery approach in a school, it will be assumed that you will have a central role in the planning and implementation of any changes. In this chapter, I will explore some of the options available to you and to a school which is planning a change. Your viewpoint and that of the resource teacher (or support, peripatetic or remedial teacher) will be taken, not because you should be the instigator of the new delivery system, but because you will probably have the day-to-day tasks of planning, implementing and monitoring the changes at the classroom level.[1] Eventually, all other members of the staff will be involved. School administrators are a crucial part of the success of any change (Fullan, 1982; Fullan and Stiegelbauer, 1991) and without their active support the likelihood of success of the project is greatly reduced. The involvement of the school's head teacher, vice-principal and department heads is therefore important at the outset of the plan.

In this chapter, a thumbnail sketch of the restorative compared to the preventive delivery approaches in an elementary school will be given, followed by a consideration of aspects that differ at the secondary level.

---

1 Many of the principles discussed will also be applicable to your role as a system-level specialist, or as a health care service provider, school psychologist or curriculum consultant. The teacher is chosen as the subject here to provide some coherence of the school-based cases and examples I will use.

Optional elements are the school-based team and its composition, the proportion of integrated or mainstreamed programmes for special-needs pupils, and the proportion of time which the resource teacher spends working directly with teachers and with pupils. How a school staff designs its delivery system will depend on many factors, including the beliefs and priorities of staff, the climate of the school and its resources and the leadership styles of the head teachers.

In order to arrive at the ideal delivery approach, the school will undergo a series of successive approximations, in which measure of the success of the project will be reviewed, and problems tackled. However, without a vision of an ideal delivery system, one in which the procedures of the school reflect the belief systems of its teachers, changes will not take place. Therefore, it is necessary for a school staff to establish its goals, to assess its current strengths and needs, and to move towards implementation of its plan with feedback from agreed-upon sources of data. All staff will not immediately subscribe to the new approach, of course. It will take push, from school policies and administrative initiatives, and pull, from the apparent successes of those who venture to try collaborative projects. By modelling the type of role which you as a resource teacher will undertake, a gradual shift will occur in the image of teachers about resource support and about their responsibilities towards all pupils, including those who are exceptional and at risk.

ELEMENTS OF THE RESTORATIVE APPROACH

The procedures for offering special education in a school in this approach have a long tradition and are typical of many of the special-education programmes in our schools.

The steps in identification, referral, assessment, placement and programming are indicated in Figure 2.1. The procedures shown here result in certain outcomes.

The educational process is divided into areas of responsibility, each of which is held by specialists, such that an exceptional pupil is the full responsibility of the special-education teachers and support staff. Regular (that is, non-exceptional) pupils are the responsibility of the classroom teacher. Hence, a child must be designated either as exceptional or non-exceptional before it is decided who should be responsible for that particular child's programme. In a variation of this, the pupil may be deemed to be 50 per cent or 25 per cent or some percentage exceptional, such that a portion of the child's school day is the responsibility of the special-education or resource teacher, and the remainder of the day is that of the classroom teacher or teachers. Alternatively, the pupil is identified as needing additional help, in the form of remedial withdrawal or resource help in addition to or as a substitute for regular classroom instruction.

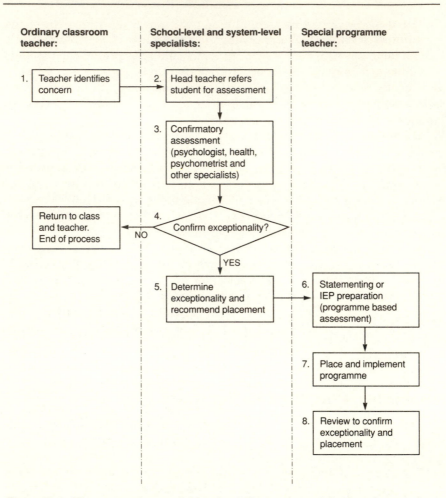

Ordinary classroom teacher:	School-level and system-level specialists:	Special programme teacher:

1. Teacher identifies concern

2. Head teacher refers student for assessment

3. Confirmatory assessment (psychologist, health, psychometrist and other specialists)

4. Confirm exceptionality?

Return to class and teacher. End of process   NO

YES

5. Determine exceptionality and recommend placement

6. Statementing or IEP preparation (programme based assessment)

7. Place and implement programme

8. Review to confirm exceptionality and placement

Figure 2.1 The restorative model for the delivery of special-education services
Source: Wilson, A. (1984). *Opening the door: The key to resource models*. Toronto: Ontario Public School Teachers' Federation.

That portion of his or her programme is the responsibility of the remedial instructor. While it may be initially chosen in conjunction with the classroom teacher, it is not directly linked to the regular curriculum.

The important point to note in this approach is that a pupil's programme is fragmented; the child comes under different areas of responsibility and usually receives quite different programmes in each context, often without any consultation or joint goal-setting between classroom and special-education teachers. This is the most familiar version of 'remedial or resource withdrawal'. Its popularity is based on several assumptions, all of which are open to question:

1 The academic or core programme in the special-education setting has different content, goals and techniques from that of the classroom. A corollary of this is that specialized curricular and programming techniques can be made available only in the special placement.
2 The special-education teacher is trained in the curriculum and programming techniques required by the pupil whereas the classroom teacher is not.
3 The small group or one-to-one instruction in the specialized placement will enhance learning of core skills significantly more than in the classroom setting.

The procedures by which a pupil is assigned to a withdrawal or special-education class under the restorative approach also contain certain assumptions; for example, it may be the classroom teacher's task to inform the head teacher or another resource person about a pupil, if the teacher believes that the pupil is having problems in school, so that the head teacher or resource person may decide whether the problem is severe enough to merit referral for assessment. If, in the opinion of the head teacher or resource person, the pupil's problem is not sufficiently severe, no further action may be taken about the teacher's expressed concern. Consequently, the child will remain in the classroom programme until the problem has increased or the child's progress has deteriorated to the point that the teacher feels justified in bringing the child once more to the attention of the head teacher or resource person. In this situation, two problems arise. First, the classroom teacher is left to cope with the child's early indications of difficulties without any formal means of help or support; if the teacher cannot cope, then the child's performance will slide, and the opportunity to prevent further difficulties will be lost.

Second, if the head teacher or resource person chooses to refer the pupil, the educational and psychological assessment specialists are usually called in. From this point on, the teacher's concerns may be secondary, if relevant at all, to the assessment and subsequent processes, and to the selection of the programme.

In the restorative approach, the task of assessment is to determine whether there is sufficient evidence to ascribe to the pupil a deficit so that the child will be formally designated or 'statemented' or alternatively will qualify for remedial assistance. Assessment must therefore be diagnostic: a search for evidence of deficits. Deficits are usually indicated by normative data; that is, test scores are collected to show that this pupil deviates sufficiently from the norm for pupils of that age and grade level to justify the child's designation as exceptional. In many North American and British school systems the 'rule of thumb' for referring a pupil for assessment is that achievement test scores must be two years behind chronological age norms.

This rule of thumb cogently illustrates the translation of the medical management philosophy into practice. It is the responsibility of students to qualify for additional help and not the responsibility of the system to prevent them from needing it. Teachers are given a clear message that children who do not fulfil the entry requirement for remedial help will not get it. Since teachers care about their pupils, they consequently believe that they are doing their professional level-best for at-risk pupils if they do *not* intervene to prevent their problems from escalating. Doing nothing will reduce the time it takes to ensure that such children qualify for help.

A multiprofessional (multidisciplinary) team or other prescribed admission committee is convened to consider the assessment information gathered about the child. In the case of the restorative approach, the predominant data to be considered are norm-based test scores of educational achievement and psychological and intellectual characteristics. This type of assessment depends heavily on some questionable assumptions.

1 Normative test data reliably and objectively measure the degree of the pupil's deficit; that is, they validly predict a cause (etiology) for which a category of deficit or label can be supplied.
2 The determination of a child's deficit category will enable that individual to be placed with similarly designated children in a specialized programme.
3 A child's performance and behavioural deficits are a part of that pupil's mental makeup such that test scores will accurately reveal them.

Normative test data, particularly IQ scores, are assumed to reflect a long-term characteristic of the child, not of his or her temporary relation to the home or school environment, or his or her past learning opportunities.

Criteria for designation as a child with special needs are understood to be absolute and not relative. As Solity and Raybould (1988) note, the presumption of the definition of a learning difficulty in the UK 1989 Education Act is that a learning difficulty is quantifiable, measurable, and of such a nature that it can be compared with other learning problems, and that differences following such comparisons can be expressed in terms of statistical significance. Solity, like others, seriously questions such presumptions (Gartner and Lipsky, 1987; Reynolds *et al.*, 1987).

Two outcomes of the assessment are possible. Either the child is found not to be exceptional and is returned to the regular classroom, or the child is deemed exceptional and withdrawal or segregated placement is recommended. Consequently, either the classroom teacher is given the full responsibility for that pupil, or the responsibility for part or all of the core academic programming is removed and placed with the special-education-trained teacher.

Under the restorative approach, the classroom teacher is not likely to receive assistance or support for attempting to cope with the needs of the pupil in the teacher's own classroom, when such needs first become evident. Further, the teacher's criterion for referring the pupil becomes that of the teacher's ability or inability to cope with the child's needs. Inability to cope results in the pupil being viewed as exceptional.

In the final stage of the restorative approach, the child is determined to have special-education needs, and a provisional statement or individual educational programme is drafted. The pupil is placed full- or part-time with a special-education-trained teacher. The task of this teacher is to implement the programme, the goal of which is to restore the pupil to an appropriate functional level for reintegration in the classroom.

Subsequent reviews will also consider to what extent the pupil still fits the designation and category of exceptional, and therefore continues to merit placement in that programme. Because 'programme' is interpreted as synonymous with placement (as in 'he's in the learning disability programme'), the review does not need to consider whether the instructional content of the programme is suitable, but rather whether the child still qualifies for the placement. The question of whether the pupil has made reasonable progress towards programme goals is therefore only of secondary importance to the question of whether he or she continues to exhibit an identifiable deficit of the kind that the placement is intended to serve. Further, in view of the severity of the child's original problem and the focus on diagnosis, there is little likelihood that the child will indeed make up sufficient ground to be returned full-time to the ordinary programme or, if he or she does, that the gains will be sustained in the absence of follow-through work with the classroom teacher.

PROBLEMS OF THE RESTORATIVE APPROACH

Several problems of the restorative approach have been mentioned to date:

- The fragmentation of the process; regular and special programmes are separated and specialists in each deal with different parts of the pupil's schedule and day. Their focus is upon the deficit assumed to reside in the pupil, rather than upon the opportunities to learn which could be provided to him or her.
- There is no formal provision for helping a pupil before his or her problems become severe. Under the restorative model, frequently a pupil cannot officially receive even a brief period of withdrawal help unless he or she has been designated as exceptional.
- There is no provision for helping a teacher who is concerned about his or her ability to meet the needs of a pupil. Indeed the teacher is encouraged not to seek help unless there is clear evidence that the pupil's progress or difficulty will merit the term 'exceptional'. In some

schools, this problem has been further aggravated by the tendency of the assessment and statementing process to designate some pupils as exceptional while at the same time recommending that the pupils be placed in the regular classroom programme.

- The expectation fostered in the ordinary classroom teacher is that he or she is responsible for non-exceptional pupils only. Those pupils who fail to reach or who exceed average class performance and behaviour levels should be the responsibility of special class teachers. 'This pupil does not fit in well in my classroom, therefore he needs special education.' The onus is on the pupil to fit the programme and not vice versa.
- Ultimately, teachers and parents are reinforced in the belief that exceptionalities reside solely within pupils, not in the suitability of the programme, students' past experiences or their home, social or learning environments. Furthermore, a learning or behaviour problem is somehow a characteristic of the pupil only. Thus, a learning disability is seen as related to mental makeup and not to learned skills, and an emotional or behavioural problem is viewed as part of the child's characteristics and not of his or her reaction to the current situation at home or school. By practising the restorative approach, teachers may find substantiation for these beliefs, even though the evidence is equivocal at best that certain categories of exceptionality are the result of organic or neurological differences.

One may profoundly disagree with these assumptions. Yet, the procedures which depend on the all-or-nothing designation of a pupil as exceptional seem to prohibit the pupil from having a mild, a temporary or a situationally based difficulty. Reluctant to have a child labelled unnecessarily, many teachers therefore avoid seeking assistance until the pupil's problems are acute.

ELEMENTS AND OPTIONS OF A SCHOOL-BASED PREVENTION APPROACH

The prevention approach differs in several important ways from more traditional models. As its name implies, resources are maximized in the early stages of the procedures for designating a child as exceptional, in the hope of preventing that child from reaching a special-education class. Consequently the formal designation process is delayed until the end of a series of pre-referral procedures, during which all possible steps are taken to accommodate the child's needs in the regular programme. Formal referral to special education takes place only when those steps have been un- successful in meeting the child's special needs.

The work of the special-education staff is not reserved solely for pupils who have already been identified as exceptional. The special-education resource staff also assist the classroom teacher with pupils who have

PRE-REFERRAL PHASE

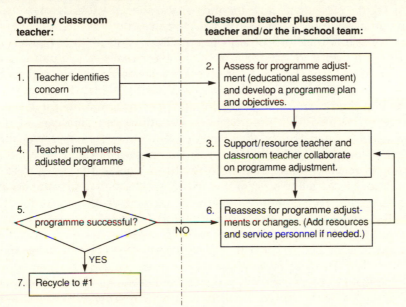

REFERRAL PHASE

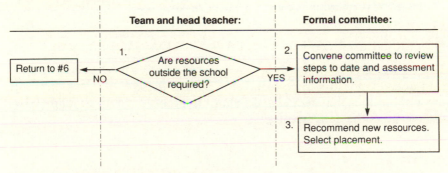

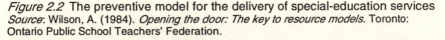

Figure 2.2 The preventive model for the delivery of special-education services
Source: Wilson, A. (1984). *Opening the door: The key to resource models*. Toronto:
Ontario Public School Teachers' Federation.

become a concern, as soon as the teacher first expresses that concern. A
pupil does not have to be severely out of step with his or her peers to
merit the assistance of the resource teacher. The criterion for requesting
assistance is that the classroom teacher feels that some suggestions or
support will be of help in meeting that pupil's needs. Support may take
many forms, including an informal observation or assessment, some

additional learning materials, a second opinion or some short-term instruction of the pupil in or out of the classroom.

As Figure 2.2 shows, the team and/or the special-education resource teacher may have several meetings with the classroom teacher to suggest programme modifications or adjustment. These will include providing resource support and materials, setting objectives and evaluating the outcome of the programme. After several cycles at the pre-referral stage (boxes 1 to 7 in Figure 2.2), the classroom and resource teachers may decide that the classroom teacher's initial concerns about the pupil have now been resolved, that programme adaptations can take care of the pupil's needs. Alternatively, they may decide that the resources available in that school are inadequate to meet the pupil's needs. If the latter decision is reached, they propose to the head teacher that referral to a more formal assessment is warranted so that an alternative placement can be sought which offers more suitable services and programmes. By this point they have accumulated a great deal of information on what programmes and techniques were tried, which were successful, how the pupil best learns and so on. This material forms the basis of the assessment on which further programme and placement selection can be made. The assessment process, then, is programme-based. As will be discussed in Chapter 4, it is structured by the questions which arise from teachers' previous attempts to meet a pupil's needs. It is not a pre-designed test battery, but formal and informal tools which have been selected in light of collective professional concerns and previous teaching attempts.

DIFFERENCES BETWEEN THE TWO APPROACHES

There are distinct differences between the preventive model and the more traditional restorative approach. Table 2.1 lists these. Differences occur in the assumptions that people make about the procedures to be followed at each stage of the delivery process.

Referral

In the preventive approach, the limitations of the school's resources eventually lead to the referral of a pupil as exceptional. In the restorative approach, it is presumed that the deficits which reside in the pupil are sufficiently deviant to merit referral.

Programming

In the preventive approach, the goal throughout is the identification of an appropriate programme for a pupil which will assist him or her within the context of the classroom curriculum. In the restorative approach,

Table 2.1 Comparison between restorative and preventive models for delivery of special-education services

Restorative model	Preventive model
Roles	
Fragmentation of pupil's schedule Each professional has defined tasks	Consultative: Pupil's schedule is designed by professionals as collaborative team
Delivery commences after pupil's problem is severe	Delivery commences before problem escalates
Statementing or designation commences process	Formal designation to special education is final resort
Classroom teacher excluded from responsibilities and resources	Classroom teacher shares responsibilities for programmes and receives resources
Resource teacher, if any, works with pupils	Resource teacher provides service to both pupil and classroom teacher
Assessment	
Assumes deficits which can be labelled	Assumes learning strengths and weaknesses requiring programme adjustments
Degree of deficit is basis for referral	Limitations of school's resources is basis for referral
Purpose of assessment is to diagnose deficit by category	Purpose of assessment is instructional: to identify programming interventions and services
Placement	
Placement is outcome Programme means placement	Programme adjustments are outcome Programme means instructional content, techniques, services, etc.
Assessment for programming follows placement	Assessment for placement delayed pending outcome of programme adjustments
Placement is location designated for each type of exceptionality	Placement is location best suited to carrying out appropriate programme
Integration/demission	
Integration refers to proportion of pupil's time in regular classroom	Integration refers to extent of modifications to pupil's regular classroom programme, including resource support
There is no provision for a bridge for pupil who is reintegrated	Resource support is available to pupil and teacher following reintegration

appropriate programmes are the responsibility of the special-education teacher in the withdrawal or segregated contained setting, at least in those subject areas where the pupil's difficulties (or gifts) are more apparent. Attention to programming only commences after the statementing or designating process and placement have been completed.

Integration

In the restorative approach, integration refers to that portion of a pupil's time in the regular classroom. In the preventive approach, on the other hand, integration refers to the extent of modifications to the pupil's programme that can be carried out in the classroom setting, in conjunction with the curriculum and objectives set by the classroom teacher. Integration or mainstreaming is also a more dynamic concept in the preventive approach, in that it includes heading off difficulties through early intervention in the regular classroom.

Demission and reintegration

The two approaches offer different possibilities for demitting pupils from special education; that is, returning them to the ordinary classroom after they have spent time in special education. The restorative approach does not provide a bridge for a pupil between the special and ordinary settings. If the pupil has not fully caught up with his or her peers by being in the small withdrawal programme or special class, or if the child has caught up but will have difficulty maintaining the gains in the large classroom, extra demands will be placed on the classroom teacher for which there is no formal means of assistance or support. On the other hand, in the preventive approach the resource teachers may assist the classroom teacher to reintegrate pupils who are leaving special education, or who are still in special education but are increasing the proportion of time spent in the ordinary classroom.

The same resources, programme adjustments and support used for the prevention of problems detected early may also be applied to assist a pupil to reintegrate and to maintain his or her gains in a regular classroom. For example, the resource teacher might work with the pupil in completing the ordinary classroom teacher's assignments, in supplementing the regular curriculum with tutorials, in conducting the class so that the classroom teacher has additional time to spend with those pupils requiring special attention and so on. The resource teacher might also conduct educational assessment and translate curriculum objectives into evaluation criteria, thereby monitoring the pupil's progress more closely than would be possible for the classroom teacher to undertake in the larger classroom.

Assessment

The nature of the assessment process differs considerably. In the restorative approach, the purpose of assessment is diagnostic. In the preventive approach, particularly in the stages prior to referral for statementing or for formal designation as exceptional, the purpose of assessment is instructional; namely to determine which skills, knowledge and behaviours have been learned and which still need to be acquired. In this way, programme adjustments can be geared to the pupil's standing and skills on a series of task-specific objectives in language arts, mathematics and so on.

The resource teacher, psychometrist or psychologist and other professionals involved in the assessment process will select informal and formal achievement tests and conduct observations and interviews so that appropriate programme objectives can be identified. Further, if the objectives prove to be inappropriate, that is, too easy, too hard, or unsuccessful in solving the classroom and resource teachers' initial concerns, these can be changed as a result of the information provided by the pupil's performance. The assessment process is cyclical, self-correcting and continuous (Chapter 4). In the restorative approach, assessment is more likely to be a once-a-year event leading to confirmation that the child is exceptional, for the purposes of assisting those who are reviewing the child to determine if his or her deficit is still present.

Record keeping

The type of records kept about a student may be quite different in the two approaches. In the restorative approach, psychological and educational norm-based assessment data may be the primary source of information on which professionals base their decisions. In the prevention approach, the records will usually describe the programme modifications, services and other interventions that have been tried, their outcome in terms of the child's progress and the observations and informal and formal assessment of the pupil's learning characteristics. The pupil's programme plan is the record, since it describes the cycle of assessment which leads to the programme objectives, the intervention used and the evaluation of progress. If the pupil is eventually referred for special education, psychological and other assessment data may be added to the records. However, the purpose of the pre-referral process in the prevention model is to find the most suitable programme interventions within the current classroom context. Therefore, the pupil's learning history is central to the information on which the teaching team makes its decision.

Much has been written in the education journals in recent years about the new role of the resource or support teacher. Whereas the more traditional approach calls for competence in educational assessment and withdrawal programming and teaching, the preventive approach requires additional consultative skills. The resource teacher has to be able to work effectively with the regular teacher, offering support, advice and resources and building upon each other's expertise while recognizing that a pupil's needs are a jointly shared responsibility of the classroom and resource teachers, and the pupil. Consultation may take place in the ordinary classroom, in the staff-room or a resource room. Administrators, ancillary and other staff members may or may not be present.

The role of the support teacher requires interpersonal skills and abilities with teaching peers that go well beyond those usually learned in training courses for special-education teachers. Furthermore, the support teacher may require other consultative skills such as coordinating and conducting meetings, interviewing parents, and presenting the child's case to the in-school team (see below) or to the statementing committee. In short, the support teacher does not simply close the resource-room door around his or her pupils during specified periods of the day but works closely with classroom teachers, attending both to the needs of the pupil and to the objectives, opinions and perspectives of the classroom teacher. The support teacher serves a client dyad; that is, at minimum, he or she serves two people: the pupil and the classroom teacher.

THE IN-SCHOOL TEAM

In many schools, the resource teacher's tasks are supplemented by an in-school team – an informal resource team that may consist of one of the following:

- Classroom teachers only, two or three of whom take turns serving on the team on a rota or duty roster.
- Classroom teachers, plus the resource teacher who collects and prepares information for the team. Other special-education teachers may be present.
- The classroom and special-education staff, plus the principal and/or vice-principal, who takes a formal role in chairing the team meeting.
- School staff, plus system-level resource staff, who may include the psychologist, psychometrist or teacher-assessor serving the school, speech and language consultants, guidance personnel (particularly in secondary school teams) and the school nurse, social worker and other specialists as appropriate.

This last team has the most formal structure. However, it should be stressed that the team is not acting in the capacity of a formal 'gate-keeping' committee to admit a child to a category of exceptionality and to make diagnostic and placement decisions. Instead, it acts as a programme and resource team to assist the classroom and resource teachers to adapt and modify programmes including those offered within the regular classroom. Further suggestions for the establishment of the in-school team will be included in Chapter 6.

THE PREVENTIVE APPROACH AT THE SECONDARY LEVEL

In schools in Canada and the United States, one often hears that prevention is suitable only for elementary schools. Secondary school staff note that consultation between staff is difficult if not impossible, in light of the rotary system of the curriculum in which pupils rotate from one subject department to another according to their timetables, the number of teachers with whom they work and the qualifying credentials for obtaining credits in their various subjects. Subject teachers argue that school-based resource consultants do not have the range and depth of subject-matter expertise to support subject teachers across the curriculum in secondary schools. Consequently, the resource staff in many secondary schools have very different roles from their elementary counterparts. They often act as advocates for pupils, negotiating reduced timetables, special criteria for examination, the use of supplementary technology for note-taking, writing, and extra time for assignments. Secondary schools in North America therefore usually have resource rooms, to which pupils either elect to seek help or are withdrawn for tutorial help. Advocacy, however, is not collaborative. One can use the tools of consultation on behalf of one's client, but the negotiations are by definition adversarial, in that the needs of the teacher and of the student are generally different. Collaborative consultation calls for the needs of the teacher-client to be focal in order for the teacher to meet the needs of the pupil.

Alternatively, in the British comprehensive school system, the prevention approach appears to be working better in the secondary schools (Green, 1989). In many comprehensive schools, all subject teachers also have a 'pastoral' responsibility. Each is assigned as home-room tutor to a group or form of students as they enter the secondary grades. The tutor is attached to the form for the duration of the students' secondary education. Tutors meet daily with their home form in order to accomplish the 'housekeeping' tasks familiar in North American schools. However, tutors must also keep track of the academic and non-academic development of each pupil. This requires coordinating progress reports from the pupil's subject teachers, preparing report cards, reporting to parents, accumulating pupil records towards the final school leaving document, the Record of Achievement (Hargreaves, 1989). In addition, study skills are taught in tutorials, and pupils receive individual help when they request it. Tutors are also responsible for communicating the child's progress and difficulties to parents.

Pupils who are at risk are referred to the tutorial team, consisting of a senior tutor and the other tutors responsible for that year's students. Through this team, decisions can be made to communicate with subject departments on behalf of a pupil, to draw upon the resources of the Learning Support Department, which consists of remedial and support teachers, or to refer to school administrators or to external support professionals such as social workers or psychologists. Of course, the pastoral system is subject to the same difficulties and variations as any support system; beliefs and attitudes of the members, leadership, tutor training, priority in the timetable, and in the school's financial allocation. Yet it demonstrates a secondary school team mechanism which facilitates the support function in the school and which conveys the message that all teachers carry responsibility for the learning outcomes of all pupils. Green (1989) describes the accomplishments of this model in one British comprehensive school. When teamed with a Learning Support staff who deliver a resource service to subject specialists, and with mixed-ability groupings in the first three years, the delivery model is a powerful mechanism for developing staff awareness of total pupil needs. In reviewing seventy British schools, Hodgson *et al.* (1984) describe the diversity of teaching strategies, modifications to curriculum and collaborative teaching arrangements that have been developed. As Bines (1986) notes, it is possible for support teachers trained in remedial techniques to work collaboratively with subject specialists, and in many settings it has been shown to benefit both under-achieving pupils and partially or totally integrated, moderately and severely disabled children.

Chapter 3

Consulting skills
Contracting

> I have defined love as the *will* to extend oneself for the purpose of nurturing one's own or another's spiritual growth . . . the principal form that the work of love takes is attention. . . . Attention is an act of will, of work against the inertia of our own minds. . . . Listening well is an exercise of attention and by necessity hard work.
>
> (M. Scott Peck, 1978, p. 121)

In this chapter, and the chapters to follow, I will suggest ways in which you might approach each stage in the consultation cycle. Examples of the phrases which you might use are also given. It is your task to adapt these to your own personal style; to find ways of expressing yourself which feel comfortable and natural for you. The 'how-to' sections of this and the following chapters are to help you make explicit and be aware of the interpersonal dynamics and skills which accompany good consultation.

For an experienced consultant, these are implicit; not part of the consultant's immediate awareness. But in order to acquire these skills, our purpose is for you to be aware of them, to practise them and then to reflect on their impact so that your approach to clients is based on your own beliefs about how to reach your objectives with the people with whom you work. Consultation skills, like other human interaction patterns, can be developed through reflection, practice and gaining the confidence to use them effectively.

GOALS OF THE COLLABORATIVE CONSULTATIVE FRAMEWORK

There are three goals to collaborative consultation:

- to solve an immediate problem;
- to assist the client to master skills and knowledge to prevent and/or respond more effectively to similar, future problems;
- to effect change – to enhance the ways in which teachers conduct their work with problem pupils.

You have a variety of ways of achieving these goals. One is to plan and implement a system for delivering instructional opportunities and services to pupils in the school. In this delivery system, all staff will assume joint responsibility for all pupils in the school regardless of their special characteristics. One of the ways in which you can affect staff change is by using your skills in collaborative consultation. By modelling your role effectively, including negotiating a balance of responsibility for each 'problem' situation, you can commence the move towards prevention and integration. Hopefully, your role will move towards your ideal role image of becoming more of a curriculum adviser and co-teacher, a staff developer rather than a 'gate-keeper' for pupil problems.

The skills of consultation are relatively simple to learn. Putting them into practice is the real challenge. In this and the next two chapters, the steps in the consultation process will be described, with examples and with exercises for you to try. The exercises will help you to plan a consultation event before it occurs, and to analyse the impact of your style of interaction after the event.

If you have studied counselling techniques, some of the following will be familiar to you. Such skills include active listening and non-intrusive responding. However, consultations and counselling techniques have quite dissimilar goals. In therapy, counselling aims to have the client know himself better. Consultation centres around a problem which is external to both the consultant and client. You hope that the client will get to know himself better as a result of your work together, but this is a byproduct of your collaboration, and neither the chief nor the ostensive purpose.

THE CLIENTS

Who are the clients? The term 'teacher' is used in these chapters to indicate the person or people with whom you consult. For the purposes of this book, this will usually be a teacher, but the principles of consultation also apply where your client is an administrator, ancillary helper or other colleague, or a parent. You may also find yourself dealing with several clients, some of whom are not readily apparent. There are nearly always one or more behind-the-scenes clients who have an interest in the issues about which you are consulting. Part of the task of the first meeting or series of meetings with your client is to determine who the behind-the-scenes clients really are. This process is *contracting* and it is the crucial first step in the consultation process. It is your first and usually best opportunity to find out what you are getting into. It will also be your only opportunity to set out before the teacher your own needs and expectations. In doing so, you set the stage for sharing the problem with the teacher, and for not carrying the whole project alone and hence the whole responsibility for its success or failure. More importantly, as previously diagrammed in Chapter 1, Figure 1.1, how you work with your colleague will establish your reputation throughout the school and parent community, and therefore each consulting event is an opportunity to model the role by which you intend to be seen in that school. The message of Figure 1.1 is that if you want to be asked to test children, be seen carrying a large and prominent test kit with you. If you want to be known as the person who distributes teaching ideas, start carrying teaching tips and good ideas for use in the classroom, and disperse them to fellow staff. If part of your role is to work collaboratively with another teacher, then you are likely to promote requests for it by modelling this role with as much publicity as possible, such as in the staff room in the lunch hour, or by discussing the role with staff in an orientation meeting. The contracting meeting is also a key opportunity for you to demonstrate your collegial role.

STAGES IN CONSULTATION

Contracting may consist of one or several meetings between you and the client. It leads to the second stage which is to develop a plan of action for you to assess the problem (Table 3.1). If the problem involves a pupil, your role will probably include working directly with the pupil through observations, interviewing and tasks in the classroom. You will most likely explore the context in which the student learns, through research of the pupil's records, background, home, and community. The third stage (reporting and feedback) will be to negotiate further with your client about the form which the intervention to the problem will take. You will plan a programme and services with the teachers, and sometimes with parents and with the pupil him- or herself and establish the criteria for evaluating the success of the interventions you designed. In the final stage, consolidation, you and your clients will evaluate the outcome.

In Table 3.1, the stages in the consultation process, from contracting with the teacher to evaluation of the outcome, are set out in the right-hand column. Each of these will be discussed under the four headings: contracting, creating a plan of action, offering feedback and consolidating the project. In the left-hand column of Table 3.1, the educational skills which are the substance of the collaboration are listed. These skills are the expertise which you bring in your support role to your client. They are familiar to you as a process for assessing the needs of a pupil and preparing an instructional plan of action to meet those needs. They are also cyclical in nature; a successful instructional intervention confirms that your original judgment about the pupil's difficulties was correct or, to use Reschly's (1980) term, 'instructionally valid'.

There has been considerable discussion in the education literature about assessment for instructional intervention. For example, Salvia and Ysseldyke (1985) have a good discussion on the process of assessment as an exploration of the pupil's needs. Little has been written about the

Table 3.1 The stages in a consulting cycle

Technical skills	*Consulting skills*
Identifying concerns	*Contracting*
evaluating outcomes of previous process identifying teacher's concerns seeking concern of parents: others (including pupil?) collecting history setting hypotheses and designing questions to be tested preliminary observations, interviews	communicating understanding of problem negotiating wants and offers – making parameters of task establishing roles – explicit responsibilities sharing concerns: vulnerability and loss of control
Assessment	*Plan of action*
matching data sources to questions data collection observation, interview, tentative judgments	surfacing layers of analysis taking account of political climate – school and system recognizing sabotage
Reporting	*Feedback*
tentative judgments team conference offering alternative interventions, modifications, services, resources	guiding team meetings dealing with forms of resistance presenting alternative solutions modelling role
Programme planning	*Consolidation*
coordinating staff responsibilities guiding the selection of a plan recording goals, objectives, time-lines and need recording evaluation criteria review – evaluation of progress of pupil, programme, services	coordinating participants guiding selection of choices/ownership recording decisions and responsibilities evaluation of outcome: • progress of consultation event • progress of role
decision to extend, recycle or terminate	decision to extend, recycle or terminate

parallel cycle of the interaction, the collaborative skills in consulting, which allow the assessment and instructional cycle to work in any given school. The relative nature of special needs and particularly of their interaction within the context of home and school, suggests that the context and the key people in the child's life are integral to understanding the child's needs. Therefore, one does not assess the child in isolation, but must negotiate with the adults and sometimes the peers around the child to seek information and to propose solutions to adjust the context in which the child lives and learns. An effective educational assessment must therefore be conducted in the manner of a collaborative consultation. Note how the two cycles of skills, technical and consulting, are listed in parallel in Table 3.1. The collaborative consultation skills provide the interpersonal dimension to make effective the technical skills of assessment for instruction.

In this and the next chapters, therefore, each of the steps in consulting will be considered. The parallel technical skills of assessment and programming will be drawn into the discussion as we proceed, since they are the substance of the collaboration. Of course, not all consulting events depend on a child's difficulties. Many consulting situations involve problems with colleagues, as conveyed in some of the problem situations in the appendix. Frequently teachers wish to discuss problems about themselves or their work situation, and raise issues about colleagues, administrators, parents or support service personnel, school policy and so forth. The same consulting procedures are used and, indeed, the same principles of assessment can be applied to solving problems in the workplace. But for the purposes of illustrating the consulting skills and assessment and programming cycles, we will continue to focus in this and the next two chapters on pupil problems referred by classroom teachers. The extension of the cyclical approach to solving other problems will then be considered subsequently.

In this chapter, the first phase in consulting, contracting, is addressed. In the next chapter, the identification of the problem and its assessment, and the consultant's skills in developing a plan of action are addressed.

THE FIRST PHASE OF THE COLLABORATION – CONTRACTING

This is the most important part of the consulting cycle, for it is here that you establish what your role will be and within what parameters you and your client will work together. If the initial contract is not clear and mutually acceptable, then the later stages of the consulting event will run the risk of falling apart.

There are two components to contracting (Figure 3.1). One is fact finding; to find out the facts about the problem situation and what the client requests and expects of you. The second component, planning, has the objective of setting out for the client what you are willing to offer and what role you will play. The contracting meeting is the opportunity for each of you to set out your 'needs' and 'expectations' and to negotiate an acceptable agreement for working together. This agreement should, at the very least, not degrade the role image by which you are aiming to be known. Your needs in modelling the support role include enhancing your image as a helpful, resourceful and successful consultant. The contracting meeting is crucial, therefore, for 'getting all the cards on the table' so that you can assess how likely you will be to succeed in solving the referral problem. If you bypass this step, you run the risk of taking on the whole programme yourself, of wearing the blame for failure if the project fails, and of creating an undesirable image of yourself in the eyes of your colleagues. At best you risk establishing a collaboration that is differently understood by you and the client. At worst, your efforts may be sub-verted or sabotaged by the apparent or hidden clients with whom you have not established a set of mutually agreed-upon procedures for work-ing together.

1 *Fact finding*

- Communicating an understanding of the problem
  - empathy
  - active listening
  - begin coping with client's vulnerability and fear of losing control
- Eliciting client's needs and offers
  - 'What can I do for you?'
  - Negotiating shared responsibility

2 *Planning the approach*

- Stating your own needs and offers
  - tone
  - explicit, short, simple statements
  - listening for agreement
- Reaching agreement
  - restating
  - recording a contract
- Giving support

Figure 3.1 Steps in the contracting meeting

The contracting meeting also establishes who is going to be responsible for making the significant decisions about the pupil's programme, for putting the programme into place, for keeping the parents involved, and for keeping track of how the programme is meeting the pupil's needs. By establishing this at the outset, clients get the opportunity to be the owners of the solution and the cause of success when the pupil makes progress. This contributes to the second goal of the consultation process; to give the client the means to solve the problem, if and when it next occurs. Further, the client is motivated by success. As Hanko (1985) notes, one of the most powerful aspects of solving a problem collaboratively can be that teachers discover that they have reassumed control of the time that the pupil had hitherto 'made' them spend unproductively with them. The positive response of most teaching professionals to the success of their pupils is one of the main tools in your tool kit (Rosenholtz, 1985). Teachers may not initially want to contract your services so that they may do a better job. You may more frequently be asked to take care of the problem yourself. However, you are working towards client ownership of problems, by giving away your expertise and providing feedback to teachers about the positive results, which occurs when they take on the responsibility for planned programming for the pupil.

Stages in the contracting phase

Fact finding

The stages of the contracting meeting or meetings are listed in Figure 3.1. The tasks of communicating an understanding of the problem and the development of an open and trusting partnership commence with how you introduce yourself to the client and to the problem expressed. The communication tasks will continue to be focal to your style of interaction throughout the consultation event, but are especially important in the contracting stage. Step 1, personal acknowledgements, and Step 2, communicating an understanding of the problem, address the first two goals of the consultation process.

You might start with an acknowledgement of the unique aspects of their situation. Your opening comments can set the stage for communicating your empathy. Here are some openers which you may be able to adapt:

> Thanks for dropping me a note about Charlie. Deciding to call me in must have been hard after the gains you've already made with Charlie. He must be worrying you.

> Tanya's problems are widely acknowledged in the school. You must be facing at least two challenges: her disruptive behaviour and the pressure from the parents of other children in your class.

Your next step will be to invite the teacher to tell you about the problem. It leads directly into the important question, 'What is it that I can do for you?' This question is crucial, for it establishes that you are not there to unleash your expertise on to this teacher, but that you intend to meet on an equal footing. This is the beginning of your negotiation of each person's tasks and responsibilities. The reciprocal phrase which you will lead your colleague towards is 'Here's what I can do for you'. These two phrases establish the main objectives of the contracting process; eliciting the teacher's needs and offers, and stating your own needs and offers in order to reach your joint objectives.

Your opening comments prompt your colleague to describe the problem situation in his or her own words. This permits you to then paraphrase back to the teacher the problem as he or she sees it. Conoley and Conoley (1982) note that in addition to paraphrasing, some verbal ways of acknowledging what you have heard include acknowledging, reflecting, summarizing, clarifying and elaborating. By doing any of these in your own words, you communicate your empathy, and your ability to centre the problem in the terms of reference of your clients. This commences the process of reassuring them that their problems are valid, that they are not incompetent in bringing them forward, and that you are not going to impose on them a predetermined solution. Accepting what you hear becomes an essential part of contracting. When you demonstrate careful listening, you indicate respect for the ability of the person who is speaking to you to perceive the nature of the problem. In this regard, Fuchs *et al.* (1990) write of the essential conditions for consulting as parity and respect. Parity, or the communication that your client is as much an expert as you, and respect for your client's concerns and perspective are the basis on which trust is established between you.

You may wish to reassure your teachers directly that the problem is valid and that you can help. The tone you wish to communicate is that you respect their experience and their efforts to date, and that you are here to work alongside them, exploring some avenues which they may not yet have had the time to follow or have not thought of trying. Your message is that two heads are better than one. It is useful to keep in mind the kinds of apprehension which the teacher might be feeling. Smith (1982) interviewed teachers about their concerns when seeking advice from a colleague. His summary of their most prevalent concerns provides a comprehensive set of notes on teachers' feelings of potential vulnerability and their fears of losing control of their teaching situation when talking to a 'consultant'.

1 Will you listen?
   Will you show awareness that you heard what I said, without blame, distortion, critical comment or expert advice?
2 Does asking for assistance imply incompetence?
   Will you assure me that you are confident in my capacity to adapt my skills to meet this need, if you'll show me some ways?
3 Will you tell the boss?
   Will my seeking advice result in jeopardy or wider publicity?
4 Does extra help mean extra work?
   Will you provide me with remedies to support and relieve me?
5 Can anything be done quickly?
   Will you shower me with expertise and resources to make me embark on a whole new career, or can you offer simple changes and immediate practical advice?

6 Can you do anything for me?
Can you understand my problem and the constraints of my job so that your suggestions will provide me with real alternatives?
- For teachers who have a strong sense of their teaching skills, this may take the form of 'Will you accept that I know best what will work for me, and I reserve the right to decide?'
- For teachers with less experience or confidence, this may be 'help me to reassume control of this situation.'

Smith's list contains several pointers for a possible hidden agenda of the contracting meeting. It emphasizes that teachers who seek support do not want to be 'taken over' as a result of their request. Teachers need to share control of the problem. They need parity; for you to request that they contribute their skills, and for your collegial support of their decisions. To a greater or lesser degree, they are risking their personal and professional privacy by their disclosure that they have been unable to resolve a professional problem. You could potentially jeopardize their confidence if you share this with others without first seeking their permission. It will take some time for you to establish a reputation within the staff as a person who preserves others' confidence, and not a pipeline to the staff-room broadcasting network, to authorities or to anyone else who might pass judgment on your teacher's competence. Becoming known as a neutral and trustworthy person who keeps confidences is an essential component of successful collaboration.

Finally, teachers seek to repossess the time, energy and resources that they have wasted in an unsuccessful attempt to solve this problem. From their perspective a solution does not involve a prolonged period of learning, or an additional investment of time and energy. Once you have explored the roots of the problem, you may conclude that the solution to the problem requires long-term commitment. You may need to work in that direction by building the learning experience in small increments, each with a reward for the teacher in terms of making specific gains towards the problem solution. It may also be in the interest of your staff development role to work effectively together particularly on the initial problems referred to you, so that the teacher will return for further collaborative support. It will be a calculated professional decision, whether to make small incremental gains with the teacher or to take drastic steps and lose the teacher, for example, because the needs of the pupil are too great to proceed incrementally. By and large, though, teachers are willing and cooperative partners who simply seek your support to find a solution to a problem with a pupil, parent or colleague. Your next step then is to reach consensus with the teacher about the nature of the problem and how to approach it.

Eliciting your client's needs and offers

By asking what it is that you can do for your client (a part of fact finding), you invite a statement of the client's expectations of the collaboration. This commences the negotiation of what tasks each of you will undertake. The kind of role that your client may expect you to play will vary at different stages of the consultation event. Initially he or she may request more information, or a temporary or interim strategy or resource to tide over the situation. Also, the client's expectations may be unrealistic or impossible to fulfil. Often your teacher's starting position may be to ask you to assume the total responsibility for the problem alone; as in 'take this child out of my classroom.' More typically, teachers will need help and encouragement to state their expectations. The isolationist culture of teaching (Hargreaves, 1989) has given few teachers the skills to ask for assistance. Helping them to make explicit their own needs and expectations begins the specification of criteria for evaluating the eventual success of the collaboration. For example, if a teacher's goal is to have a child 'feel better about himself', you might ask what the child does that leads the teacher to believe that the child currently does not feel good about himself. By asking what the child *does*, you begin to establish objectives which you both can observe, and which can tell you if your work together is successful.

Planning the approach: the consultant's needs and offers

You next set out what you are prepared to do for the client, and what you expect him or her to do in return. The purpose is to establish explicitly what each is willing to contribute and what each must derive from the collaboration in order for both parties to judge it to be successful. By making explicit the contributions of each party, you will be assisted by recording notes that will be shared with your client and which will form a written contract. Success, therefore, has two components: observable changes in the problem situation as a result of the collaboration, and consultant and client each accomplishing their own goals in sharing responsibility for the outcome.

It is sometimes helpful, ahead of the contracting meeting, to make a list of the parameters in which you are prepared to work. In the planning and review sheet (Appendix 1), you will note that a distinction is made between your essential and desirable needs. One practical way to plan your role is to make lists of your essential and desirable needs in your workplace. You may wish to review it at intervals to see how you are progressing. Essential needs are those which are crucial for the maintenance of your role image, and for vesting the ownership of the problem squarely with your client. Your essential needs may include access to information and to key people, a commitment from the teacher to meet you part-way, commitment from the head teacher to proceed, priority for this activity in the school, a high profile among the staff, time and basic resources to do the task. Your desirable needs are those which would ensure your success and enhance your image as collaborator, but which would be negotiated away for this particular client if the outcome was acceptable. Some desirable but not essential needs may include someone to work with you, more time, working indirectly not directly with the pupil, a meeting of the staff in which the head teacher explains the project and declares support for it, more resources, the commitment of the administration, being the first person alerted when problems arise, feedback on what has happened after you left, the authority of the head teacher to support you, and so forth. Your list will depend on your own school situation, the nature of the problem and of the clients.

Listing your essential and desirable needs is part of both short- and long-term planning. As your role develops in the school, you will move towards your long-term goal, and your needs list will change. Desirable needs which you initially may give away should gradually become essentials, as you establish others' expectations of how to collaborate. Bound in with this list is your plan for the development of assessment and instructional skills in the staff; it may include changes in expectations about the role which assessment will play, a move towards assuming responsibility for all pupils in the school, a shift in staff attitudes from blaming pupils for failure to seeking instructional solutions, and so forth. These outcomes are long-term. Don't be discouraged by contracting meetings that fail to reach an agreement or by consultations that are not fully successful. Indeed, it is sometimes necessary to know when to walk away, but more will be said about this and about maintaining your own sense of self and purpose in later chapters.

Here are several pointers for expressing your own needs and offers and for encouraging your clients to do the same. If possible, avoid lengthy verbal elaborations of your needs that beat around the bush. This communicates your uncertainty in asking for what you deserve. The alternative is to use simple, active declarative sentences; 'I would like to . . .', 'In dealing with this issue, I need to be able to. . .'. Having stated your needs in a low-key, simple way, it is helpful to pause quietly and let the statements register with your client. If the client raises a question, you may answer as simply and briefly as possible. Then be quiet and listen for a 'yes' or 'no'.

For many people, being quietly and gently assertive takes some skill and practice. You may wish to prepare for a contracting meeting by listing your needs and offers, then finding ways to express them in simple, declarative terms. After a while, the skill will become easier, and seem to flow from your fingertips instead of from the knot in your stomach.

A second pointer is to be careful not to overstate your own offers, or the potential results of your collaborative venture. Rather than promise that you will be able to solve the teacher's concern in no time, make your offers cautiously realistic. If you are more successful than this, then your colleagues will consider you to have delivered a bonus to your joint project. Do not offer guarantees such as 'I'll be able to get John back on track in no time.' Make allowances, instead, for the possibility that either the teacher or the pupil will renege on his or her responsibilities; 'I can work with John to use the tape-recorder for next Wednesday, to help him hear what he's writing. Then he'll be ready for you to direct his use of it for your history class, as you agreed to do.'

Reaching agreement

Even at the risk of appearing pedantic, it is useful to restate or summarize once again what each has agreed to do. At this point, you may take this opportunity to consolidate your notes with the teacher. You could leave a carbon copy as your preliminary contract (see the section on keeping notes before you design this, though, since it should include clear objectives and time-lines for your next tasks).

Client feedback

Finally, give the teachers the opportunity to say how they feel about working with you. Ask them if they are satisfied with what you've agreed to do so far. In more difficult negotiations your questions might be more direct, as in asking how they feel about working with you. You may not get a completely truthful answer, but you are putting the client on the spot. If the client foregoes this eleventh-hour opportunity to object to your agreement, he or she will have made at least a partial commitment to it. A statement from you of appreciation for the teacher's willingness to work together with you will complete this stage of your collaboration.

An exercise in preparing for a contracting meeting

The following exercise illustrates the analysis of a problem situation into its component parts. By analysing the consulting event ahead of the contracting or subsequent meeting, you allow yourself the advantage of being ready to state your needs and offers, and of not being caught off guard by the requests of the client. This becomes essential when you meet resistance from the client, as will be discussed in the next chapter.

On a sheet of paper, prepare four headings: the players, the needs and objectives of the primary client(s), your needs and offers, and the potential reasons why the situation has arisen. This fourth heading is to contain a list of your tentative hypotheses about the possible reaons which might explain the client's motives and behaviour. As will be discussed in this chapter, your hypotheses about this case should be tentative, subject to review and revision as you learn more about the case, and there should be as many alternative hypotheses as you can generate, to allow you to keep your options open for deciding how to proceed.

In Appendix 3 the case of Hubert Brown is offered as an exercise in problem analysis. It is used here to illustrate the proposed format. Hubert, a principal, wants you, the board-level consultant, to ratify his idea of sending a group of staff to a course on integration. There are problems among the staff associated with the attempts of one teacher, Mrs Jones, to integrate a class of students with behavioural problems into another teacher's music programme. The analysis of this case migh proceed as follows:

The players:
Hubert, the principal.
Mrs. Jones, the special-class teacher.
The music teacher.
The rest of the teaching staff.

Hubert's potential needs and objectives:
He wants you to solve the integration problem.
He is keen to preserve a good image of himself with staff.
He seeks a fast solution.
He genuinely believes that the course on integration will solve the concerns of the staff.
Etc.

Your possible wants and offers:
You need access to the staff to hear their opinions.
You want to meet with Mrs Jones.
You want to meet with the music teacher.
You would like Hubert to be willing to delay a decision until you are ready to propose it.
You want Hubert's support.

Possible hypotheses:
Hubert's pressing for a solution may be masking some serious problems among the staff.
Hubert may be unwilling to take responsibility for dealing with the situation.
Hubert may be trying to put you on the firing line to solve the problem.
Hubert may not want the problem solved, but may want to play for time.
Etc.

Having completed this preliminary partitioning of the alternative motives which might be giving rise to Hubert's request, you are in a better position to meet with him and to listen for confirmation of your various hypotheses. How he responds to your request to delay a decision, and to talk members of the staff, will give you an indication of which of the hypotheses seem to best represent the situation in the school.

Active listening

As suggested by the quotation from M. Scott Peck's writing at the beginning of this chapter, listening to another person is not an easy task, and one which many people do poorly. Active listening involves a conscious attempt to hear another person without slotting what he or she is saying into one's own predilections and understandings. Two ways to approach the task of hearing your client are, first, to ask for their viewpoint, and second to restate in your own words what you hear them say. The first is a simple step often overlooked by people who are caught up in their own expertise. Yet keeping an open mind on another's perspective is at the heart, not only of good consultation, but of good diagnostic skills, particularly when one is seeking a solution to a child's learning difficulties. The second, restating skill, serves several purposes. First, it clarifies for both the consultant and client that they are similarly understanding the task. Second, it provides assurance to the client that the consultant views their perspective as equally as important as their own in tackling the problem. Third, restating contains a message to the client that helps to establish trust between client and consultant. The message is that the client is not losing control of the situation to the expertise of the consultant, and that the consultant is not degrading the client's concerns by failing to take account of them in defining the task. The client therefore feels less vulnerable and less open to criticism. Restating can sound less than genuine, however, if it is overdone. Ways to try to communicate that you are attending to the teacher's concern include summarizing, asking for clarification, and asking for confirmation that you've understood the main points (Conoley and Conoley, 1982).

Listening actively and restating what is said, simply and non-judgmentally, are essential consulting skills. They are seldom easy to acquire; the gamesmanship of human interaction often has different rules for giving away confidences and retaining power. As Peck says, listening well is hard work requiring self-discipline, but the respect which it conveys is an important compliment to give to one's client.

Use and tone of language during consulting

Throughout the contracting meeting, and indeed in subsequent meetings, the consultants will discover that their use of language has a significant impact on how the meeting progresses. For example, when stating one's own needs and offers, it is often tempting to 'beat around the bush', to provide too much justification for the proposed role, and thus to convey subtly to the client one's own insecurity. Simple, active, declarative statements are the most effective means of communicating, provided that they are assertive but not confrontational, low key but not apologetic.

The ability to name your own and others' needs in a quiet, simple way is disarming to the client. In cases where the client may have mixed motivations or agendas for the collaboration, naming them conveys that you are fully aware of the nuances of your interaction, and that you are not prepared to engage in mutual hoodwinking in order to achieve your objectives. Yet your tone of voice conveys that you are simply stating your parameters for working together and are not critical of your client's style of responding. Wolfgang (1984: 73) has a useful acronym for the non-verbal qualities of a communication that convey credibility to the client, and that are persuasive. These characteristics have been shown to be effective in the Anglophone community; however, there are considerable cultural and social differences in how people interpret them.

S Stance – face the client, lean forward slightly.
O Openness – hands upturned, arms and body relaxed. Don't cross arms.
F Facial expression – positive, relaxed, a smile if appropriate.
T Touch – e.g., a handshake but *caution* as people accept touch to differing degrees.
N Nod – shows you understand and are listening.
E Eye contact – fairly steady eye contact indicates that you are listening, but avoid staring.
S Speech – moderate pace, warm, positive voice tone.
S Space – interpersonal distance. Arm's length is the neutral zone.

As will be discussed in the next section, the consultant's use of language and the tone of his or her statements is central to dealing with a resistant client, and with the hidden agendas of a consultation.

Chapter 4

Consulting skills
Assessment, feedback and developing a plan of action

I will consider what you have said,
  what you have to say
I will with patience hear, and find a time
Both meet to hear and answer such high things.

(Julius Caesar, I.ii. 1168)

By the end of the contracting meeting or series of meetings, you will have the following:

(a) Technical information:
- the teacher's statement of the problem.
- other information supplied by the teacher, and perhaps from sources such as school records, and samples of pupils' work. You may have talked to other people about the problem also.
- your own hunches, or hypotheses about the areas of the problem which need further exploration.

(b) Consultative:
- an understanding of your teacher's and your roles, expectations, and offers and a feel for the teacher's willingness to work with you.
- hunches about others who have an interest in the project or who may be part of a hidden agenda.
- an understanding of some of the context in which your teacher is required to work; its pace, the school's climate, policies, interpersonal issues, etc.

The contracting meeting may not have taken place as a single event, but may have been conducted piece by piece at different times. Both you and the teacher may have sought further information between meetings. In fact, the assessment cycle has begun; the very act of interviewing your client has communicated your ideas and priorities which you have now drawn to his or her attention. Contracting with another person has already affected the way in which that person is looking at the problem and has therefore had an impact on the child's needs.

It is not useful to seek solutions and recommend interventions at this stage, however, since you know that there is more to be uncovered with respect to the child's needs, and to the effects on his or her functioning of the home, school and teacher. You commence an assessment cycle, based on the teacher's statement of the problem, but you will be aware that there are layers to the problem to be uncovered. In Table 4.1, the technical process for conducting an assessment is highlighted, and will be the subject of the next section. Then the consulting cycle counterparts will be considered, in which you will develop a plan of action and provide feedback and consolidation to your clients.

ASSESSMENT OF THE CONSULTING EVENT

In the assessment process you have two (or more) clients; your primary client, the pupil, and your immediate client, the teacher. A group of teachers and the school administrators may also play the part of clients. You have explored in the contracting phase the expectations and needs of the teacher or teachers. Your focus on the pupil will be tempered by the teacher's needs, by his or her response to your question 'What is it that I can do for you?' This will be particularly important when you come to the point of giving feedback to the teacher about the pupil's needs. It is quite possible that the referral concerns expressed to you by the teacher differ from the outcome of your assessment of the pupil. One purpose of an assessment therefore is to conduct a comprehensive exploration for yourself of the set of factors, both internal and external to the child, which are giving rise to the teacher's concern.

The other purpose of assessment is to prescribe an intervention; to find how the child best learns, and what his or her current learning capabilities and difficulties are, so that a remedial or enrichment programme can be designed. In order to do this, you will want to take into account the specific learning difficulties that the teacher has identified. For example, if the teacher believes that the child has difficulty in comprehending what he or she reads, then an assessment must look at the specific skills of reading comprehension. This might be done through a standardized test of reading comprehension or through a more informal method such as listening to the child read and analysing his 'miscues' (Goodman, 1976). However, by interviewing the teacher further, you may discover something of the teacher's understanding of the child's comprehension problem. Perhaps it is specific to mathematics problems, or the teacher considers that the child is too focused on accurate word identification to the detriment of searching for meaning. The first point to be made is that assessment, in the consultative framework, seeks to probe the teacher-client's knowledge of the problem situation, so that the consultant can collaborate with the teacher in finding a solution that is relevant from the

Table 4.1 The assessment, planning and feedback stages in the consulting cycle

Technical skills	*Consulting skills*
Identifying concerns	*Contracting*

evaluating outcomes of previous process
identifying teacher's concerns
seeking concerns of parents:
    others (including pupil?)
collecting history
setting hypotheses and designing
    questions to be tested
preliminary observations, interviews
	communicating understanding of the problem
	negotiating wants and offers – making
	    parameters of task
	establishing roles – explicit
	    responsibilities
	sharing concerns:
	    vulnerability and loss of control

Assessment	***Plan of action***

matching data sources to questions
data collection
observation, interview,
    tentative judgments
	surfacing layers of analysis
	taking account of political climate –
	    school and system
	recognizing sabotage

Reporting	***Feedback***
tentative judgments team conference offering alternative interventions, modifications, services resources	**guiding team meetings** **dealing with forms of resistance** **presenting alternative solutions** **modelling role**
Programme planning	*Consolidation*

coordinating staff responsibilities
guiding the selection of a plan
recording goals, objectives,
    time-lines, and need
recording evaluation criteria
review – evaluation of progress
    of pupil, programme, services
	coordinating participants
	guiding selection of choices/ownership
	recording decisions and responsibilities
	evaluation of outcome:
	• progress of consultation event
	• progress of role

decision to extend, recycle or
    terminate
	decision to extend, recycle or
	    terminate

teacher's perspective. Assessment in this context is not the application of a pre-selected battery of tests which assume that the deficit is reflected by the deviation of the child from the performance norms of age peers. It is a

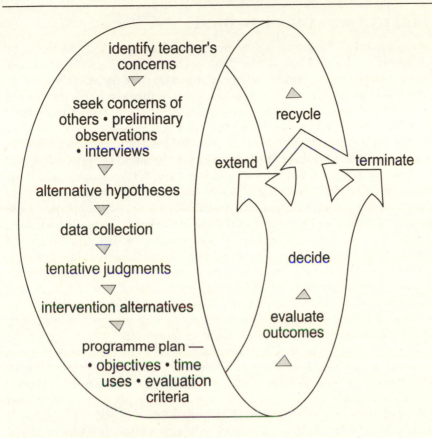

Figure 4.1 Assessment as a cycle

treasure hunt, in which you look for clues in the referral process, and follow them up. But not only do you explore in depth the issues which the teacher raises, you wish to go beyond them. If the child is considered to have a reading comprehension problem, you may want to explore both the child's skills and the situational characteristics of the learning environment. The second point, then, is that assessment is a dynamic process, along the lines of scientific enquiry. Its goal is to explore all the factors contributing to a problem situation so that the consultant can display, as feedback to the teacher, a picture of those factors, and how they interact to create the problem. At that point, solutions or interventions can then be discussed.

One final point to be made about the assessment process is that it is cyclical. As mentioned in Chapter 3, the steps lead to a programme plan and intervention which is self-correcting or self-validating. In Figure 4.1, the steps in assessment are highlighted in the steps for the technical skills cycle. These will now be discussed, and their cyclical relationship explored.

STAGE 1: GENERATING HYPOTHESES

As a result of the contracting meeting, you have learned the teacher's perception of the problem of the child, and some inklings about causes. The next step is to expand the scope of your search to include the perceptions of others, including the parents and often the pupil him- or herself. The child's learning history and background, medical history and school records are reviewed. By observing the child in the classroom and in other settings in the school, you add to your information base not only an indication of the child's abilities, behaviour and style of responding but also how he or she interacts with teachers and fellow pupils, and how they in turn interact with the child. Throughout this preliminary data collection, the assessor generates hypotheses or tentative questions about the pupil's needs and strengths and their underlying causes. As a rule of thumb, there are seldom less than six possible factors that might be contributing to a problem situation. For example, a reading comprehension problem may stem from past learning experiences, current or previous vision or hearing problems, early delays in language, inappropriate instructional level, task- or domain-specific difficulties as for example in failing to comprehend mathematical problems but not literature, general learning difficulties, lack of emphasis on reading as an activity in the home, lack of opportunity to read, poor concept of self as a reader, and very occasionally, a neurologically based impairment. These 'hypotheses' or preliminary questions clearly reflect a set of factors internal and external to the child. There is a very good possibility that several of them may be confirmed as interacting to produce the child's current difficulties. In any case, they will probably extend beyond the concerns initially expressed by your client, the referring teacher, since the teacher cannot explore all the possible contributing factors within the context of the classroom. We will return to this topic when addressing the development of an action plan, a phase which parallels the technical cycle phases of identification of concerns and of assessment.

STAGE 2: DATA COLLECTION

In the second stage, these hypotheses or questions are explored through systematic data collection; formal and informal testing, trial teaching, further observations and interviews with parents, teachers and the pupil and so forth. The scope and nature of this second stage depend on the preliminary hypotheses or questions about the factors contributing to the problem. The purpose is to rule out some questions as not relevant and to confirm that others are probably central to the pupil's characteristics. This leads to tentative explanations about the underlying causes, the environment conditions which are contributing to the problem, and their interplay in the pupil's current profile of achievement.

The assessment tools are chosen to answer the initial hypotheses. They may include standardized achievement or ability tests if the hypotheses warrant their use. However, they are likely to be centered upon the classroom concerns of the teacher, and therefore will be more task-specific and situational than some standardized test data can easily and efficiently provide. Since the objective of the assessment process is to prescribe one or more instructional interventions, it may be more relevant and efficient to use curriculum-based assessment, guided observation and dynamic assessment methods. Guided observation of the child includes participating in the classroom to check on hypotheses about the child's everyday behaviour and learning. For example, if the concern is lack of reading comprehension skills, it may be informative to listen to the flow of the child's reading aloud, whether the child backtracks to early parts of the text when meaning breaks down, what the child does when encountering a difficult phrase, and whether the comprehension deteriorates over passages with stories of high or low familiarity. How does the child use the information in pictures? How fluent is the child's language in retelling the story?

Dynamic assessment is a term coined by Feuerstein (1979). Feuerstein assessed where a child was lacking a specific skill in the context of a learning objective. He first measured the child's skill, then, after endeavouring to teach the child, would measure any improvement in his or her skill level, immediately after instruction and often after a delay to assess the child's mastery and retention. This test-teach-test method provides insights not simply into the lack of a classroom-based skill, but into the potential for the child to benefit from instruction to remedy the deficit. Hence the term 'dynamic assessment' is used in contrast to the static assessment given by single measure test scores. In recent developments, assessors have measured how skills that are taught for the mastery of specific tasks have been transferred or generalized by the pupil to closely related tasks, or even applied to more distantly related performance objectives. Humphries and Wilson (1986) and Campione (1989) discuss how the training itself in the test-teach-test sequence may be geared to weak, general hints and proceed through much more directive instruction. This allows the assessor to gauge the extent and depth of assistive instruction which the child may need to reach a given performance criterion, as well as to compare the effect of differing levels of support on the transfer of new skills. The test-teach-test routines can also be used to compare the relative effect of different kinds of instruction on the child's learning needs and style. Does the child benefit from discussion of a metacognitive or strategic level of skill, such as 'when you are losing track of the meaning, it helps to go back to the last paragraph and pick up the threads again'? Or will the child need the task to be broken down into a series of specific skills, such as 'read the passage into a tape recorder.

When you have finished, rewind the tape and listen to yourself reading. As you listen, try to answer this comprehension question'?

STAGE 3: ARRIVING AT TENTATIVE JUDGMENTS

From the process of assessment, the consultant is able to rule out some of the initial hypotheses, and to arrive at a tentative judgment about those hypotheses which appear to hold. The judgments are themselves hypotheses; they are tentative and temporary pending the outcome of the interventions that are designed to address them. The tentative judgments are translated into an action plan. In conjunction with those who will put the plan into action, programming alternatives, services and techniques are considered, and from among them, a programme plan is chosen. At the heart of this process is what Reschly (1980) has called 'instructional validation'. If the tentative explanations are indeed correct, then the action plan which is designed to alleviate the difficulties or change the outcomes will be successful. At this point, in order to judge the validity of the chosen plan, the expected changes and a reasonable time-line need to be selected, and provisions made to review the results.

Assessors, like scientists, build a picture of the child and his or her learning context in increments. Tentative hypotheses lead to observations or to instruction which in turn leads to new questions; to the unexpected. The assessor may revise the plan of action as things develop. New issues may emerge so that the assessment and programming cycle is never tidily linear. It is open-ended and exploratory; indeed, like a treasure hunt. More important, the outcome in the child's learning and behaviour is the key to evaluating how accurate was the assessment. The process is cyclical and self-correcting. If the problem still exists at the end of the process, the assessor must return to the drawing board. He or she must consider what might have been overlooked, or what other hypotheses or explanations and sources of information should be considered.

STAGES 4 AND 5: REPORTING AND INSTRUCTIONAL PLANNING

Having arrived at an understanding of the child's difficulties, and developed some tentative conclusions and possible interventions to deal with them, your task is to communicate them to the teacher in a manner that will facilitate the selection of an intervention, and motivate the client to put it in place. You now have a professional responsibility to balance the needs of your client with your judgment about the needs of the child. Most of the time these will be compatible, but occasionally you will be faced with a compromise between two differing sets of needs. There will also be occasions when you feel that the teacher is not acting in the best interests of the child, and you are obliged to confront the teacher with the child's difficulties and strengths, as you see them, and perhaps – at the risk of losing the teacher as a client – with a statement of how the teacher is contributing to the problem. More will be said about working with the teacher as a contributor to the problem when we consider the consulting cycle counterpart to this phase of assessment in the next section.

It is helpful to plan your presentation to the teacher ahead of the meeting. By thinking through your options ahead of time, and by anticipating the teacher's response to your report, you make best use of your time, as well as improving the likelihood that your work together will be collaborative.

You arrive at a picture of the situation, and a set of tentative conclusions or judgments about its causes and possible solutions. It is your current 'best guess' about the problem situation, and it has been explored with the purpose of identifying instructional possibilities and resources which will maximize the child's progress and prevent further escalation of difficulties.

Reporting to your clients is a consulting event. However, preparation of your information and opinions for this meeting should be considered. How might your analysis of the situation and your assessment results be conveyed in simple, non-technical language? How can the findings be geared to address the teacher's initial referral concern, without excluding other important results?

For instructional ideas, are there several alternatives which you can present to allow your teacher to make a selection? Similarly, if other professionals should be involved, if services such as modifications to the physical space or changes to timetable, to presentation of materials and to examination methods are needed, can these be presented with sufficient options to allow the teacher to have some choice in their selection?

Consider also the subsequent stage of implementing the selected programmes and services. It is important to have thought ahead of time about the role which you are prepared to play. How much direct involvement with the child will you have? How do you plan to work with the client, and for how long? What support services will you offer?

Finally, you will need to plan the evaluation component of the programme. The criteria for evaluating the effectiveness of your assessment and the intervention follow directly from the initial statement of the teacher's concerns and from your judgment of the situation. Evaluation criteria complete the assessment cycle. If you have correctly diagnosed the situation and have devised an appropriate intervention and support system, the child's performance and behaviour should improve. Just as important, the teacher's actions should change and his or her satisfaction with the outcome of those actions should increase. How might you gauge such changes? What is a reasonable time-line for both of you to judge that gains have been made? What observable changes in both the pupil's and the client's behaviour will satisfy you and the client that the gains have occurred? Such changes need to be observable so that you both can share in evaluating the success of your project. The criteria do not have to be formal, complex and time-consuming, however. They simply have to be appropriate to the original purpose of your collaboration. For example, a teacher might identify a concern about a pupil's sense of self-worth. He or she initially reports that the pupil 'needs to feel better about himself'. In your early interviews with the teacher you ascertain what behaviour by the pupil gives rise to the teacher's belief that the pupil 'feels bad' about himself. You ask what the pupil does that indicates to the teacher that the pupil feels this way. Perhaps the pupil arrives late in a morning, wears a frown and slumps into his seat, fails to complete assignments and complains that the work is too hard. The evaluation criteria, following your proposed intervention, should then be reflected in these behaviours. Does the pupil seem more willing to arrive for class, look more interested, complete work and receive better marks and, when asked, state that the work seems to be more manageable? You don't need a high calibre measuring instrument for you both to agree that gains are being made.

The point to be made is that programme objectives need to be chosen to be specific to the instructional goals which you have both agreed on. They should not be so vague as to be unmeasurable, nor yet so specific that they become onerous to monitor. In some cases, for example when you design a behaviour modification programme, you may wish to contribute to your work together some very specific measures of behaviour; number of instances per minute at baseline and at intervals during intervention, etc. More frequently, though, your objectives do not need to be so specific. As Campione (1989) and Salvia and Ysseldyke (1985) note, current thinking about programme-focused assessment has retreated from the formal scientific criteria of assessment of earlier decades, and seeks to address the situational needs of pupils and teachers in ways that are operationally definable, contextually relevant, and instructionally valid, rather than statistically sound.

You have now developed a working contract together which includes the objectives, time-lines and evaluation criteria, upon which you and the teacher have both agreed. When recorded as notes of your reporting meeting, they are the contract for the final stage of your collaboration, which will be the implementation of your instructional plan.

They may include the following:

- tentative judgments and conclusions about both the initial concerns of the teacher, and your analysis of the situation;
- proposed solutions including interventions and modifications, possible services, resources and support to assist the teacher to implement those solutions that he or she selects;

- evaluation criteria or objectives relevant to each proposed solution, including proposed time-lines for evaluating each outcome.

You are now prepared for the feedback meeting itself.

THE FEEDBACK MEETING

Following the assessment and the plans which you make for reporting and proposing a programme and interventions, you schedule a reporting meeting with the teacher. Many of the principles of consultation which you encountered in the contracting meetings will resurface at this stage; communicating an understanding of the problem plus its possible causes and solutions, negotiating the responsibility for implementing the solution, dealing with your teacher's concerns, and even dealing with his or her resistance to implementing the programme.

Preparing an action plan

In Chapter 3 we noted that teachers who refer a problem feel varying degrees of vulnerability and loss of control about the situation. They may have tried to cope with their problem, and are likely to see their failure to solve it as due to factors beyond their control. They may attribute the problem to the child's mental makeup, to his or her inability to fit into the class or group, to the child's environment. Some teachers may suspect that there may be a part of the solution within their control, but by asking for help they often express their inability to cope.

While your task is to offer support and to assure them that help is at hand, your ultimate goal is to have them see themselves as responsible for meeting pupil needs and for working with you collaboratively to achieve this. When working for the first time with some teachers, this objective may be far from realized. The referral problem may sound like blaming the child, or may be a demand for you to remove the child from the classroom, and take care of the problem yourself. Even after probing the teacher about the nature of his or her concern, you know that you want to look well beyond those issues. There are three layers or levels to address:

- What is the child doing?
- What is the home and general school environment contributing?
- What is the teacher contributing?

Depending on the teacher's perspective, aspects of all of these levels may be addressed. But beyond these will be further concerns.

To illustrate the steps, let us consider an example. The head teacher asks you to work with a classroom teacher who is reticent about integrating a deaf child into his or her classroom. In an interview with you, the teacher states that the pupil is not functioning at an appropriate level for the class. While the functional level of the child is one issue to be explored, other factors may be contributing:

- The teacher is concerned about his communication skills with the child.
- The teacher believes that the withdrawal of the child from the class for speech therapy and for instructional support will fragment the child's programme, or will cause the teacher to gear his or her teaching around that child's schedule.
- The teacher is doubtful whether he or she has sufficient information or skills to know how to proceed.
- The head teacher did not consult the teacher about the placement, and the teacher feels that, as in the past, he is never involved in decisions.
- This child has a behaviour problem.
- This child's parents are difficult to satisfy.
- And so forth.

The layers below the referral problem may thus reflect aspects of the organizational climate of the school or the home, and factors associated with the client him- or herself. It is not always possible to bring these issues to the surface in the contracting meeting, but placeholders should be kept for them in the formulation of concerns about the problem and the pupil's needs. As you explore the problem situation by observing the child in the classroom and by talking to other teachers and education professionals, the parents, and the child, these factors may surface. They are usually related to the organization and managerial climate of the school, to the struggle for power and precedence in the school community, to the politics of parent organizations, and simply to interpersonal conflicts, allegiances and attitudes.

In collaborative consultation, you are responding to two clients; the teacher and the pupil. The needs of both must be considered in the context of the school, if you are to find solutions which are acceptable to the teacher and which will be successfully implemented in the context of that school. Recommendations that focus exclusively on the pupil frequently fail to take account of the school context, and their adoption is therefore jeopardized. Only your hunches will allow you to be sensitive to those elements which have an impact on the teaching staff, such as the character of the school and its staff dynamics, the leadership, the organizational climate, the support or pressure brought by parents, and the self-images of your clients themselves. There are no reliable and valid assessment instruments to tap these, beyond careful observation and active listening. In Chapter 5, some techniques will be discussed, however, for dealing with clients' resistance and for other organizational and interpersonal obstacles frequently encountered by the consultant.

To summarize, the assessment phase of the consulting skills cycle involves bringing layers of the initial concern to the surface, not only in terms of the child's difficulties but in terms of the teacher's ability and willingness to deal with the problem. Factors which contribute to the latter may exist in the organization of the school, in its leadership, communication channels and social fabric. They may also exist in the home and community of the child and of the school. Their impact will be felt by both the teacher and the child, and an effective solution must take account of them. Some pointers are:

- to explore the teacher's personal role in the development of the problem;
- to explore the factors in the school which contribute to the teacher's involvement in the problem;
- to cast your hypothesis net widely, trying not to overlook any factors, at any level of analysis which might be contributing to the problem;
- to seek a solution that is appropriate to that context, and compatible with the structure of that school as well as relevant to the teacher's concerns and ability to implement it – and when implemented, it will address the total profile of the pupil's needs.

In contextually based assessment of the type we have been considering, the consultant is not taking the role of a researcher, in the way this is usually understood. A researcher must maintain neutrality, be comprehensive in the collection of data, derive conclusions only within the parameters of the data, and eliminate personal bias. The consultant is far from neutral, but takes account of client's needs and feelings, and guides their activities. Although I have used the term 'hypotheses' to name the possible questions and concerns which arise in the consultant's mind, these are often hunches, derived from gut reactions towards people, and from sheer experience. The process is made objective not by the collection of validating information, but by the outcomes which result from a good plan of action if such 'hunches' and hypotheses are correct. Furthermore, you have been placed in a consulting role because of your 'biases', if this is the term for your experience and expertise, and the intuitions which you derive from observing an instructional episode. These are not hard facts, but the basis for your judgments. As Salvia and Ysseldyke (1985) note, judgments are both the best and the worst of educational assessment. They are akin to clinical inference. They expand well beyond the data, and are invaluable when correct. Should they be incorrect, the only safeguard is that they be viewed as temporary, tentative and disposable.

The feedback meeting

In this final step in the consulting cycle, some pointers for conducting the feedback meeting are considered. Frequently, the meeting might involve a team of people, a topic to be discussed in Chapter 7. If the consulting problem is with members of management or concerns a sticky social situation, it may hardly be one in which feedback is offered but rather a continuation of negotiating your role. Here, however, we continue the theme of a teacher and consultant working together to solve a pupil's problem.

Reporting

As a result of your preparation, which you have completed prior to this meeting (Appendix 1), you have in hand the following:

- A picture of the total situation, as you see it, commencing with the referral problem, but extended and redefined by your subsequent observations, interviews, and work with the pupil.
- A set of judgments, including hunches, conclusions, contributions from your experience and expertise, which are your best guess about the causes of the problem.
- A number of recommended solutions which take several forms and address different aspects of the problem. These will include possible instructional ideas and suggested services to the child, such as modifications to physical plant, augmented timetables, etc. But they may also include suggestions for the organization and activities of the classroom, the school and perhaps the home. You will know if further professional expertise should be enlisted in the school or in the community at large. You will also have suggestions for tackling problems in meeting the child's needs as seen from the perspective of your client and other staff.

The meeting in which you report this information to the teacher has two components; a presentation of your findings to him or her and a collaborative session with him or her to devise a plan of action. Plan to take no more than one-third of your time together to present your findings. The development of an action plan requires more time, and will draw upon those skills which you previously applied in the contracting meeting.

Your presentation

Some of the issues which you might plan to include in your presentation are the following. Commence with a brief reminder about your initial contract and your joint objectives. Then describe four or five issues which you consider to be the major contributors to the referral problem. If there are many factors, select only those which the teacher is in a position to respond to. For example, it may be true that a relative of the child could benefit from psychotherapy, but this concern is beyond the control of the average teacher. The child's reaction in the classroom to this person's appearance might be discussed, however.

It is helpful to present your findings and judgments about the problem in simple, clear language. Your purpose is to convey an understanding to the client and not to overwhelm him or her with technical expertise. If in doubt, check whether the client understands by asking if he or she is comfortable with this kind of terminology, and by inviting your client to interrupt at any time for clarification.

It is also helpul to let your client know what the consequences will be, in your opinion, if the problem is left without attention. You may also want to ask the teacher for a reaction to your analysis by asking if what you say matches their own perceptions, or if your analysis rings true for them.

The presentation of findings and judgments should take only a small portion of your meeting time. In typical school meetings, your total reporting session may be condensed into a ten- or twenty-minute session. Condense the presentation of your findings accordingly, to leave as much time as possible for the development of an action plan.

Two rules-of-thumb are worth keeping in mind while making your presentation. It is easy to fall into the trap of blaming the child or other people not present with you for the problems, and so colluding with your client to avoid naming that part of the responsibility which rests with the teacher. For example, a behaviour management problem may stem from the child's successful attempt to control the teacher's attention. It can be presented in this manner. 'Robert certainly knows how to divert you from what you're doing by starting his clowning routine. He must have been getting away with doing that routine for years.' On the other hand, the teacher carries some of the responsibility for relinquishing control. The purpose of an intervention may be to assert the teacher's responsibility and recommend ways in which he or she can reassume control. The same behaviour management problem could therefore be stated in terms such as 'By responding to Robert's clowning around, you may have allowed him to demand your attention whenever he chooses. Let's see if we can take back control from Robert.'

A second pointer is to use language that will have maximum effectiveness. Two groups of phrases, enabling and temporary state, are worth remembering. In enabling phrases, you are offering support to your client:

You've dealt with this problem for long enough.

You were right when you thought that . . .

I'm glad you raise this point; it's an important part of the puzzle.

Temporary state phrases avoid implying a long-term deficit or problem, and instead imply that a solution is possible:

Chris hasn't yet learned to . . . (instead of 'Chris can't . . .')

Angela has yet to master . . .

Once we've distinguished the wood from the trees . . .

and so forth.

As in the contracting meeting the tone of your presentation should be assertive but not confrontational, and descriptive rather than evaluative. A client will more readily accept a statement of fact than your evaluation of that fact. Compare these two statements; for example, 'Your questioning technique came unglued at about the start of the lesson on differential equations' and 'Your selection of questions for the lesson on differential equations failed to do justice to your instructional plan.'

Many words have emotional connotations or become ' red flags' and should be used with caution: 'lazy', 'stupid', 'awkward', or 'tense'. It is helpful to create your own 'hit list' of red flag words and phrases, and to search for alternatives that feel comfortable for you, and are acceptable to your client.

Collaborating on the action plan

If possible, try to allow two-thirds of the meeting time for this phase, and be prepared to add more time if necessary. The goal of this phase is to establish the teacher's ownership of the chosen solution, and to agree upon reasonable criteria for evaluating the teacher's eventual success.

First set out the proposed instructional alternatives and the offer of other proposed solutions to the problems which you have just outlined. Presenting alternative resources and other routes to the goals allows your client to participate in selecting a means to achieve the goal. For example, if the child has a poor self-concept in the classroom, suggest two or three interventions that might tackle the problem. Might the child benefit from tutoring a younger child? Is a class buddy feasible? Is a systematic reinforcement schedule appropriate? And so forth.

Ask for your teacher's reaction. Permit him or her to say what might work within the context of the curriculum and classroom, and what he or she is willing to tackle. As will be discussed in Chapter 7, a reporting meeting which involves a team can be a useful collaborative setting in which fellow members of staff 'brainstorm' ideas and, by pooling their expertise, are able to offer a wide variety of resource materials and techniques from which the teacher may choose.

Thirdly, you may be willing to negotiate a further role for yourself. You may offer to support the selection of an approach by providing materials, or documenting changes in the pupil's behaviour and performance. It is essential to check that the teacher is willing to undertake a commitment to the plan, however, and not to leave it to you. Checking for the teacher's commitment may include questions about his or her reaction to selecting an action plan: 'How do you feel about moving in this direction?' 'Do you feel you'll be in control of the programme if you embark on this?' 'I will set up the taped teaching sessions for Patti, provided you can direct me about the content of each lesson, and can then build them into your overall teaching. Does this fulfil your view of how it's going to go?' 'I will keep track of Jason's progress by looking through the task sheets that you collect from him at the end of each week. Do you think this will be sufficient to help you meet his needs?'

When you have agreed upon the programme plan, the responsibilities for implementing it and the evaluation criteria, you may once again summarize your agreement. This may be an appropriate time to make notes, which you can then develop and provide to the teacher. As in the beginning of the consultation process, this is a contract; a plan that does more than outline the objectives and means to an instructional end. Ideally, it is a working document, containing a statement of purpose, responsibilities, and tentative outcomes that can be modified or abandoned if the results merit it. It can be viewed as a contract between you and the teacher, but more importantly between the teacher and the pupil. It will hopefully contain monitoring and evaluation criteria. Because the teacher has a primary role in selecting and designing it, it is likely to be compatible with the context in which the teacher and child must interact.

In the final step, give support to your client: 'I like the route you have selected. I'll do whatever I can to help you succeed.'

Chapter 5

Difficult consulting situations

The plan is always perfect, . . . until the enemy
makes the first move.

(Attributed to General Helmuth Von Moltke)

In this chapter, the difficulties of the relationship between the consultant
and teacher-client are considered. The context of pupil problems at the
pre-referral and integration stages of service delivery will continue to be
used, to explore the relationship between the consultant and classroom
teacher. The first issue to be explored is how to deal with a client's
resistance. Then we will consider consulting situations beyond the classroom. These include negotiating with one's supervisor, and dealing with
conflicts of interest. What should one do when there is a conflict of
interest between the needs of the teacher and the needs of the pupil? How
does one resolve the conflict when the views of the teacher about the
needs of the child seem to be in opposition to the views held by the school
administrator, or the policies of the school system?

Finally, we will briefly touch upon some broader issues of conflict and
conflict resolution, including dealing with the threat of legal action.

We have assumed throughout the discussion of the consulting process
in Chapters 3 and 4 that teachers feel positively about collaborating with
you and are comfortable with the consequences. While many collaborative efforts will result in a pleasant exchange of expertise and a rewarding
outcome, this will not always be the case. As Margolis and McGettigan
(1988) state, classroom teachers are reluctant to integrate students for
whom they must alter instruction. They are expected to change their ways
of working with children. Change requires new ways of thinking and
doing. It is difficult, and demands excessive effort and self-analysis
(Fullan, 1985). Clients therefore predictably exhibit resistance to the
possibility of having to change what they are comfortable with doing. In
Chapter 3, Smith's (1982) list of teachers' reactions to working
collaboratively provides some reasons why teachers are resistant.
Margolis and McGettigan (1988) also list reasons. From the teacher's

perspective, solving a student's problems may be dependent on more support than is available, may be contingent on special training they lack, may be incompatible with the needs of other students, may be unlikely to produce the desired result in time, may not be fully understood, and may be imposed without due regard for their opinions. In this list, we see the elements of people's resistance: fear of being vulnerable to others' opinions and fear of losing control of the situation.

The important principle underlying resistance is that no matter what the client is actually saying or doing, it is not a personal attack on you, the consultant. At times, resistance may certainly feel and sound like a personal attack. The key to managing resistance, however, is not to take it personally, but to see it as an opportunity to address the underlying concerns which are inhibiting the client from participating in finding a mutually acceptable procedure to follow. Managing resistance, then, is viewing it as an expression of underlying concerns which need to be acknowledged and given a voice.

Most of us feel conflicting emotions, and exhibit a degree of 'approach vs. avoidance' behaviour when we face a challenge. In Chapter 1, a distinction was made between educators who hold restorative beliefs, in which they tend to view learning problems as arising from factors within the pupil and therefore beyond their control, and those with preventive beliefs, who take a share of the responsibility for the pupil's difficulty by looking for ways to amend the learning context for the pupil. Beneath these different belief systems are learned ways for the teachers to maintain control of their pupils. Restorative teachers may attribute the failure to the pupil's lack of effort or ability over which they feel they have little control. Preventive teachers, on the other hand, may attribute the failure to the pupil's environment over which they have some control. In both cases, their desire to 'explain' the problem arises from their need to understand the cause of the problem so that they can delimit their own responsibilities should the pupil fail to improve. Indeed teachers' attributions may change from one child to the next depending on factors such as the degree to which the class will be disrupted by a reaction, the degree to which the problem is in a domain in which the teacher feels he or she has some experience, and so forth. In our research, we have found that teachers who hold more preventive beliefs also have a higher regard for their abilities as teachers than those who hold more restorative beliefs (Jordan, Kircaali-Iftar and Diamond, 1993). The restorative teachers who had lower self-efficacy also expressed a preference for the withdrawal of problem children from their classrooms, over assistance in the classroom to solve the problems. Gray and Richer (1988) distinguished two groups of teachers of adolescent pupils with behaviour problems. One group attributed the behaviour to characteristics of their pupils; laziness, a bad attitude, etc. The other group attributed the behaviour in part to the pupils' environments; their economic situations, their peers, etc. Interestingly, the pupils of teachers in the latter group held higher academic self-concepts than those in the former group. Teachers' beliefs about the causes of problems were somehow communicated to their pupils and had an impact on the pupils' own attributions of their ability to achieve.

There is a considerable literature to support Gray and Richer's findings (Levine and Wang, 1983; Cooper and Croyle, 1984). While teachers' attributions of pupils' problems are complex and vary according to situational characteristics, there is a general finding that teachers offer explanations of pupils' behaviours which are consonant with the likelihood that they will be able to control the outcome. Not surprisingly, pupil failure is most often attributed to pupil characteristics, while pupil achievement is most often attributed to successful teaching. Part of the task of the consultant, then, is to assist the teacher in contributing to a pupil's success, or to use Thomas's (1986) term, to 'enable' the teacher.

MANAGING RESISTANCE

The purpose of exploring aspects of teachers' attributions is to highlight the fact that resistance is a very natural and almost inevitable outcome of being placed in a client role. While the resistance may range from mild caution to extreme anger, it does not usually stem directly from the consultant's role, skills or expertise, but from factors that are internally triggered within the client. It is an expression of underlying fears of loss of control and vulnerability. For the consultant to respond to it defensively is to miss the opportunity to help the teacher to deal with the fears. Yet most of us are likely to react as if we were personally under attack.

The art of dealing with teacher resistance, therefore, is to: (1) pause long enough to recognize the resistance; (2) resist the temptation to counter the resistance by defending yourself; (3) consider why the teacher is exhibiting the resistance and what he or she is feeling, and then, (4) support the teacher in expressing those feelings.

Step 1: Recognizing resistance

This can be the most difficult part of the task, since you may find yourself reacting emotionally before you have recognized that the discomfort you feel is a response to your client's resistance. The way to counter this is to become familiar with your own 'early warning' signs that this consultation is stalling and that you are feeling defensive. Early signs include hearing yourself state something for the third time, paying attention to the uneasy feeling in your stomach or the shuffle on your seat. You might watch for the body language in the teacher which counters what he or she is saying. When a teacher is expressing pleasure but is hunched in her chair, or is telling you that the problem has now solved itself but is clenching his hands, these are early warning signs of resistance.

Step 2: Resisting a defensive response

An early detection of a problem gives you the opportunity to undertake steps 2 and 3. The lead time allows you to reflect briefly on your own reaction, and to take a deep breath, lower the tone and pitch of your voice and speak slowly. This will help counter the normally defensive reaction signs in your language of shrillness and volume. If necessary, seek a reason to ask for a short break in the meeting. Propose taking a coffee break, or ask for a minute or two to take a stretch. If necessary, adjourn the meeting for a while. This will allow you time to grapple with your reactions and to work at understanding the expression of discomfort which the teacher is exhibiting.

Step 3: Asking yourself 'why is the teacher resisting?'

By searching for possible causes for the teacher's resistance, you avoid reacting to the symptoms. As discussed above, the client may be feeling helpless, a victim of the setting, or alienated and rejected by the group with whom he or she works, or confused and swamped by too much information. In any case the client is feeling vulnerable and/or out of control of the problem, and as Block (1981) says, 'control is the coin of the realm in organizations'. To feel out of control is to feel exposed and vulnerable to the unknown. Your task, then, is to give yourself the opportunity to assess whether the resistance is directed at you personally (and it seldom is), or whether you just happen to be in the line of fire because of your role.

In Chapter 3, I proposed a framework for analysing a problem situation. The case of Hubert Brown (Appendix 3) was used to illustrate the analysis. This case can be extended to illustrate how a consultant might manage resistance. Let us suppose that, at your meeting with Hubert, he presses you for a solution, and resists your attempts to delay a decision until you have fully assessed the situation. He is persistent in returning several times to his proposed solution, that is, to have you recommend that his teachers be required to take a course. Your personal feelings about his manner might include annoyance, frustration at being unable to get him to see your perspective, or even a little anxiety that he may not respect your opinion or be trying to use you. Although these feelings may be justified, focusing on your own state will not assist you to solve the impasse. By focusing on why Hubert may be resisting your message, you may be able to address his underlying anxiety about this situation. Once you have partitioned the apparent and potentially hidden needs of your client, you are in a better position to plan your own list of needs and offers, and to tackle the issues which are causing resistance. You may also note some areas for discussions with your client(s) in which more information and opinions need to be sought.

Step 4: Expressing the underlying concern

To help the teacher to come to terms with the resistance, you might begin by acknowledging that you are aware of it, and that, by allowing it to be part of your discussion, the teacher has your support in dealing with it. The need to resist your joint negotiations is reduced.

Block (1981) offers a four-step 'recipe' for this process:

(a) Name the form which the resistance is taking.
(b) Be quiet, let it register with the client.
(c) Offer a 'good faith' statement of support.
(d) Return to negotiating needs and offers.

Naming the resistance takes a little skill. The statement you choose should begin with 'you', the teacher, and not 'I' (think, feel or believe). To state that 'I believe that you are . . .' is to offer your client the right of denial. Instead state how the teacher is resisting as an assertion. Your statement should be straightforward, simple and short. You should not feel obliged to justify your statement, or your impact will be lost. If you beat around the bush or offer explanations of your perceptions, you let the client 'off the hook' and give him or her an opening to dispute your statement. The support you offer is not in the words you speak, but rather in your tone of voice, and facial expression. You should sound and look empathetic, quiet and low key. The way to do this effectively is to feel empathy, and to have distinguished between the facts of the problem which you both face, and the emotional reaction which the teacher is grappling with.

To a client who insists on flooding you with details:

You are giving me a great deal of useful information, but let's focus on the main problem.

To a client for whom resistance takes the form of humour:

You're really amusing, but let's get serious about this problem.

In Appendix 2, you will find an exercise to assist you in naming resistance. For each type of resistance, write down what your one-line response would be. The key to the exercise is for you to develop statements which are in a style of language that is natural and comfortable for you. This exercise might be something you want to keep handy, to add to it when you have new ideas. When you have completed the exercise, check it to see if each assertion fulfils the following criteria, discussed above:

- Does it solely address the client, and not your reaction to the client? ('You' not 'I'?)
- Does it convey assertion but not confrontation? Can it be stated in a low-key, empathetic way?
- Is the language simple, the statement short and to the point?

In Appendix 2, some suggestions are made for naming resistance for each of the forms listed there. How do these match your own ideas? Can you borrow or modify any of them to incorporate into your own repertoire of consulting skills? You may find it useful to have some personal phrases on hand when faced with some forms of resistance. For example, when someone is becoming angry or agitated, it is useful to be able to detach from the topic under discussion and say 'This issue is upsetting for you.'

Dealing with a client's resistance can be intimidating and difficult. Many of us respond by avoidance. If we view the resistance as a personal attack, we might retaliate, thus damaging the collaborative relationship beyond repair. Each of you is familiar with your own reaction patterns. It is worthwhile reflecting on these, to develop a personal strategy for handling difficult consulting situations. In a training session, a very shy, quiet resource teacher, engaged in thinking about her style of handling resistance, suddenly remarked to our group that she realized she must be more assertive. Over the next few weeks, she kept a personal diary of her attempts, both in and out of the training course, until she felt relatively comfortable about holding her ground. Another consulting teacher recognized her tendency to react to a painful situation by delivering a verbal stab to her client. The client's anxiety level therefore increased, and the resistance intensified. It is preferable to restate the teacher's concerns in a relatively low-key manner. Offer a statement of support before naming the form which the resistance is taking. In Hubert Brown's case, his pressing for a solution might lead you to respond, 'You seem to have gained a great deal from the course and I appreciate that you are keen to have your teachers attend it. But you seem to want us to agree to that decision before we have fully understood the problem.' You then pause, which allows your client to verbalize the difficulties that he or she may be having. By acknowledging the resistance and providing an opportunity for the client to be open about it, you are respecting the client's concerns. This is one form of active listening which conveys the importance that you place on your client's needs. While responding to resistance in this way may at first appear to be confrontational, if it is carried out with empathy, in a low-key tone, it can offer an opportunity to your client to unload his or her concern, and to break the log-jam in your interaction.

In Appendix 3, there are two exercises to try with a colleague, for those readers who wish to practise their skills in dealing with resistance. Rules of Order is a problem situation, which could have several strategic resolutions. Brad the Brat is a simulation task in which you and a colleague conduct a role-play of a classroom conflict. You may choose to make the conflict more or less resolvable by varying your role-play. At the end of this exercise, there is a series of review questions to assist you in analysing the outcome.

CONTRACTING WITH SUPERIORS

There are situations where it is particularly difficult to express one's needs and offers. One such situation is when you are negotiating your own job with a supervisor or manager. Thus in the case of the support teacher, it is usually necessary to define the role, and the delivery approach towards which you are working and to negotiate these with your immediate supervisors. Yet with prior planning and a comprehensive list of your needs and offers, this contracting event could set the stage for the success of all the projects which you will subsequently undertake in the school. Some of the needs which you, the consultant, might wish to include in your contract are a clear role definition, adequate resources to do the job, the freedom to negotiate with each member of staff based on their individual characteristics, priority for your role in the tasks of the school, and your supervisor's demonstrated support of it. Some of these needs might be set out in a written proposal. For example, rather than ask the administrator to prepare a job description, you may wish to write a draft of it to use as the basis for your negotiation. Similarly, a brief plan for the future delivery model within your school and your role in it might become a working draft for the head teacher and later for discussion with staff. Developing a blueprint for your role within a school delivery model is the topic of Chapter 8, and will be expanded there.

The strategies for contracting apply in the situation where you are bringing the problem or proposal to the consulting meeting, as well as in the case where you are responding to it. As Fullan (1982; Fullan and Stiegelbauer, 1991) and others have noted, creating change in a school system requires that participants in the change see a clear need for it, from the outset. You are therefore in the position of selling your idea, and a part of selling is to gather the information to support your contention that the change is needed. Such information might include a survey of staff concerns, an event which has triggered the need to re-examine the whole process of delivering instruction to special-needs pupils, or a policy statement from head office. Prior planning and the collection of information to support your position are simply the data collection and reporting phases already discussed, and reversed to occur prior to the phase of contracting with the staff of the school.

CONFLICTS OF INTEREST

In Chapter 4, the difficulties were raised of resolving a difference between your analysis of a problem and the views of the teacher-client. You may be faced with a difficult professional decision of choosing between the needs of the pupil and the risk of alienating the teacher, or compromising your opinions about the child's needs in order to maintain your relationship with the teacher. This decision may depend on many factors; the severity of the disagreement, the consequences to the child, the possibility that the teacher may come to see your viewpoint, the time left in the school year, the possibility of other solutions. If you reach the decision that the conflict is sufficiently severe that you are ethically bound to respond to the needs of the child, you must consider the consequences on your relationship with the teacher and on your role image in the school. You will wish to minimize your losses in terms of the consulting role. Knowing when to minimize your losses and withdraw from a consulting event is a skill. It can be informative to keep a diary or logbook of your consulting activities, in order to reflect upon your progress. The diary allows you to keep track of your plans for meetings, the process that evolved and its outcomes. A *post hoc* analysis of the process which you considered during a consulting event is often a useful tool for furthering your personal professional development. It will also remind you that you are undertaking a pioneer role as an adult educator, and have made progress, even though the truism still holds that you can't win them all.

In Appendix 4 there are worksheets to assist you to plan an upcoming consulting event and to reflect upon its success afterwards. The framework of the worksheet could be developed into diary or journal form, to assist you in working through a difficult consulting situation over a series of meetings.

A very useful resource is Fisher and Ury's (1981) book, *Getting to Yes*. The authors describe approaches to negotiation, and identify some of the traps which people fall into when they take an entrenched or a compliant position in a negotiation session. Fisher and Ury offer a strategy which can avoid these traps.

LEGAL AND OTHER ADVERSARIAL SITUATIONS

Occasionally teacher-clients or parents will present such a difficult problem that the consultant will wonder whether he or she should disengage from it. One rule of thumb is to know where your sources of support are located. These include knowing where to contact other professionals who have expertise in specific problems, and who are responsible for them. Here are two examples; facing the threat of a legal action from a colleague or parent, and recognizing that your client is on the verge of an emotional breakdown.

It is important to recognize the boundaries of your own expertise and that some problems require that you refer the client elsewhere, and that you seek administrative and legal support for yourself. To illustrate, the move to integrate severely handicapped pupils into Ontario local schools raised the issue of who should administer medication, and conduct medical procedures such as injections and catheterization. Until recently, most school boards lacked a policy for this. Few teachers initially realized that this was a policy issue, and felt bound to respond when faced with a child who had just delivered a bottle of medicine with his name on it, or a child with an antedote kit for a bee-sting allergy. In these and similar cases, the problem is not one of professional responsibility, but part of a much larger issue of administrative and legal complexity. Without written parental permission and adequate training, a teacher could be liable for the consequences of misusing the medical treatment, even if acting in good faith. Support teachers and consultants are faced almost daily with similar problems, in which the special needs of the child are apparently of educational concern, but are to some degree the responsibility of medical and health personnel, or which have administrative or legal consequences. If in doubt about your responsibility, check with your supervisor or with the administrator in charge.

REWARDING YOURSELF

Most resource consultants are working in a lonely environment. They are frequently the only person in the school or group of schools who is engaged in the task of staff development. Their mandate is to give away their expertise and skills, and to allow the recipients of their support to earn the rewards for success. It is therefore important for consultants to seek their own forum for support. Because there will be less-than-successful consultation projects, and downright failures, it is crucial to establish a network of colleagues who are similarly struggling. Consultants need their own consultation time to share consulting skills tips and to tackle mutual difficulties.

Other on-the-job factors can also contribute to mental well-being. In an excellent article, Menlo (1986) stresses the importance of being oneself; of being genuine and authentic in the consulting situation. He discusses the stresses which result when the consultant feels 'constrained to sustain a viable image of himself in the eyes of others . . . constantly employing little schticks to keep himself in some sort of defensive position' (p. 64). Do not, Menlo suggests, engage in interpersonal politics of favour-giving as a means of gaining influence. Instead, he counsels, trust your own efforts at helpfulness to gain legitimate power as a change agent. Authenticity and genuineness with others are basic conditions for one's mental health and for a meaningful connection with other people.

Consultants also need a personal life, a time and place where they can remove themselves from the stresses of their working environment. They need a self-administered reward system for a job well done. Finally, they need some tangible index of their successes in the workplace. Such evaluation criteria serve not only to convince management that the project is worthwhile, but also to remind the consultant that this is the case. Planning, implementing and evaluating such a project are the subjects of Chapter 8.

Chapter 6

Working with parents, administrators and service professionals

At this time it is still almost axiomatic that quite a number of peda-
gogues are inclined to consider the world as their classroom and will
continue to examine and give homework to any casual strangers that
happen to cross their paths.

(Alexander King, *Rich Man, Poor Man, Freud and Fruit*, Chapter 8)

Up to this point in this book, the role of the resource teacher has been
considered solely from the perspective of the classroom-based resource or
support teacher. Classroom and subject teachers are but one group of
clients with whom the consulting teacher works. In this chapter, the focus
of the consulting role is broadened to include other professionals and
parents as clients and as partners. Difficult consulting projects frequently
involve other staff members, either as clients or as partners in working
with other clients. Further bonuses or complications to consulting are
added by the role played by the head or principal. The administrative
leader may indeed be crucial to the success of a collaborative model in a
school. Finally, service professionals, that is, psychologists, social
workers, speech therapists, and others who have a collegial role with
consulting teachers, will also be discussed.

PARENTS, TEACHERS AND THE RESOURCE ROLE

If one accepts that special needs can only be meaningfully interpreted in
the context of children's total lives, then clearly parents are a crucial
component in ensuring children's educational and social success. Yet
consulting with parents presents special problems for many educators.
Conversely, parents are often nervous about meeting teachers, and particularly so when their children are experiencing difficulty.

Such reticence on both sides has its justification. Parents and teachers
do not always agree about the sharing of responsibilities for children.
Davies and Davies (1985) describe conflicts arising from the different
perceptions of each group. In general, teachers see the most important

tasks with their pupils as cognitive and social training, while parents view social training as their own responsibility. Parents may also believe that the school should play a role in children's moral development, while few teachers see moral development as their mandate.

A considerable number of parents believe that schools exclude them from important decisions, while educators believe that they make great efforts to involve parents. In a review of the first sixty-eight special-education appeal board hearings in Ontario, in which parents and school boards sought a legal resolution to their disputes, the majority of parents felt that schools had communicated poorly with them, while school board personnel rated their own efforts with parents as satisfactory or good (Metcalf, 1987). The breakdown in communication between home and school at a very early stage in the conflict led to mistrust and finally to a legal solution to an adversarial conflict. Active, empathetic listening on both sides in the initial stages of the dispute would have gone some way to preventing the escalation of mistrust. Even relatively late in the conflict, a mediator might have assisted both parties to find a resolution, particularly if the mediator was viewed by both as objective and uninvolved and able to make each listen to the other's grievances. A sad outcome of the Metcalf study was that, when asked in whose best interest the dispute was resolved, all parties agreed; the lengthy, emotional, and financially draining hearings resulted primarily in benefits to educators, to parents, and least of all in benefits to the child in dispute.

Winkley (1985) discussed the 'formidable one per cent' of parents who are viewed with fear by teachers. The response of some teachers is to 'hold parents at bay' in the belief that the less parents know, the less opportunity there will be for dispute. Some of the reluctance of teachers to discuss problems with parents may also stem from their lack of confidence in expressing their professional actions in verbal terms. As Hargreaves (1989), Little (1985) and others have noted, teaching is an isolated profession, conducted within the confines of a classroom, with a sparse tradition for verbal exchange of ideas and concepts. Thus some of the factors which inhibit teachers' acceptance of a consulting relationship with a professional colleague may also arise when teachers talk to parents about the substantive aspects of their profession. At a different level, teachers may also feel unsure about their own ability to communicate with parents.

Consulting teachers can assist parent–teacher communications in two ways; by participating in meetings directly, and by assisting classroom teachers to acquire communication skills with parents. In both cases, consultants may frequently find themselves involved as a party in parent interviews, both to provide support for their colleagues and to offer their expertise to a parent.

PARENT–TEACHER CONFERENCES

Let us consider aspects of working with parents. Resource teachers are in a unique position to develop a different type of partnership with parents. As Davies and Davies (1985) note, parents bring a perspective which is unavailable in the context of school. In most cases, they desire and are able to assume an equal role with educators in planning programmes and monitoring their children's success. Early and ongoing parental involvement ensures not only mutual trust, but also the benefits of carrying over decisions made collaboratively about children into the home. Consistency in a child's life enhances the benefits of the chosen interventions, and permits everyone to assess their impact. In addition, one of the contributions of the resource role most valued by teachers is assistance with parent involvement (Speece and Mandel, 1980).

As a participant in parent–teacher conferences, there are several points to keep in mind. The quality of interaction with parents depends on many of the factors already discussed in consultation skills and in conducting a meeting. These include establishing objectives, demonstrating a genuine concern for parents' perspectives, respecting parents' concerns, actively listening, responding with empathy, and interpreting by restating what parents say. Where parents have a role to play in implementing interventions at home, the interaction becomes a contracting meeting. If you have a hand in guiding the organization of the conference, the principles for contracting (Chapter 3) might serve as your guide. Both teacher and parents are encouraged to express the parameters of their involvement, to state the commitment they are prepared to make, to learn what each expects to receive from the other, and to agree upon methods to monitor and report the outcomes to one another.

Because of the emotional salience of parent–teacher meetings, there are several specific issues to keep in mind. Teachers should focus on teaching procedures and classroom behaviours. They should avoid raising topics which do not directly bear upon the child's school problems. Kelly (1974) suggests that teachers should distinguish between problems that have an educational solution and those that do not. In the case of the latter, teachers should know what resources are available to offer a parent, but should not attempt to solve such problems themselves. Problems involving medical concerns, chronic poverty, and emotional and marital breakdown are beyond the expertise of most teachers, and without specialized training, teachers should be attentive listeners, but should not venture advice or opinions about such problems. While it is beneficial to refer parents to experts within or outside the school system, the parents themselves should be allowed to make the decision to follow through. It is not wise to specifically name experts such as psychiatrists, counsellors and others unless there is only one available. If the parent requests you to recommend a specialist, consider naming the service, and providing its telephone numbers and the name of a contact person, if appropriate.

Use of specific diagnostic labels to identify children's problems should also be avoided, particularly where such labels are ill-defined and open to interpretation. Instead of referring to a child as 'learning disabled', a 'slow learner' or a 'behavioural problem', it is preferable to specify a number of common problem areas and specific instances that are observed in the classroom. IQ scores similarly carry too many differing and potentially harmful connotations to convey an accurate picture of a child's level of function. Instead, you might describe how this child performs on tasks that are usually accomplished in the classroom at this age or level. Take care not to draw your examples of problems or characteristics of individuals that do not relate to this family and also take care when you share your own experiences. An empathetic statement can be an ice breaker, but avoid describing your or your family's experiences in detail, since cases are seldom the same. In Chapter 4, the discussion of the emotive impact of 'red flag' phrases is particularly germane in parent–teacher conferencing. 'Deficits' and 'deviance' imply an insoluble state in the child, one that will never be remedied or outgrown. Unless all parties already know and can accept that this is the case, such characteristics of children can usually be discussed in other ways. For example, the value of 'temporary state' phrases was raised in Chapter 4. Rather than telling parents that their child is unable to achieve an objective, it is often useful to convey the same information in terms of the achievements yet to be mastered. Parents want and expect a realistic appraisal of their child. They want to know how their child's performance or behaviour compares to that of other children. Give the information clearly in non-technical terms, keeping in mind that comparisons with other children and with norms are seldom accurate. Then pause and listen to the parent's response. The meeting might be completed on a positive note by your giving them support and offering them the opportunity for further discussion.

THE HEAD TEACHER AND THE RESOURCE ROLE

The impact of the head teacher on a school's approach to offering services to special-needs pupils is discussed in three sections of this book. In Chapter 5, the special case of negotiating one's role with one's boss was briefly considered. In this section, the essential role played by the head teacher (or other supervisors) in consultation events in the school is explored. In Chapter 8, we examine the leadership styles of head teachers and the effect of such styles upon efforts to bring about changes in the delivery of services in a school. In this chapter, the relationship of the head teacher to the in-school consultant is considered.

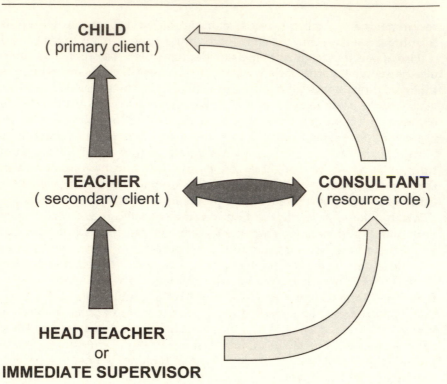

School-based consultation is most effective when it is an indirect service to children. The support or resource teacher (consultant) works directly with other professionals (clients) to equip them to provide services to their pupils. Two of the goals of a consulting event are to solve an immediate problem and to equip the client to solve similar problems in the future.

In order to maintain this relationship with one's teacher-client, the consultant must establish a working relationship with the client which eventually must be fully understood by all the staff in the school. The head teacher's endorsement of this relationship, early in its introduction to the staff, is essential. Conoley and Conoley (1982) list the important aspects of school consultation. By taking each of these, we explore how the head teacher can support or sabotage the relationship between consultant and client.

1 The consultant and client hold a peer–professional collaborative relationship

In Figure 6.1 the relative status of the head, the consultant, client and child are represented. The teacher must be able to view him- or herself as lateral to the consultant, otherwise the purpose of vesting the primary responsibility for the child with the teacher is lost.

Head teachers have a major role in protecting the laterality of the relationship. By providing resources to the consultant role such as time and materials, a budget, and office space, they indicate their endorsement of it. Other forms of endorsement may also be required; time to discuss the role in staff and parent orientation meetings and on INSET or professional development days. Lack of support may also provide a clear message to staff about the value of the consulting service. A head teacher who preempts the consultant's schedule to assign him or her another task is communicating that it holds little prestige. Resource teachers report being used to organize fund-raising events or sports days, to step in when a teacher is absent, and to monitor examinations, all of which can potentially sabotage the image which the consultant is attempting to establish as lateral and equal to colleagues. Block (1981) talks about this form of sabotage as using the consultant as 'a pair of hands'. He indicates this by representing the consultant as stooping over rather than meeting face-to-face with colleagues and peers.

2 The consultant is viewed in a facilitative rather than an expert role

Certainly, the consulting teacher brings to the consulting relationship a unique expertise and training. However, the task of the consultant is to hand this to the client, rather than to take personal credit for it. This is seldom easy. As a result, the lack of a head teacher's support may jeopardize this fragile understanding. Frequently, the consulting function role is a part-time position or is carried out as overload, to be shared with a deputy or department headship, a teaching role or another position of responsibility. In this case, the facilitative role of the consultant may be hard to establish, since positions of responsibility call for a more subordinate relationship to staff. In part, the personality and collegial style of the consultant will determine the success of the combination of tasks, together with his or her ability to move among the roles of mentor, coach, facilitator and evaluator. Block (1981) terms this subversion of the consulting role 'the expert'. He represents it by portraying the consultant running wildly about, attempting to be all things to all people.

A client will experience greater reluctance to ask for help of a colleague who is seen as the authority or expert, not only because this increases the client's vulnerability, but also because of the following.

3 Consultation is in essence confidential

Perhaps the most important component of the consultant–client relationship is that the client must feel secure. He or she will not become the subject of the administrator's evaluation through the consultant's reports, nor the topic of public scrutiny in the teacher's room. Yet head teachers or other supervisory personnel will from time to time ask you for evaluative comments about a teacher, placing you in a very difficult position. You feel accountable for your work, yet you are being elevated to the level of evaluator which will violate your client's trust. Ways to respond to this request from a supervisor may include inviting him or her to talk directly to the person they wish evaluated, or to observe your work together, provided your client agrees, or to invite him or her to ask the client directly how your collaborative project is progressing.

4 The client has the freedom to terminate the project at any point

Head teachers can greatly assist the consulting role by accepting that some consulting projects will not be successful, and by not asking for a detailed report of those that fail. Again, it is helpful for the consultant to use a colleague as a 'sounding board' for analysing both successful and failed projects. The head teacher may be just such a person if he or she truly has the skills of mentor, provided that the confidentiality of specific teachers is not encroached. At the end of Chapter 5, we discussed how consultants also need to establish their own 'sounding boards', as well as people to whom they can vent their frustrations outside their daily working environments.

5 Consultation is concerned with work-related problems only

In those cases where personal, cultural and societal problems impinge on the consulting relationship, the head teacher can provide significant support to both the consultant and teacher. Provided the teacher consents, the head teacher may be able to assume responsibility for handling crises or for securing external resources for the staff member, parent or child. Within a school, it is usually the head teacher alone who has the authority to take the problem and its solution beyond the scope of the school's immediate resources.

6 Consulting is primarily preventive in nature, though acute and chronic problems are dealt with as well

In many schools the consultant and head teacher have established a partnership. On the one hand, the head has a supportive but 'hands-off' relationship with the resource teacher, while the latter reports work-related issues to the head and keeps the head informed about the progress of the consulting role and the service delivery process. Information takes two forms. Specific cases are reported of children who the head should be informed about.

Some head teachers wish to be informed about all the children the staff consult about. Second, information is provided about the number and type of referrals, the progress of the project and so forth. Project implementation and evaluation will be examined once more in Chapter 8.

In Appendix 3 we offer two case studies based on real school problems involving supervisors. In 'the Hubert Brown case', a consultant is requested by a head teacher to organize some staff development activities. But the head may have more than one agenda. In 'sharing ownership' the resource teacher feels caught between different interests.

WORKING WITH SERVICE PROFESSIONALS

In the last two decades, there has been what Reschly (1988) terms 'predictions of a revolution' in the literature of each of the educational specialists who provide support services to schools. In particular, school psychology has witnessed a dramatic shift away from the classification of students and determining their eligibility for services, towards the design, implementation and evaluation of interventions (Medway and Updyke, 1985; Reschly, 1988). Other professionals, such as those in speech and language pathology (Frassinelli, Superior, and Meyers, 1983), social work (Zischka and Fox, 1985), and counselling and guidance (Ehly and Dustin, 1989) remark on the shift from diagnostic and confirmatory assessment towards the design of collaboratively developed interventions in natural settings. In Great Britain, Wolfendale (1988) notes that the 'whole school approach', in which all professionals take a collective responsibility for remedial and special-needs children, is at the heart of the 1979 Warnock report, the 1981 Education Act and the subsequent decade of debate on special needs.

Like the teaching profession, service professionals are involved in a revision of attitudes and beliefs about their roles and responsibilities, not only towards children with special needs but all the children in their schools. Working collectively with other professionals to provide indirect service to children requires some organization and planning which was unnecessary in the previous era of direct service based on short-term withdrawal of children from classrooms.

To date, little has been written about the impact of the simultaneous evolution of roles upon the organization and function of the school. The multidisciplinary team is one attempt to bring a variety of expertise to support the classroom teacher. In Chapter 7, the school-based team is discussed, although the addition to the team of multidisciplinary service professionals adds new dimensions and complexities to the objectives and conduct of the team. In discussing the notion of establishing collective responsibility within a school staff, Wolfendale (1988) alludes to a 'network of services' in which specialists are committed to providing continuously available support to a school. From the perspective of the consulting teacher, however, the task is to establish a working relationship with a variety of professionals who are usually short-term visitors, who frequently have no more than a tangential acquaintance with the philosophical and social structure and ecology of the school, and who have very large caseloads of children. Further, the assumptions and perspectives of each service professional may be more or less consistent with those of the consulting teacher. Often, the most critical test of one's consulting skills is to establish the territory and rules of order of the collaboration.

In Appendix 3, the exercise 'rules of order' is an example of a situation in which the lines of formal authority are not clear. As a case study, it illustrates some of the complexities of consulting within the formal hierarchies of school systems. The need to establish both clarity of roles (Fullan and Stiegelbauer, 1991), and guidelines for coordinating services to individual children, are demonstrated by this case.

It is probably essential that, at an early stage in a school's adoption of a preventive delivery approach, the service professionals assigned to the school are invited to meet with resident staff. Their reaction to the approach towards special-needs pupils taken by the school and to suggestions for their own participation in the delivery process may guide the way in which working relationships evolve and their involvement in it. If possible, service professionals should volunteer to become associated with the project, or have the opportunity to select alternative schools. Similarly, their involvement with in-school team meetings might be optional; an opportunity to work preventively with teachers, and to practise an in-service role is viewed as uncomfortable and time-consuming by some, and as exhilarating by others. Once again, the willingness of individuals to go along with the project and their participation in the context of the school will be governed by individual and local factors, and by the particular needs of each child.

On this last point, Wolfendale (1988) raises the notion of the 'key staff member' or 'reference point'. Each member of a multidisciplinary or school-based team has the potential to play a key role for each special-needs child. This person assumes the responsibility for coordinating the curriculum and services for the child, and for monitoring and evaluating the outcome. This notion, which is paralleled by that of the pastoral tutor in the comprehensive school, is intriguing, for it permits one professional to take a leadership role for a pupil for whom he or she has particular expertise. To this task are added the requirements of coordinating the contribution of other experts, educational and parental, and of monitoring the resulting outcome in order to evaluate the instructional effect of decisions collectively made. At a practical level, Wolfendale gives examples of cases in which pupils receive instruction in more than one setting, or have provisions and services both in and out of school, justifying the appointment of an itinerant, outreach or advisory teacher, or a school psychologist, social worker or health professional as the key worker for that child. The support teacher's role would then become that of in-school liaison, and overall team leader.

In Chapter 7, the secondary school model of pastoral care is described. In this secondary level adaptation of a pooled-resources model, the key worker is usually the form tutor. However, if the case merits it, the tutor may be replaced by the resource teaching staff, or by a centrally deployed service professional who becomes the case worker and service co-ordinator for that child. One difficulty with the secondary model is that tutorial responsibilities draw upon skills which subject teachers may not have acquired; monitoring both academic and personal development of pupils across the spectrum of secondary curricula, sharing information and collaborating to solve problems with other staff members, and reporting to parents. One source of in-service expertise in many secondary schools is the staff of the guidance and counselling department. Often, the counselling training of such staff can complement the consultation skills of the support teachers in providing professional development within the school.

Chapter 7

Drawing on the resources of the school

> It is perfectly possible to say what schools and teachers need to do in principle in order to accommodate pupils with special needs. It is quite another matter to develop realistic strategies for change which are based on and grow out of existing classroom practice.
>
> (Hodgson *et al.*, 1984, p. 165)

Since the inception of the mainstreaming movement, there has been considerable debate about the adaptation of the ordinary classroom in order to meet the needs of a wide range of pupils. Both in Great Britain and in North America, researchers have explored how teachers cope with integration and the creative ways in which they solve difficulties. In Great Britain, the 'whole school approach' prescribes the involvement of all members of staff in sharing the responsibility for students with special needs. One component of this approach is the release of staff resource time to allow teachers to work collaboratively. At the secondary level, resource staff are 'designated' to provide curriculum and teaching support to subject teachers. The effectiveness of this type of support has received mixed reviews (Clark *et al.*, 1982; Hodgson *et al.*, 1984; Bines, 1986, 1989). In North America, the extensive analyses of the efficacy of mainstreaming also report equivocal results (Carlberg and Kavale, 1980; Wang and Baker, 1985–6; Epps and Tindal, 1987).

It is apparently not a simple task to prescribe the school- and classroom-based practices which lead to successful prevention and integration. In this chapter, an attempt is made to synthesize some of the methods reported in the literature. As Hodgson *et al.* imply in the quotation which opens this chapter, how these are implemented in a particular school will largely depend on existing practices in that school, the context for change and the personnel involved.

THE IN-SCHOOL TEAM

We begin this chapter with a description of a school-based teacher support team, which will be termed the 'in-school team'. This team also has many names on both sides of the Atlantic; for example the Teacher Assistance Team (Chalfant *et al.*, 1979; Chalfant and Pysh, 1989), the School Support or Appraisal Team (Pugach and Johnson, 1989), and the Teacher Advisory Committee (Postlethwaite *et al.*, 1986). An in-school team is now becoming a standard feature of North American elementary schools, under such labels as the Special Education Resource Team (SERT) and School-based Support Team (SBST).

Graden (1989b) and Pugach and Johnson (1989) discuss the characteristics of team meetings which form part of the pre-referral stage of service delivery. While a formal team meeting might be convened following a referral, in order to comply with the legal procedure for identifying a child as having special needs, the pre-referral team has, as its focus, supplying alternative resources to the teacher to assist him or her to prevent the student from needing to be formally referred. Therefore, the purpose of the in-school team is to provide a pool of ideas and resources to which any member of the teaching staff may have access in order to solve a classroom problem. Thus, the primary client of the team is the teacher, and the topic discussed is the teacher's concerns, actions and resource needs for one or more pupils. In the prevention approach to delivering school resources (Chapter 2), the team plays a central role in the pre-referral process, for it is at the team meeting that the resource teacher and other team members engage in a 'brainstorming' of ideas to assist the referring teacher to meet the needs of one or more pupils. In order to be effective, the team must be informal and collegial, and its purpose must be to assist the teacher. Formal teams, aimed at reaching a decision about the designation status of a pupil, do not qualify for this task. Yet frequently, in-school teams are unsuccessful because their purpose, which is to solve instructional problems and to provide staff support, is not clear to the members. Consequently the team regresses to a more traditional committee which focuses largely upon the deficits of children, and on the mechanics of admitting them to special services, to 'statement' them or to review their designation as special, with little reference to the needs of each child as viewed from the perspective of their current teachers.

The membership of the team has some part to play in the direction that the team will take. The composition will vary from one school to another. However, the core members of a pre-referral and support-focused team usually consist of teachers who are respected by fellow staff for their pedagogical wisdom. Sometimes team members may serve on a rota. In secondary schools, the team may be drawn from a variety of subject departments or chosen to represent different ability groupings. In elementary schools, team members may represent the divisions of the primary school (primary, junior and intermediate levels), or core curriculum areas such as language arts, reading and mathematics. Expertise in remedial education is provided by the resource teacher, and frequently by a special-education class teacher who is located in the school. Support service representatives, such as the psychologist, the speech therapist and the school nurse may also be present, but in order to preserve the focus of the team upon locally appropriate instructional interventions, such specialists may be held in reserve for later stages in the prevention cycles should the resources in the local school be insufficient to solve the student's problem.

A building administrator, that is, the head teacher or deputy head, or a department head may also be a member of the team. Again, this will vary from school to school, depending on such factors as the leadership style and personality of the head and the ethos or climate of the school, the role which the administrator is perceived to take and the available resources, as discussed in Chapter 6. The presence of the head teacher has several advantages. It emphasizes the importance given to the team's function. The head is also in a unique position to exercise authority and can therefore assume responsibilities and draw upon resources which may be beyond the reach of the resource teacher. As an example, the author participated in a team meeting, chaired by a principal, at which an elementary school teacher referred her entire lowest reading group of five children for 'removal from the class for remedial instruction because they don't fit into the three reading groups I already have'. The principal, acting as a model consultant, explored with the team and teacher each one of the referred pupils in turn. By the end of the meeting no decision to withdraw any child had been made, but two had been identified for further assessment, one had a serious home situation on which the principal agreed to act, and the teacher had decided to change her approach to the remaining two children, with assistance from the resource teacher. The teacher declared her satisfaction with the process, possibly without realizing how far the outcome varied from her initial request. The gentle authority of the principal and the empathy and attention of the team fulfilled the teacher's need for support. Further, work had begun on the teacher's real need, for a classroom organization for reading which she could control.

The most frequently asked question about the in-school team is how to find the time to meet. From the point-of-view of the members of the team, it is advantageous to free members during instructional time. Frisby (1987) boldly proposes that the head or deputy head could serve as a relief teacher for such meetings. He also looks at creative uses of ancillary and parent helpers. In Canada, the author has witnessed some creative solutions to freeing staff time; grouping two classes together and adding teaching assistants, using time when special-class pupils are integrated to free the special-education teacher for the team or as a relief teacher, using the mandatory French period and library time for meetings, and scheduling a twenty-minute silent reading period for the entire school. Chalfant and Pysh (1989) trained teachers in 71 schools in Pennsylvania to establish Teacher Assistance Teams, freeing the staff by introducing sustained-reading periods in the children's timetables.

Chalfant and Pysh also recommend that team meetings should be scheduled regularly and consistently rather than on an *ad hoc* basis, and preferably once a week. Their training programme is designed to accomplish a total cycle of referral, diagnosis and plan of action in a thirty-minute session. In order to keep the meeting within the time-lines, they propose that referral information be circulated to team members ahead of the meeting. This includes the teacher's statement of the problem and the resource teacher's preliminary assessment notes. Members are asked to generate preliminary questions or hypotheses. If needed, requests for further information can be met before the meeting. An alternative format was developed in one group of schools in which the in-school team has been a central component of the delivery model for several years. The teachers devised a different type of referral form, also with the intention of minimizing the length of the team meeting. The referring teachers specify the problem, the type of help they need, and the people whom they would like to attend the team meeting. This form is then sent to the team members ahead of the meeting, to help them to organize their thinking, and to plan and collect possible resource materials to share with the referring teachers.

At the meeting itself, a structured allocation of time is often helpful to ensure that time is well spent and that the team is satisfied with the outcome. Chalfant and Pysh (1989) propose the following breakdown for a thirty-minute case conference:

1 Reaching consensus on factors contributing to the
   problem.   5 minutes

2 Permitting the referring teacher to identify the problems
   which he or she sees as central.   5 minutes

3 Brainstorming instructional and intervention ideas,
   offers of help, etc. As a rule, twenty-five or more ideas
   should be generated.   10 minutes

4 Referring teacher selects four or five useful ideas.
   The means of implementing them and criteria for
   evaluating them are selected. The meeting to
   review progress is scheduled.   10 minutes

   Total time   30 minutes

Chalfant notes that the team meeting should be conducted as a high-profile activity in the school. It should be fun and snacks should be provided. The meeting will be more favourably received if it begins and ends on time.

One of the main advantages to the team meeting is its power to influence reluctant members of staff to join in the collaborative process. As Frisby (1987) notes, it can serve as a 'drip-feed' for in-service staff development. It can do so in a more relevant and appropriate way than many INSET or professional development activities, because it addresses teachers' individual needs within their current working environment. A high-profile location, such as the staff-room, the apparent enjoyment of members, and the expressed satisfaction of the teacher-client all contribute to its appeal. If the resource teacher is in charge of the team, its scheduling would be important in bringing about changes within the plan to implement a preventive and integrative delivery system in the school.

How the team is managed is therefore an important part of its success. It is a group setting in which to use many of the consulting skills discussed in previous chapters. As with contracting, the role of team members is to listen and respond to the needs of the referring teacher. While the pupil's needs are discussed, the team members invite the referring teacher to corroborate their analyses, and to select two or three ideas from among the many resource options that they offer. The leader of the team will need to establish such ground rules clearly with team members. An in-service session at the beginning of the year would help to serve this purpose. Also helpful are intermittent reviews with the team about how well it is functioning. Any paperwork such as the pre-meeting data, and jotting sheets for the meeting itself may be designed to reflect the objectives and stages of the process. Some schools have even developed a team logo to identify their paperwork.

The interpersonal dynamics of team meetings are the subject of much research both in education and in industry and business. A few pointers to handling the dynamics of the team will suffice here, although there are other sources, for example examining the effective organization of the multidisciplinary teams in the United States (Fleming and Fleming, 1984; Courtnage and Smith-Davis, 1987).

A team is a group of separate individuals who do not necessarily concur with each other. Each should therefore be given the opportunity to express him- or herself. The leader may therefore need to involve the quiet people and quell the most vocal. Statements which might do this are:

> We've had several good ideas from John and Wendy. Let's hear what the rest of you are thinking?

> Charles, you always have some unique suggestions. Do you have anything in mind on this topic?

The team leader should attempt to get the members of the group to listen to each other. If this is not happening, you might consider asking some of them to react to what the others are saying, or ask an individual to summarize the sense of the meeting to date.

Block (1981) suggests that team leaders should be sensitive to the real power in the group. The power of influencing others does not always reside with those designated as having formal authority, or with those most vocal. Keen attention to the quieter members of the group will sometimes reveal those who, with a few words, can sway the group. The leader must also pay attention to the goal of the team; that is, to permit the referring teacher to have ownership of the solutions.

When an in-school team meeting is carefully planned and regularly conducted, it is a powerful tool for providing teacher support. Teachers bring a wealth of tacit knowledge to such meetings, knowledge which they have acquired in the isolated context of their classrooms, and which has seldom been shared. Some may have difficulty finding the words to express their understanding, and may need time and encouragement to do so. However, the insights, ideas and creative assistance that result are surprising and gratifying. In twenty or thirty minutes, considerably more resource ideas and support for a staff member can be generated than several hours of the work of a solo resource teacher.

The team at the secondary school level

The team approach operating in many British comprehensive schools, associated with the pastoral system (Green, 1989), provides a structure which allows all members of the teaching staff to contribute expertise to solve pupil difficulties. It also provides a vehicle to release teachers from an identity associated solely with their subject specialization, and a forum to experience the concerns and strengths of the staff members.

In this delivery approach the progress of every pupil in the school is monitored by assignment to a form tutor. Teams of tutors meet regularly under the leadership of a senior tutor to discuss problems and needs. The tutorial team attempts to supply a corrective intervention at the first level, that is, within the resources of the teachers involved, or as a result of the tutorial team meetings. If this is unsuccessful, the team may move to the second level in which the support teacher, sometimes associated with a Learning Centre, provides additional help across subject areas. The support teacher will contact subject teachers and parents, and may engage in one-to-one tutorials with the pupil. The main task of the support teacher, however, will be to conduct an instructionally based assessment of the pupil across several core subject areas, by interviewing the pupil, teachers and parents and by observation in classes. This results in a team meeting in which solutions are considered through interventions and support to the pupil, parents and teachers. If further professional expertise is needed, the third level is begun, involving service professionals, such as psychologists, speech and health specialists and community resources. The resource teacher becomes the case coordinator at this stage, and replaces the form tutor as the key worker until additional support is no longer necessary.

In discussion, the head of one such school programme told the author that the main difficulty with implementing this programme is that form tutors, like key workers, have seldom had training to undertake a prevention role in the school. Such skills, including collegial collaboration, must not only be developed through in-service support for those involved, but it is helpful to accord the position as one of responsibility with financial or other incentives, to establish its importance among the staff. Other characteristics of the programme in this school which make it a model for its jurisdiction are as follows: children placed in mixed-ability groups are assigned to each form for the first three years, ensuring a diversity of pupils; children are assigned to a form tutor for their entire secondary school years, which permits the tutor to get to know each pupil in depth; the head teacher provides resources to the tutorial process so that it is seen to have priority in the school as the means whereby academic and personal achievement are monitored.

Curriculum-support teams

When in-school teams have been successfully operating for a number of years, as is the case in some Canadian primary schools, an interesting transformation occurs. In the first years following their formation, in-school teams respond to teachers' requests about students' learning and behaviour problems. Indeed, in the first year of team operation in Chalfant and Pysh's (1989) study, 57 per cent of the intervention goals written by the teams were directed towards management of student behaviours, by improving work habits, classroom conduct, interpersonal behaviour and attention problems. Only 22 per cent of referrals were directed towards academic learning problems. When the teams have been successfully operating for a number of years, a transformation occurs in the type of requests which teachers make, away from behaviour management and help to complete formal procedures towards more subtle aspects of instruction, such as developing pupils' problem-solving and memory strategies and learning styles. Eventually, the staff may request support for the development of adaptive curriculum units; teams become curriculum designers who modify existing curricula or develop new material which will satisfy the teachers' needs to cope with the diversity of all pupils in their classrooms. One elementary school principal in rural Ontario has considerable cultural and economic diversity in the population of her school. She establishes curriculum support teams, providing release time for teachers early in the school year. The needs of individual children become built into the curriculum goals and design. Thus, teachers learn from colleagues about the characteristics of special-need and at-risk children and how to meet their instructional needs, although the ostensive purpose of the team is to develop adaptive curricula. Administrators of a large suburban school board, less than half of whose pupils are from English-speaking homes, have combined the special-needs units with the multicultural units to provide personnel who are able to pool their experience to recommend culturally appropriate modifications to curriculum, classroom practices and materials.

The natural emergence of curriculum and programme concerns from the practices of a preventive delivery approach in a school could be viewed as justifying the current trend to merge special and regular education. However, rather than being imposed by fiat, the merger evolves as the product of changes in staff thinking. Concommitantly, there is a spontaneous decrease in the rate of referrals of pupils for withdrawal and segregated class placements, as teachers discover ways to cope with the diversity of their classes.

IN-CLASS SUPPORT

In her book on the redefinition of remedial education Bines (1986, p. 28) soberly remarks that 'the diagnostic skills of assessment of special needs children should be extended to assess and remediate the curriculum and organization of the school'. Bines's remark highlights the lack of appropriate framework and procedures in schools to support the successful integration of remedial and special-needs students. Bines further reflects upon the prevailing attitudes towards such students, distinguishing the psychometric perspective, in which 'the child is viewed as an empty vessel to be filled with approved knowledge', and the epistemological–phenomenological perspective which considers the interaction between the learner and environment and which is 'crudely represented by progressive primary school education'. Rejecting both stances, Bines proposes the 'experimental approach' in which the process rather than the product of education is focal. This approach seems to reflect elements of what we have termed 'the preventive–integrative' approach, stressing collaborative consultation to support the instructional process. In British secondary schools, there is a move towards this latter perspective. Both Bines (1986) and Hodgson et al. (1984) note how few secondary school teachers regard remedial teachers as needing greater subject specialist knowledge in order to work together collaboratively. Indeed, some subject teachers considered it to be an advantage for their resource colleagues not to have specialist knowledge in the subject concerned, since this might help the resource teacher more fully to understand the difficulties experienced by pupils. Suggestions for adapting the organizational structure of secondary schools, in particular, seem to come largely from Great Britain. The pastoral system in the comprehensive schools permits student groupings to be innovative. At the same time, pastoral tutors have the responsibility in their home room, a responsibility which draws them away from an exclusive commitment to subject material (e.g., Green, 1989; Postlethwaite *et al.*, 1986). Unfortunately, secondary school staff in North America are more inclined to express concern about compromising the integrity of standards in subject areas, often to the detriment of being open to collaborative and experimental approaches to remediation.

How effective schools are organized

Grouping

In a large cross-sectional study of seventy secondary schools, the NFER team (Hodgson *et al.*, 1984) documented the ways in which teachers had engineered the integration of special-needs pupils, often by developing innovative collaborative relationships across the staff. These included grouping pupils for mixed ability, and by ability bands which varied from one subject to the next. Some vertically organized groups developed a family atmosphere in which peer support, peer tutoring or 'buddy systems' could be enhanced. Team teaching was frequently used, sometimes with regrouping of pupils. A designated (resource or support) teacher might conduct the class to free the subject teacher to work with remedial groups. This is an example of what is called 'reverse integration' in North American schools.

Grouping of pupils can have significant benefits on their achievement. Slavin (1988) notes that there is still much to understand about the effects of grouping. While certain forms of grouping can be an effective part of adapting instruction to individual needs, the costs and benefits of different forms of grouping must be carefully weighed. McKenzie (1983) also reviews characteristics of classroom organization which lead to effective instruction. In the Hodgson study, flexible use of staff time was noted; timetables were developed for groups rather than classes. Traditional remedial periods were redesigned to include co-teaching periods.

Room management

Thomas (1985, 1986, 1987) explores room management procedures which enhance the teacher's direct instructional time with individual pupils. Thomas claims that, by coordinating ancillary and volunteer personnel to help individual groups of pupils, and to manage the flow of activities, the engaged time of pupils on learning materials can effectively be doubled. Further, teachers feel empowered by such structural changes to control the learning of the total class by directing the activity and flow managers, while being freed to devote more time to individual pupils. Peripatetic and remedial teachers are also able to fit into the organizational structure of the classroom, without disrupting the control of the resident teacher.

Time

Teachers usually make the decisions about how time is allocated in the classroom. Berliner (1983) distinguished between allocated time, engaged time and academic learning time. Allocated time is the proportion of the time within the constraints of their timetables that teachers choose to devote to aspects of subject matter; reading comprehension vs. decoding skills, or discussion vs. seatwork. Engaged time or time-on-task refers to the behaviour of students within allocated instructional time, and academic learning time is a subunit of engaged time in which the students are directly engaged in instructional tasks that relate to the teacher's objectives for the instruction. The proportion of time spent on each of these time components, and the nature of the activities within each, directly affect student achievement.

In the United States, the Minnesota group (Thurlow *et al.*, 1983) have also explored the length of time in which exceptional pupils are actively engaged in learning in integrated and segregated classes. Their findings are startling. In an average primary classroom, about half of the school day (180 of 390 minutes) is allocated to academic instruction. Of this, pupils will spend an average of 25 per cent or forty-seven minutes actively engaged in instructional events. Less than eleven of these minutes will, on average be spent reading; ten silently and less than one reading aloud. Apparently less than 3 per cent of the school day is spent on tasks which give rise to the most significant academic problems in the middle school years.

Pupils in segregated learning disability classes fare little better, although the smaller ratio of pupils to teachers should in theory result in enhanced learning time. After two months in a class with a pupil–teacher ratio of 8:1, the learning-disabled pupils had managed to reduce their academic engaged time to approximately that of similar pupils integrated in regular classrooms. Thomas (1986) may be correct that we have yet to come to terms with the challenge of the classroom environment in assisting class teachers to adequately meet the needs of children who are experiencing difficulty. Ironically, however, the same conclusion may also be true for the small-group remedial and special-needs class. Organizational strategies are notably lacking from texts on instructional methods in special education.

How instruction is adapted

Teaching methods and adaptations of the curriculum to meet the needs of exceptional learners are the subjects of numerous books, articles, conferences and professional development packages. There are, however, some excellent summaries for the resource teacher to explore. For example, in Canada, Bachor and Crealock (1986) provide a comprehensive manual of instructional topics, their assessment in pupils, and strategies for programming. In Great Britain the NFER study (Hodgson *et al.*, 1984) describes teaching strategies such as pre- and post-lesson teaching, and modifications to curriculum content.

In the United States, there has been a plethora of reports of what is now termed 'effective instruction'. Rosenshine (1983) reviews the work on the instructional behaviours of teachers which were shown to have the greatest benefits for student achievement and self-concept:

1 reviewing (a) the previous day's work, and reteaching, if needed, and (b) weekly, monthly and end-of-unit reviews;
2 presenting new content and skills;
3 allowing students to practise (a) in class, and using this to monitor students' understanding, and (b) independently, out of class;
4 giving frequent, detailed feedback and corrections. Feedback or reinforcers are most effective when they are positive and therefore enhance the pupil's self-concept.

Effective teachers take an active role in the management of their classes, structuring the learning process in response to students' production, and creating an orderly climate to which students are expected to contribute.

As Bickel and Bickel (1986) conclude, much of the discussion of the efficacy of mainstreaming has been focused upon the pupil within the placement. Little emphasis has been given to the variations in the instructional procedures within such placements. For a resource teacher who is attempting to bring about change in the school setting, the instructional process becomes the primary focus. As discussed earlier, the in-school consultant is attempting to empower or enable teachers, to convey to teachers that they are craftspeople with skills to encompass a range of instructional levels and variations. Further, such skills are as relevant to children with exceptional needs as to children who perform within the classroom norm. To convey this understanding, tacit permission must be given by the support providers and by the school system at large for the teacher to experiment with instructional adaptations. Support must also be given to monitor the outcome of such explorations upon student performance.

Huberman (1992) notes that, in the later stages of their careers, teachers who experience career satisfaction turn away from in-service initiatives and professional development. They generally prefer to close their classroom doors and 'tinker' with instructional variations. They frequently excel in adaptive instruction, but their senior expertise is neither accorded credit nor well used in the coaching of more junior colleagues. Coaching, defined by Joyce and Showers (1980) as providing companionship, giving technical assistance to a colleague and analysing the outcomes, can also have a profound impact on the instructional productivity of teachers and on the instructional time they spend with students. In many respects, the ideal support teacher is the staff coach in the school. Ironically, the components which are essential if the support person is to create change in a school through staff development initiatives, such as coaching, are identical to those which make a classroom teacher effective with students:

- providing information and rationale;
- demonstrating the target performance;
- allowing teachers to practise;
- providing feedback.

The resource consultant, as coach, is really practising a variation of good teaching with the adult staff in the school. The notion will be developed in Chapter 8 that by doing so the consultant is creating change within the school.

Chapter 8

Overview of classroom consultation
Getting ready to begin

There is a certain relief in change . . .; as I have found in travelling in a
stage-coach, that it is often a comfort to shift one's position and be
bruised in a new place.

(Washington Irving, *Tales of a Traveller*, Preface)

THE GOAL

The premise was made at the outset of this book that the resource teacher,
acting as a consultant to fellow teachers, is performing the role of a
change agent in the school. You will recall that I contend that consultation
techniques are merely tools, to be used in the reform of a school's delivery
approach. The purpose of the approach is to establish a collective responsibility among the staff for all of the pupils in the school. While many
school staffs are currently well versed in this approach, there is considerable variation from one school to the next, and even from one staff
member to the next within the school. There are curious differences in the
approaches taken on each side of the Atlantic to the reformation of special
education. The British literature is promoting the concept of collective
responsibility through the 'whole school' movement, under the urging of
government reports (DES [Warnock], 1978; DES, 1984, 1985) and policy
analysis (Brennan, 1979; Solity and Raybould, 1988). Because a relatively
small proportion of children are eligible for the specialized resources of
special education, concern has been raised about the 15–20 per cent of
pupils for whom there is little or no likelihood of specialized help outside
of the ordinary classroom. In North America, the pressure for change has
developed from a quite different source. Consultation, as a vehicle for
enhancing collective responsibility, is seen as a pre-referral process
(Johnson *et al.*, 1988; Pugach and Johnson, 1989; Graden, 1989b) aimed at
reducing the escalating number of pupils who are referred for special
education and in particular for the category of learning disabled.
Currently, 15–20 per cent of children in the school systems of the United
States and Canada are formally designated as exceptional, and provisions

are made which parallel the statementing process in Great Britain. The irony is that ordinary classroom and subject teachers have for several decades been given the implicit message that any such underachieving children should not be held indefinitely in the ordinary classroom but should be referred to specialized personnel and services. Legal and political mechanisms, including strong, vocal parent organizations, have added to the weight of this belief.

Despite the differences in policy and context, both North American and British school systems see the consultation process as an antidote to the fragmentation of service delivery. Its aim is to redistribute resources towards immediate problem solving in the ordinary classroom, and to create a dialogue between regular and special education. It is seen as providing informal alternatives to assist teachers within the current organizational constraints of the classroom, and to avoid the expensive and cumbersome mechanisms and the psychological fallout of formally admitting pupils to special education. The provision of a consultation service also acknowledges that ordinary teachers are faced with complex problems from the requirements of mainstreaming, not only to integrate children who were previously served in contained schools or classrooms, but also to meet the needs of the diversity of children within a classroom for whom referral is not an option. Today's classes are indeed complex, reflecting the cultural, social and economic diversity of modern society. As previously noted, most teachers feel that they are running just to keep up. The challenge of further innovations, including prevention mechanisms and integration, are therefore daunting.

Barriers

While the notion of collaborative consultation among staff has intuitive appeal and pragmatic benefits, all is not rosy in practice. Ordinary classroom teachers have many reasons to mistrust the injection of a collaborator into their traditionally isolated work setting. For several reasons, they are justifiably cautious. While they have understood that they are not qualified to cope with the 'specialness' of children's needs, they are told that, morally and ethically, such children have a right to a 'normalized' educational setting among their peers. Teachers frequently view these conflicting messages as a managerial ploy to save money and resources at their professional expense. When teachers are offered resources to assist them to meet the needs of such pupils, they are suspicious of the relevance and usefulness of such resources. Support teachers drawn from special-education backgrounds may bring a knowledge base and understanding of the ordinary classroom which diverges from that of ordinary teachers (Furey and Strauch, 1983; Johnson *et al.*, 1988; Thomas, 1986). Support teachers have training in individualized instruction, behaviour modification techniques and the application of specialized materials to specific handicaps, while classroom teachers develop expertise in large group management, time-sharing between individual and group needs, classroom organization and group behavioural management. Classroom teachers occasionally remark about support consultants that 'they don't know what it's like to have thirty children all day long'. Support teachers, in turn, feel that they lack credibility with teaching staff because they are seen as tutorial supports to children and not as colleagues with relevant expertise. Indeed, Furey and Strauch (1983) note that special-education-trained teachers rated substantially higher than ordinary teachers their ability to teach mildly-handicapped children, while ordinary teachers conversely rated their own ability as higher than that of special education teachers. Rosenfield (1985) describes the lack of match between teachers' organization and management of classrooms, and the intervention strategies which resource teachers frequently recommend for individual children. The predominant instructional methodology of teachers trained to remediate learning disabilities is individual behavioural techniques, while classroom teachers are increasingly exploring cooperative learning and metacognitive strategies as part of their classroom practice (Pugach and Whitten, 1987). Johnson et al. (1988) list those factors which in their opinion create the main obstacles to effective special-education consultation; lack of clarity about who is responsible for what, skepticism about consultants' expertise about regular class pupils, the overwhelming schedule and lack of time of the classroom teacher and teachers' lack of training for the collaborative process.

One might argue that the credibility gap which exists between ordinary and support teachers is the reality of modern schools and is the raw material with which a staff must wrestle. Yet, Graden (1989a) argues that such barriers are not as impenetrable as they may appear. In a collaborative, consultative framework which offers truly reciprocal sharing of expertise, which is supported appropriately by administrators, and which is adaptable to the ecology of each school, a workable delivery approach can be developed. The NFER study (Hodgson et al., 1984) and Bines (1986) offer clear evidence that secondary school staffs develop innovative and mutually supportive ways of collaborating. In the remainder of this chapter, we review how a plan to implement a consultation service within a school's delivery approach may be devised which has the potential to overcome both attitudinal and practical barriers. However, as Goodlad (1975) notes, depending on the readiness factor in the school, the implementation could take an average of three years to accomplish, and may take considerably longer.

TAKING STOCK OF YOUR SCHOOL

No matter who has initiated changes in a school's delivery model, nor what form the plan takes, the success of the project will be affected by the resources and particularly the strengths of the staff in the school. It will also be affected by the climate or ethos of the school, and the context of the school set by the school system and the community. Taking stock of these helps to design a plan which may eventually be implemented. Before commencing a new project, a careful stock-taking, or needs-assessment phase should be planned, in which staff participate in identifying their strength and needs.

The support role

Time

What time allocations are available for the support role?

Tindal and Taylor-Pendergast (1989) provide descriptions of the break-down of a support teacher's time, corroborating the findings of Evans (1980) that the time actually spent in consultation with teachers was less than 5 per cent of the working week. The consultation activities them-selves consisted of preparing written material (14.6 per cent), communi-cation with teachers, parents, students, administrators and other professionals (22.4 per cent), non-interactive observation, usually as part of an assessment (22 per cent), interactive testing time using formal testing techniques (32 per cent) and reviewing records and preparing materials (9 per cent). Of the time that the consultant spent interacting with others, interacting with a teacher accounted for only 14 per cent.

Evidently the support consultant spends very little time actually engaged in consultation. One way to take stock of the resource role is to establish a baseline for the time spent on various activities. Not only will this furnish data to support your negotiations for changing the role, but it will also provide the basis for later comparative evaluation.

Technical capability

What is the technical capability of the resource teacher and how is it viewed by the staff?

Feedback from staff about the credibility of your consulting role may be difficult to elicit. Yet the credibility issues discussed above may be crucial to the success of the implementation. The staff may be forth-coming about their views in a general staff meeting, or at the inception of an in-school team, particularly if the proposed approach is about to be implemented. It is more likely, however, that you will have the confidence of one or more members of the teaching staff with whom you will work, while other members of staff will hold their reactions in reserve until they see how the project evolves. One way to commence the project, therefore, is to start small, with one or two like-minded colleagues.

Credibility

Does the support teacher have the ability and credibility to be a 'master teacher' on staff?

In some schools, the support role is designed not because of the special education expertise of an individual, but because he or she is acknowledged as the senior statesman on the staff. Johnson *et al.* (1988) are cautious about the promotion of expertise over peer acceptance. They suggest that schools are inherently hierarchical organizations, and that the support teacher is, de facto, an expert on the staff. They state 'as long as consultation exists within a hierarchical organizational structure in which classroom teachers are perceived to have less expertise, its success is likely to be limited, especially in a climate of reform in which classroom teachers are rightfully searching for the professional respect due them in view of the difficult job they perform' (p. 44). Potts (1985) is more pessimistic about the inhibiting effects of perceived expertise. She recommends that professionals must 'relinquish some of their mystique, and bring their arcane jargon and rituals to public scrutiny'. While this may be an overly negative reaction to the hierarchical nature of the teaching profession, the point must be made that mutual respect and 'genuine teamwork which does not imply one-way support' (Quicke, 1985, p. 22) are crucial to the resource role. The reciprocity of the teacher's and resource roles will take time to establish and as Little (1985) recommends, skillful pairs build trust by acknowledging and deferring to one another's knowledge and skill, by talking to each other in ways that preserve individual dignity, and by giving their work together a full measure of energy, thought and attention. In a reciprocal relationship, the resource teacher demonstrates respect by attending seriously to the client, and demonstrates a degree of modesty by being willing to learn from the other and to improve. In Kilbourn's (1990) text of the two-week interaction of a science teacher and his department head, Oliver and Taylor, the reciprocity of roles, and the ability of each to learn from the other is delightfully portrayed.

Perhaps the final word on the issue of credibility is that each person brings expertise to the consulting event. You most certainly need to be recognized as credible if not an expert in aspects of each case. However, to portray yourself as an expert at the expense of the client is to endanger the collegial relationship. Trust will be established when each views the other as contributing knowledge and experience in a complementary way. As discussed in Chapter 3, fear of vulnerability and loss of control will endanger the trust of the client, leading to sabotage of the contract between you and the client.

While Pugach and Johnson (1989) may view the consulting role as confounded by its own mandate, it is the consultant's delivery of the role, rather than his or her substantive knowledge and technical expertise, which will ultimately ensure that the role is successful.

Personal skills

Is the support teacher sufficiently confident to be able to give away his or her expertise and resources to fellow staff, without desiring to take credit for their impact?

Few analyses of consultants' skills remark on the importance of personal self-confidence. Yet, in working with consultants-in-training, I have come to believe that this attribute is central to success in the role. When training resource teachers to assume a consulting role, I favour a programme that permits participants to opt out of the role without personal or professional consequences if they so desire. Miles *et al.* (1988) list the eighteen characteristics of successful 'change agents' which span personal, educational and interpersonal domains. West and Cannon (1988) surveyed seventy-one competencies of consultants in nine categories, and found that forty-seven were rated as 'essential'. The highest ratings were given to interactive communication skills and to personal characteristics, values and beliefs. These are tall orders for any person who is undertaking a new role, and doing so with relatively little historical context or current training to guide him or her. Today's resource teachers are pioneers in a new field, and pioneering is not for everyone. The right to opt out of the role applies as much to the resource teacher as to his or her clients.

The school administrator

Will the head teacher, deputy head and department head, set the tone for the project and give you support to allow you to move ahead?

The literature of leadership styles identifies many factors which assist or detract from the effectiveness of the head teacher (Leithwood, 1988). It has been concluded that the role and style of the head teacher has a central influence on the nature and success of a policy implementation and its acceptance by the staff (Fullan, 1982; Fullan and Stiegelbauer, 1991). In our research (Wilson and Silverman, 1990, 1991), the restorative or preventive beliefs of a school staff regarding the school's delivery model correlated significantly with those of the principal. Indeed, in evaluation of the most important contributions to an effective consulting role, school staff most frequently mentioned the support of the head teacher.

Support includes tangible resources such as time to consult, financial resources such as supply or relief teachers to conduct in-service meetings to orient the staff, and incentive systems for participants. While creative systems may include time, recognition through positions of responsibility, space and materials, much of the literature on school reform demonstrates that the incentives most valued by teachers are intrinsic to the teaching process. Crandall *et al.* (1986) and Rosenholtz (1985) both note the power of administrators who buffer teachers to ensure that teachers' efforts may be devoted to raising student achievement. Teachers see increases in their own effectiveness with students as the single most important factor in their job satisfaction. Effective school heads are able to enhance teacher effectiveness by providing technical assistance and opportunities to establish strategies which achieve instructional goals. Heads also take a personal and active role in monitoring the academic progress of pupils in the school, and reflecting their findings back to the teachers whose efforts contributed to the progress. Thus a detailed monitoring system for pupil achievement, as proposed in the assessment and programming cycle and in reporting to teachers (Chapter 4), is an integral tool in developing staff participation and incentives. To be effective, the head teacher must demonstrate that the monitoring and reporting process has high priority in the function of the project and the school.

The head teacher's involvement in monitoring the progress of the delivery approach itself is also important to the perception of staff about its value in the school culture, an issue to be considered in monitoring project outcomes, below. As Idol and West (1987, p. 490) note, 'Lack of administrative support is a major barrier in initiating and implementing teacher consultation programs between regular and special educators.'

Relationships among staff

While the barriers to the role of the resource teacher were discussed, other obstacles may arise from the climate or ethos of the school, and from the relationships of staff with each other. For example, Hargreaves (1991) discusses four types of school climate:

1 isolated, in which each member of staff works alone, with little collaborative interaction, and a high degree of mistrust;
2 balkanized, in which staff align themselves into factions, often on the basis of common interests or out-of-school social interactions. Groups may see other groups as oppositional;
3 contrived collegiality, in which the staff of the school appear to be working collaboratively, but they feel little commitment to the process, and their morale may be low; and
4 collaborative, in which true interaction for support and assistance occurs.

Obviously, one hopes to work in the latter type of school, particularly if introducing new ideas involving support teams and resource distribution. If your school staff are not truly collegial, are there groups of people with whom you might start to work collaboratively?

Strengths of the staff

How open are the staff to new ideas? In plotting the career development of teachers in Switzerland, Huberman (1992) found that there was greater receptivity to change in the first seventeen or so years of a teacher's career, after which they depended on their own past experiences to guide their work. Who, on the school staff, has experience and ideas that are respected by others? What areas of curriculum expertise exist? Who demonstrates clinical ability and insights with problem pupils? How can their expertise be drawn into a collaborative process?

External resources

What are the strengths of resource people deployed from the central school board office? Who might be willing to work in a more preventive mode? How could this be done? Are there resources upon which to draw in the community, for example, in local mental health units and social agencies? What resources could parents provide? In previous chapters, the potential roles of service professionals and parents were discussed. It was proposed that, by involving these people early in the project, their later involvement would be accepting of it. Fullan (1985) notes that teachers and other professionals are most receptive to an innovation when they have a hand in its planning, design, implementation and evaluation.

Central office press

Fullan (1985) also notes that successful change involves pressure, both through the interaction of staff and leaders within the school, and from priorities set by the school system. In one school system, the central office administrator presented the heads of his schools with a choice. A pool of resource teaching and supply or relief staff was made available. Any school who referred and had accepted eight children for a special-education class would receive a contained class teacher, as required by policy. This teacher was drawn from the pool. Support staff not allocated in this manner could be shared between the schools. These teachers would be available to work directly with staff and students for 75 per cent of their time, in ways jointly selected by the head, resource teacher and staff in that school. The administrator then permitted his head teachers to negotiate with the staff, to assess the amount of special-education staffing which would be needed. The heads then met to divide the pool collectively, all aware of the benefits of being able to staff their schools over formula with additional support teachers if they minimized their demand for teachers of self-contained units.

Administrative press, in this example, created by resource allocation rather than in the form of written policy, has a direct impact on the perceived need for local changes.

THE STRENGTHS AND PROBLEMS OF THE SCHOOL'S CURRENT DELIVERY SYSTEM

By taking stock of the internal and external resource base, the support teacher is equipped to consider the feasibility of proposed changes. Before making proposals, however, a survey of the staff will reveal what they see as the advantages and shortcomings of the current delivery procedures. This survey could be conducted by a questionnaire, but a staff meeting allows discussion of the concerns which each person raises, and it provides an opportunity for the staff to become aware of ideas and perspectives which are new to them. Any survey should be completed with a reporting back of the opinions and issues which predominate, and an identification of the problems to be solved. Staff are now able to be involved in the dialogue from which the plan will be drawn. There will probably be resistance and disagreement. The proposed plan, with changes to the existing role of the resource teacher, the creation or modification of an in-school team, and any policy directives to support it, may begin as a pilot endeavour, with a few staff members who are supportive of the ideas and direction. Alternatively, consider a full staff meeting at the beginning of the year. As Idol and West (1987, p. 490) write, 'Those consultants, who perceived themselves as successful in their consultation roles, sold consultation as a valuable support service to colleagues through in-service or written descriptions of their consultative services.'

After the first meetings, every consultation event becomes an opportunity to develop the model and to demonstrate its potential effectiveness in meeting its goals.

GETTING STARTED

The subject of implementing innovations in the schools is too large to tackle within the confines of this chapter. Fullan and Stiegelbauer (1991), among others, provide an excellent synopsis of critical issues in educational change. As stated earlier, however, a support teacher or consultant cannot practise consultation skills and engage in collaborative work with fellow teachers without a clear sense of the intended outcome of this work. In order to abandon the relative safety of a closed classroom or resource area away from the mainstream, and to become a staff developer on the heels of a career in working with children, you will need to have a vision of what you hope to achieve, and some criteria to help you see the results of your efforts.

Involving the staff

By sharing the development of the proposed delivery model with colleagues and by inviting them to participate in modifying and evaluating it, their long-term commitment is established. However, the participant role is not simply a matter of involvement. A school innovation is most likely to succeed if:

1 The participants see a clear need for change at the outset. This may require hard evidence that the problem exists. Fullan (1982) states that breakthroughs occur when people cognitively understand the conception and rationale for why the new way works better.
2 The project leaders design the proposed solution to the problem, in this case a revised model of how the school delivers services and instruction which focus on prevention and integration, but participants are able to respond to it, and to modify how it will be implemented.
3 Implementation is flexible, allowing changes to occur as the evidence is collected. Peer norms, leadership and in- and out-of-school factors affect whether change will occur.
4 Evaluation is based upon criteria that were mutually established at the outset, and which are objectively applied and openly available.

As an example, in one school a staff meeting commenced with the flow-chart diagrams of the restorative and prevention models (Figures 2.1 and 2.2 in Chapter 2). Staff were invited to specify the strengths and weaknesses of each model, and to compare them with their own. The staff collectively redesigned the flow chart to incorporate the multicultural consultant and English-as-a-second-language (ESL) teacher assigned to their school, since multicultural and first-language issues were viewed as central to their resource needs. From this base, the staff began to examine proposals for reorganizing their tasks to include an in-school team to promote collegial support, and classroom consultations with both the resource and ESL teachers. This case illustrates the flexibility and collaborative involvement in the design of this school's project.

The teachers' dance

Little (1985) describes the first faltering steps of a project to promote teacher advisors as in-class consultants. She describes the delicacy of the interaction between consultant and client. In response to the invitation to 'use me', teachers were hesitant to request advice from the consultant for fear of being perceived as using them as assistants, or as Little remarked, as 'gofering' ('go for this . . .'). Simultaneously, the consultants feared stepping on the toes of the teachers, by intruding in their routines; hence, the teachers' dance. The interactions became successful when consultants commenced their invitations with a specific description of their services; 'this is what I've done before, and this is what I can do for you'. Little followed six technical principles in training successful teacher advisors; principles which serve as a review of points developed in the discussion in this book on consulting skills:

1 Establish a common language, minimizing jargon and maximizing understanding about the curriculum, programme and problem.
2 Focus on one or two key issues, with depth, persistence, imagination and humour.
3 Select observational methods, together with the teacher, that provide hard evidence of the success of your joint efforts.
4 Demonstrate a lively, mutual respect.
5 Establish trust by bringing predictability to the relationship, one in which each could rely on a known set of topics, criteria and methods to be used.
6 Reciprocity: acknowledge each other's expertise, and talk to one another in ways that preserve each other's dignity and professionalism.

EVALUATION MEASURES

What objective and mutually observable measures of progress are available to monitor the project?

Evaluation measures can be considered in three categories:

1 *input measures* – the number of referrals, the attitude of staff, their expression of satisfaction with the planned changes, etc.
2 *process variables* – the conduct of team meetings, the rate of requests for consultation, how well the consulting event progresses, changes in time allocations to consulting functions, etc.
3 *outcome variables* – changes in student achievement, staff and parents' attitudes to the process and to children's achievements, changes in the number and types of request for assistance, cost effectiveness measures, indices of willingness to collaborate.

Input variables might include a reduction in teachers' requests for confirmatory assessment, and a change in the nature of the requests directed to the resource teacher which reflect a change in teachers' expectations. In one school system in which a training programme for resource teachers was implemented, the people responsible for formal admission meetings to special education noted a far greater amount of documented programme modifications for pupils being presented by teachers than prior to the training programme. The nature of the teachers' under-standings about pupils' needs had also changed, from 'X appears to have a learning problem and needs more specialized help than I can provide' to 'I have run out of ideas for helping X and I want your help to find new ways to tackle his difficulties'. The messages were similar but the locus of responsibility had shifted from the pupil to a joint one between pupil and teacher. One evaluation criterion, then, might be a change in the attri-bution of causes of problems made by teachers.

Process variables might include a sample of the time allocations of the resource teacher's week at the beginning of the project, and at subsequent intervals. Tindal and Taylor-Pendergast (1989) have developed an instru-ment for observing the engaged time of consultants who are actively engaged in consultations. One might also expect the total time engaged in consultation to increase relative to the time spent in direct activities with pupils, as teachers increasingly draw upon the service. Time with pupils, also, might reveal a decrease in confirmatory assessments and remedial instruction in proportion to the time spent in instructionally based assessments.

Outcome variables might include both quantitative and qualitative measures. Quantitative indices include pupil achievement records, the number of referrals to special education, the number of integrated pupils successfully accommodated, and the return rate to special education or to more restrictive settings of pupils who are integrated. Qualitative indices are equally persuasive; teachers' expressed satisfaction or lack of it with the delivery approach, teachers' views about their own ability to meet the needs of pupils, and about the type and value of the support they receive. Parents, service professionals, administrators and the pupils themselves will furnish opinions about the success of the approach and their involve-ment in it.

Idol (1988) proposes that a project team be established at the outset to design evaluation criteria appropriate to the context of that school. The team should select criteria which indicate the impact of the overall programme, as well as the effects of specific consultation services upon staff, students, parents, administrators and the support teachers themselves. They may consider using comparison schools or classes not involved in the project. As a team, they would be involved in the synthesis and interpretation of the data, and have the major responsibility for reporting it to staff.

Will a collaborative consultative process among staff have an impact on the achievement and well-being of students? This is the ultimate question posed by school governors and trustees, by academics and by parents. The issue is that of accountability. However, the research evidence to date is sparse and equivocal. In attempts to prove the efficacy of one whole-school approach, the Assisted Learning Model (ALEM), Wang and colleagues (1985) have come under considerable criticism for lack of scientific rigour in their measures and bias in their evaluation (Fuchs and Fuchs, 1988). Pugach and Johnson (1989) and Fuchs *et al.* (1990) also articulate the problems of the pre-referral process and the lack of evidence by which one might evaluate its impact upon students. This research has yet to be done. Other writers (Heron and Kimball, 1988; Lloyd *et al.*, 1988) suggest the need for a great deal more research evidence before issues of integration, pre-referral intervention and consultative support can be viewed as valid solutions to current learning problems.

One must ask, however, if lack of empirical support is reason to forsake the approach. There is mounting evidence that segregated class programmes and the alternative traditional delivery procedures are not effective overall in enhancing student achievement (Carlberg and Kavale, 1980; Epps and Tindal, 1987; Wang and Baker, 1985–6). From a pragmatic point of view, therefore, alternative approaches must be found, and their impact assessed. Of course, research data which are used to compare the overall efficacy of integration versus segregation explore the trends in populations of students, without focusing on those single individuals who benefit greatly from any specific approach. I claimed at the outset of this book that, in the absence of unequivocal evidence, one should not align a consultative approach with total integration, or any other prescribed delivery model. A range of alternative placements will currently serve the interests of all students. The task of consultation is to ensure that the best selections are made.

Collaborative consultation is a process that aims to assist people in maximizing resources for pupils on a case-by-case basis. It should not be considered as a policy for placement, or as a mandate for the delivery of resources. Teamed with a viable approach to delivery instruction and services to special-needs pupils, such as a prevention-based process, it could have the potential of redeploying staff in order to optimize the use of local resources and expertise, and therefore, over the long term, of enhancing the performance of both teachers and students. This case has yet to be proved, however. It is therefore essential that evaluation criteria are established upon which all participants agree. The criteria should reflect both philosophical and organizational goals of each school system. Qualitatively and quantitatively defined outcomes are essential components in any implementation plan.

The goals of consultation are to solve an immediate problem, to enhance the ability of the client to solve similar problems in the future and to effect change. Accountability for the last two goals is difficult to establish, and requires evidence over sustained periods. Consultation involves giving one's skills away to others, enabling them to succeed, and systematically reflecting back to them that they have achieved success. To enable teachers to be successful through day-to-day interaction on a case-by-case basis is a slow process, and particularly difficult since teachers have held long-term beliefs about their own lack of expertise, and their low status in a hierarchy of authority and specialized skills. The school-based consultant therefore begins with a negative ledger sheet; he or she must establish trust, mutual respect and a common language with the teaching staff, before consultation can begin to be effective.

Nevertheless, the potential for a consultative approach is enormous. It has both logical and intuitive appeal. In the face of escalating referrals for special education in North American schools, and the increase in concern expressed about Warnock's 18 per cent (Gipps *et al.*, 1987) in Great Britain, there is clearly something amiss with the current system. Also, preliminary evidence from school systems which are in the process of implementing a preventive approach is that there are significant decreases in referral rates of children to special education, and changes in the nature of the services which are requested. There has yet to be a critical test of the complementary component of the prevention approach, that students demitted from an intensive remedial withdrawal or segregated special class placement into the ordinary classroom are able to sustain gains in achievement when consultative support is provided to their teachers.

Thomas (1985, 1986) identifies perhaps the major factor that will determine the eventual success or failure of collaborative consultation. Thomas distinguishes between the organizational priorities of classroom and remedial teachers. Classroom teachers manage the structure and flow of work of groups of children, while remedial teachers conduct diagnostic procedures and design interventions for individuals. In effect, there are differences in organizational priorities and the definition by which each group considers itself to be successful. The culture of the classroom, to use Hargreaves's (1991) term, differs from that of traditional special-education settings. In a profession in which organizational priorities require alignment, dialogue and therefore collaboration are essential. Consultative events, therefore, require a context of commonly held beliefs, about the needs for the dialogue and about the benefits to be accrued.

Consultation embedded in a school-based delivery procedure that emphasizes the impact of the classroom context on children's characteristics has considerable potential as a staff development tool. Although the case has been made in this book for serving the needs of exceptional and at-risk children, the process is measured in many initiatives in other areas of staff training: mentoring, coaching, teacher training. We may be witnessing the start of a quiet revolution. It begins in the fragmentation of current professional roles, and in the isolated culture of one teacher per classroom. It may end in new organizational structures in schools, in collegial interaction and school-based professional development.

Appendix 1

The contracting meeting and identifying needs and offers

PLANNING AND REVIEW SHEETS FOR CONTRACTING

Planning the contracting meeting

A 1 What do you think the client's problem is?

2 What will the client need/want?

3 What do you need/want?

4 What can you offer?

5 What will you want the client to offer?

B 1 In what ways might the client be feeling vulnerable? Out of control?

2   What resistance can you anticipate?

_____

_____

3   How might you be able to deal with the resistance?

_____

_____

4   What support might you be able to give?

_____

_____

C 1   What other agendas, clients might be involved, if any?

_____

_____

2   What other information can you anticipate needing to access?

_____

_____

3   What time and reporting schedules do you need?

_____

_____

4   How will you monitor/evaluate your joint involvement?

_____

_____

Reviewing the meeting

This exercise provides a framework to help you to review your role in a meeting.

1   How did you feel about the meeting?

_____

_____

2 Who controlled the steps?

_____

_____

3 How were initial needs and offers exchanged?

_____

_____

4 What form did resistance take?

_____

_____

5 What form did support take?

_____

_____

6 What would you do differently next time?

_____

_____

7 What are your next steps?

_____

_____

LISTING YOUR ESSENTIAL AND DESIRABLE NEEDS AND OFFERS

Planning activity

1 In developing your role and plan in your work setting, list

   (a) your essential needs and offers

(b) your desirable needs and offers

2 In collaborating with a client, list

    (a) your essential needs and offers

    (b) your desirable needs and offers

Responding to resistance

Write down how you could name each of these forms of resistance (Chapter 5, p. 66)
Avoiding responsibility

Flooding you with detail

Attacking you

Compliance

Humour

Changing the subject

Silence

_____

_____

Pressing for solutions

_____

_____

Other?

_____

_____

RESPONDING TO RESISTANCE – SOME SUGGESTIONS

Suggestions for how you could name each of these forms of resistance (Chapter 5, p. 67)

Avoiding responsibility

You're unwilling to carry out your end of the bargain.

_____

Flooding you with detail

You have a lot of knowledge about this problem. What do you see as the

two essential issues?

_____

Attacking you

This issue really upsets you.

This situation makes you very angry.

Compliance

You're willing to do what I want, but what do you want?

_____

Humour

You're very funny, but let's take a serious look at the problem.

_____

Changing the subject
You keep changing the subject.

Silence
We won't solve this unless I hear from you.

Pressing for solutions
You want to contribute some possible solutions, but first let's get the measure of the problem.

Other?
Acting helpless. 'You're unwilling to take the first step.'

Appendix 3

Exercises and case studies

BRAD THE BRAT

Ask a partner to assume the role of Mrs Higgins. Give him or her the case description below, but *do not read it yourself*. Read the consultant case description.

Try your hand at simulating a contracting meeting with 'Mrs Higgins.' Give yourselves ten minutes to follow the steps for contracting (p. 68).

After ten minutes ask the review questions of each other which are listed beneath the cases.

Mrs Higgins

TO BE READ ONLY BY THE CLIENT

You have been teaching for twenty-one years, and in this school for the last fifteen. You have seen head teachers and administrators, board policies and curriculum trends come and go. You have developed a sound, structured approach to teaching. You expect discipline, compliance and hard work of your pupils and colleagues.

Brad has perplexed you. He arrived lacking even basic grade 2 reading skills. His previous teacher was inexperienced and inconsistent in her discipline. Your whole grade 3 is suffering from this, but Brad particularly has 'run ragged'. Your real agenda is to get the consultant to do something about the grade 2 teacher before you get another group like this one. Your principal, Mr Gentle, sympathizes but hasn't acted, and, since it's January, you're getting worried. You think you can handle Brad by stomping hard on him, but you're not going to reveal this to the consultant until Brad has served your purpose to bring about a change in the grade 2 situation.

Consultant

TO BE READ ONLY BY THE CONSULTANT

Mrs Higgins, a grade 3 teacher, is having trouble with Brad, and asks you to drop by to discuss him. Brad is extremely disruptive in class, kicking and fighting with other children, failing to stay in his seat, loud and abrasive towards Mrs Higgins, never completing assignments. On checking the school's record you establish the following:

- Brad has not had negative reports prior to grade 3.
- Brad's academic functioning is slightly below grade level in core areas; but not enough to prevent his promotion into grade 3.

Your knowledge of Mrs Higgins is that she has been at the school for a long time, she is seen by her colleagues as 'senior statesman' in the school. You have not yet met her.

Review questions

1 Did you reach a mutually acceptable outcome?
2 What did the client particularly like about the consultant's approach?
3 How did the client feel about the process?
4 How did the consultant feel about the process?
5 What might the next steps in the contract be?
6 What other outcomes and possible next steps might have been negotiated?

HUBERT BROWN – A CASE STUDY

A meeting is about to take place between you (a board-level consultant in special education) and Mr Hubert Brown (principal of an elementary school). Hubert has concerns about the attempts of Mrs Jones to integrate her behavioural class into the music programme. The staff are responding negatively to any attempts to integrate Mrs Jones's exceptional pupils.

You've done some work in the past with the staff in this school; that's how Hubert came to know you. Over the phone, Hubert mentions a growing interest in integration, derived from a course he has taken at the local university. He proposes to send six of his teachers to the course, and wants you to stop by to discuss this idea.

Since you have been with the board for some time, you've met Hubert casually, but you've never worked with him directly. From what you've heard, you tend to doubt whether Hubert understands the implications of integration within his school, his role in it, and the understanding needed by the school unit, but you're not really sure.

Plan your next meeting with Hubert. Consider your wants and offers, his wants and offers and the possible reasons, stated and otherwise, for his request to you. Use the steps in contracting in Figure 3.1 to guide you.

SHARING OWNERSHIP – A CASE STUDY

At the beginning of the year, Ballard School has an unexpected influx of students, leading to some reorganization of classes. The grade 4 class becomes a mixed grade 3/4 group, with some objection on the part of the teacher. The superintendent suggests to her that the class can be taught as a grade 4 programme with some modifications to accommodate the five grade 3 students.

The class teacher meets with the resource teacher, requesting that the resource teacher assume responsibility for the grade 3 students during the grade 4 French period. The resource teacher replies that she will be happy to help with the programme. To her surprise, the class teacher replies that she doesn't want help; she wants the resource teacher to take over the programme. The resource teacher explains that she doesn't have the needed forty minutes per day, but she is willing to help the teacher find some suitable materials for these students. She suggests thematic units, research skills, and a variety of activities relating to an individualized reading programme. The class teacher rejects the ideas. She wants activities which the students can do without supervision so that she can spend the French time preparing for her other students. She notes that preparation time is a union issue and she is willing to call upon her union to support her. The resource teacher proposes that they meet with the principal to discuss the issue.

At the meeting the principal decides that the grade 3 students should be in a reading programme, and promptly orders a commercial kit, consisting of four readers, a tape and workbooks. No decision is made about who will introduce the programme, although the principal claims that it will not need much supervision once it has been introduced.

The kit arrives. The class teacher brings it to the resource teacher so that she 'can prepare the first few sessions'. On reviewing the material, the resource teacher sees that:

(a) the workbook activities are too advanced for these particular grade 3 students; and
(b) the material cannot be used without considerable assistance from the teacher.

The resource teacher begins to tell the principal that the kit is unsuitable, but at the mention of it, the principal makes it clear that this programme is to be used throughout the primary grades, and that this is the solution to all reading difficulties.

The resource teacher feels that she is getting deeper into hot water at every move:

(a) she has precipitated the introduction to an approach to reading instruction which she, and other teachers, dislike;
(b) she has failed to help the grade 3 students receive a suitable programme; and
(c) she has not resolved her own role with the teacher.

Questions

1 What options does the resource teacher have?
2 With whom should she meet?
3 What contingency plans should she have?
4 What plans should she make for the next meeting?

RULES OF ORDER – A CASE STUDY

Discuss the following case with a friend or colleague.

Mrs Flavan, an occupational therapist, is employed by a Local Education Authority to work with developmentally delayed pupils. Mrs Flavan works with a school psychologist, who, like her, is assigned to a group of schools to provide resource support to the teachers of the pupils. In each school, the occupational therapist, the psychologist and teacher work as a team to establish a programme for each pupil. Neville is a 14-year-old pupil with multiple handicaps including severe developmental delay. He is disruptive and occasionally violent. The psychologist has drafted a proposal for a behaviour modification programme which includes an aversive response when Neville is acting out. The aversive technique involves the teacher standing behind Neville and drawing back his arms and shoulders in a locked position until he quietens. At a team meeting, Mrs Flavan looks over the programme and states that she cannot approve its use because Neville has yet to be given an x-ray to determine whether his growth has changed his skeletal structure. The proposed aversive technique could be physically damaging to Neville. The meeting adjourns without resolution.

The chief psychologist, her colleague's boss, writes a strongly worded letter to Mrs Flavan which states that she has no business interfering with psychological procedures. Mrs Flavan telephones the chief psychologist to ask for an appointment to discuss the case. The chief psychologist replies that he has no time or inclination to talk to her, and he hangs up the telephone. She is angry.

Since her psychology colleague evidently believed it was important to talk to his boss, in order to convey the information that she has blocked the proposed programme, she confronts him. However, he is contrite and apologetic and insists that it was a chance remark to his boss, to which he expected no such reaction. He states that his loyalty is to the team and that he is prepared to continue to look for a solution to Neville's needs.

What should Mrs Flavan do? She considers her options:

1 Since her local psychology colleague has expressed agreement with her about the problems in administering aversive techniques to Neville, there are no negative consequences at the school level if she chooses not to take further action. She could let the matter drop.
2 She could discuss the issue with the superintendent to whom the chief psychologist reports. However, since there are no consequences at the school, that is, further discussion about Neville's programme is not in jeopardy, she might create the appearance of being a tell-tale (tattle tale).
3 She could speak to her own boss who is at the same rank as the chief psychologist but reporting to a different superintendent.

Review questions

1 Is there further information which Mrs Flavan needs? How would she go about getting it?
2 Are there other options?
3 Which options do you prefer, and why?
4 What are Mrs Flavan's wants and needs in this situation?
5 What outcomes, if any, would you seek?

References

Bachor, D. G., and Crealock, C. (1986). *Instructional strategies for students with special needs*. Scarborough, Ont.: Prentice-Hall.

Berliner, D. C. (1983). Developing conceptions of classroom environments: Some light on the T in classroom studies of ATI. *Educational Psychologist, 18*(1), 1–13.

Bickel, W. E., and Bickel, D. D. (1986). Effective schools, classrooms and instruction: Implications for special education. *Exceptional Children, 52*(6), 489–500.

Bines, H. (1986). *Redefining remedial education*. Beckenham, Kent: Croom Helm.

—— (1989). Whole school policies at primary level. *British Journal of Special Education, 16*(2), 80–2.

Block, P. (1981). *Flawless consulting: A guide to getting your expertise used*. San Diego, CA: University Associates.

Brennan, W. K. (1979). Curricular needs of slow learners. *Schools Council Working Paper, 63*. London: Evans/Methuen.

Campione, J. C. (1989). Assisted assessment: A taxonomy of approaches and an outline of strengths and weaknesses. *Journal of Learning Disabilities, 22*(3), 151–65.

Carlberg, C., and Kavale, K. (1980). The efficacy of special vs. regular class placement for exceptional children: A meta-analysis. *Journal of Special Education, 14*, 295–309.

Chalfant, J. C. (1987). Providing services to all students with learning problems: Implications for policy and programmes. In S. B. Vaughn and C. S. Bos (eds), *Research in learning disabilities: Issues and future directions* (pp. 239–51). London: Taylor & Francis.

Chalfant, J. C., and Pysh, M. V. D. (1989). Teacher assistance teams: Five descriptive studies on 96 teams. *Remedial and Special Education, 10*(6), 49–58.

Chalfant, J. C., Pysh, M. V. D., and Moultrie, R. (1979). Teacher assistance teams: A model for within-building problem solving. *Learning Disabilities Quarterly, 2*, 85–96.

Clark, M. M., Barr, J., and McKee, F. (1982). *Pupils with learning difficulties in the secondary schools: Progress and problems in developing a whole school policy*. University of Birmingham/Scottish Council for Research in Education, Edinburgh.

Connelly, F. M., and Clandinin, D. J. (1988). *Teachers and curriculum planners: Narratives of experience*. Toronto: OISE Press; New York: Teachers' College Press.

Conoley, J. C., and Conoley, C. W. (1982). *School consultation: A guide to practice and training*. New York: Pergamon Press, Inc.

Cooper, J., and Croyle, R. T. (1984). Attitudes and attitude change. In M. R. Rosenzweig and L. W. Porter (eds), *Annual Review of Psychology, 35*. Palo Alto, CA: Annual Reviews, Inc., 395–426.

Courtnage, L., and Smith-Davis, J. (1987). Interdisciplinary team training: A national survey of special education teacher training programmes. *Exceptional Children*, 53(5), 451–8.

Crandall, D. P., Eiseman, J. W., and Louis, K. S. (1986). Strategic planning issues that bear on the success of school improvement efforts. *Educational Administration Quarterly*, 22(3), 21–53.

Davies, J. D., and Davies, P. A. (1985). Parents, teachers and children with special needs. In C. Cullingford (ed.), *Parents, teachers and schools* (pp. 117–29). London: Robert Royce Ltd.

DES [Department of Education & Science] (1978). *Special educational needs* (Warnock Report). London: HMSO.

—— (1984). *Slow learning and less successful pupils in secondary schools*. London: HMSO.

—— (1985). *Report of Her Majesty's Inspectors on the effects of Local Authority expenditure policies on educational provisions in England*. London: HMSO.

Dessent, T. (1987). *Making the ordinary school special*. Lewes, Sussex: The Falmer Press.

Diamond, C. T. P. (1988). Construing a career: A developmental view of teacher education and the teacher educator. *Curriculum Studies*, 20(2), 133–40.

Edgar, E., and Hayden, A. H. (1984–5). Who are the children special education should serve and how many children are there? *The Journal of Special Education*, 18(4), 523–39.

Ehly, S. W., and Dustin, E. R. (1989). *Individual and group counselling in schools*. New York: Guilford Press.

Epps, S. and Tindal, G. (1987). The effectiveness of differential programming in serving students with mild handicaps: Placement options and instructional programming. In M. C. Wang, M. C. Reynolds, and H. J. Walberg (eds), *Handbook of special education: Research and practice* (Vol. I). Oxford: Pergamon Press.

Evans, S. (1980). The consultant role of the resource teacher. *Exceptional Children*, 46, 402–4.

Feuerstein, R. (1979). *The dynamic assessment of retarded performers: The learning potential assessment device: Theory, instruments and techniques*. Baltimore: University Park Press.

Fisher, R., and Ury, W. (1981). *Getting to yes: Negotiating agreement without giving in*. Boston: Houghton Mifflin.

Fleming, D. C., and Fleming, E. R. (1984). Consultation to improve a special education teacher's participation in annual review multidisciplinary team meetings. *Special Services in the Schools*, 1(2), 59.

Frassinelli, L., Superior, K., and Meyers, J. (1983). A consultation model for speech and language intervention. *ASHA*, 25(11), 25–30.

Frisby, C. (1987). The role of the subject coordinator. In I. Craig (ed.), *Primary school management in action*. Harlow: Longman.

Fuchs, D., and Fuchs, L. S. (1988). Evaluation of the Adaptive Learning Environments Model. *Exceptional Children*, 55(2), 115–27.

Fuchs, D., Fuchs, L. S., Bahr, M. W., Fernstrom, P., and Sticker, P.M. (1990). Prereferral intervention: A prescriptive approach. *Exceptional Children*, 56(6), 493–514.

Fullan, M. (1982). *The meaning of educational change*. Toronto: OISE Press.

—— (1985). Change processes and strategies at the local level. *The Elementary School Journal*, 85(3), 391–421.

Fullan, M., and Stiegelbauer, S. (1991). *The new meaning of educational change*, 2nd edn. Toronto: OISE Press; New York: Teachers' College Press.

Furey, E. M., and Strauch, J. D. (1983). The perception of teacher skills and knowledge by regular and special educators of mildly handicapped students. *Teacher Education and Special Education*, 6(1), 46–50.

Garnett, J. (1988). Support teaching: Taking a closer look. *British Journal of Special Education*, 15(1), 15–18.

Gartner, A., and Lipsky, D. K. (1987). Beyond special education: Toward a quality system for all students. *Harvard Educational Review*, 57(4), 367–94.

Gazda, G. M. (1973). *Human relations development: A manual for educators*. Boston: Allyn & Bacon.

Gipps, C., Gross, H., and Goldstein, H. (1987). *Warnock's eighteen per cent*. Lewes: Falmer Press.

Goodlad, J. (1975). *The dynamics of educational change*. Toronto: McGraw-Hill.

Goodman, K. S. (1976). *Miscue analysis: Applications to reading instruction*. Urbana, IL: RRIC Clearinghouse.

Graden, J. L. (1989a). Reactions to school consultation: Some considerations from a problem-solving perspective. *Professional School Psychology*, 4(1), 29–35.

—— (1989b). Redefining 'prereferral' interventions as intervention assistance: Collaboration between general and special education. *Exceptional Children*, 56(3), 227–31.

Gray, J., and Richer, S. (1988). *Classroom responses to disruptive behaviour*. Oxford: OUDES.

Green, H. (1989). Curriculum and pupil behaviour: One school's experience. In N. Jones (ed.), *School management and pupil behaviour*. Lewes: Falmer Press.

Hanko, G. (1985). *Special needs in ordinary classrooms*. Oxford: Blackwell.

Hargreaves, A. (1989). *Curriculum and assessment reform*. Milton Keynes: Open University Press; Toronto: OISE Press.

—— (1991). Cultures of teaching. In A. Hargreaves and M. Fullan (eds), *Understanding teacher development*. London: Cassells; Columbia: Teachers' College Press.

Heron, T. E., and Kimball, W. H. (1988). Gaining perspective with the educational consultation research base: Ecological considerations and further recommendations. *RASE: Remedial and Special Education*, 9(6), 21–8.

Heward, C., and Lloyd-Smith, M. (1990). Assessing the impact of legislation on special education policy: An historical analysis. *Journal of Education Policy*, 5(1), 21–36.

Hodgson, A., Clunies-Ross, L., and Hegarty, S. (1984). *Learning together: Teaching pupils with special education needs in the ordinary school*. Windsor: NFER-Nelson.

Huberman, M. (1992). Teacher development and instructional mastery. In A. Hargreaves and M. Fullan (eds), *Teacher development and educational change*. London and New York: Falmer Press.

Humphries, T., and Wilson, A. J. (1986). An instructional-based model for assessing learning disabilities. *Canadian Journal of Special Education*, 2, 55–66.

Hunt, D. E. (1989). Teachers' centres: Professional renewal for the nineties. *Orbit*, 20(4), 1.

Idol, L. (1988). A rationale and guidelines for establishing special education consultation programmes. *R.A.S.E.*, 9(6), 48–58.

Idol, L., and West, J. F. (1987). Consultation in special education, Part II. *Journal of Learning Disabilities*, 20, 474–97.

Johnson, L. J., Pugach, M. C., and Hammitte, D. J. (1988). Barriers to effective special education consultation. *Remedial and Special Education*, 9(6), 41–7.

Jordan, A., Kircaali-Iftar, G., and Diamond, C. T. P. (1993). Who has the problem, the student or the teacher? Differences in teachers' beliefs about their work with at-risk and integrated exceptional students. *International Journal of Disability, Development and Education, 40*(1), 45–62.

Joyce, B., and Showers, B. (1980). Improving inservice training: A message from research. *Educational Leadership, 37,* 379–85.

—— (1988). *Student achievement through staff development.* New York: Longmans.

Kelly, E. J. (1974). *Parent–teacher interaction.* Seattle, WA: Special Child Publications.

Kilbourn, B. (1990). *Constructive feedback: Learning the art.* Toronto: OISE Press; Cambridge, MA: Brookline Books.

Leithwood, K. A. (1988). *A review of research on the school principalship.* Washington, DC: The World Bank.

Levine, J. M., and Wang, M. C. (1983) (eds). *Teacher and student perceptions: Implications for learning.* Hillsdale, NJ: Erlbaum.

Little, J. W. (1985). Teachers as teacher advisors: The delicacy of collegial leadership. *Educational Leadership, 43*(3), 34–6.

Lloyd, J. W., Crowley, P., Kohler, F. W., and Strain, P. S. (1988). Redefining the applied research agenda: Cooperative learning, prereferral teacher consultation and peer-mediated interventions. *Journal of Learning Disabilities, 21*(1), 43–52.

McKenzie, D. E. (1983). Research for school improvement: An appraisal of some recent trends. *Educational Researcher, 12,* 5–17.

Margolis, H., and McGettigan, J. (1988). Managing resistance to instructional modifications in mainstreamed environments. *Remedial and Special Education, 9*(4), 15–21.

Medway, F. J., and Updyke, J. F. (1985). Meta-analysis of consultation outcome studies. *American Journal of Community Psychology, 13,* 489–505.

Menlo, A. (1986). Consultant beliefs which make a significant difference in consultation. In C. L. Warger and L. E. Aldinger (eds), *Preparing special educators for teacher consultation.* Toledo University: Ohio College of Education and Allied Professions.

Metcalf, L. (1987). Special Education Appeal Boards in Ontario. Unpublished dissertation. University of Toronto, OISE, Toronto.

Miles, M. B., Saxl, E. R., and Lieberman, A. (1988). What skills do educational 'change agents' need? An empirical view. *Curriculum Inquiry, 18*(2) 157–93.

Miller, J.L. (1990). *Creating spaces and finding voices: Teachers collaborating for empowerment.* SUNY series on teacher preparation and development. New York: SUNY Press.

Peck, M. S. (1978). *The road less travelled: A new psychology of love, traditional values and spiritual growth.* New York: Touchstone.

Postlethwaite, K., Hackney, A., and Raban, B. (1986). Provision for pupils aged 11–13 with special educational needs in Oxfordshire. Unpublished manuscript. Oxford Educational Research Group: OUDES.

Potts, P. (1985). Training for teamwork. In J. Sayer and N. Jones (eds), *Teacher training and special education needs.* Beckenham: Croom Helm.

Pugach, M. C., and Johnson, L. J. (1989). Prereferred interventions: Progress, problems and challenges. *Exceptional Children, 56*(3), 217–26.

Pugach, M., and Whitten, M. E. (1987). The methodological content of teacher education for learning disabilities: A problem of duplication. *Learning Disabilities Quarterly, 10,* 291–300.

Quicke, J. (1985). Initial teacher education and in role support agencies. In J. Sayer and N. Jones (eds), *Teacher training and special education needs.* Beckenham: Croom Helm.

Reschly, D. J. (1980). School psychologists and assessment in the future. *Professional Psychologist, 11*, 841–8.

—— (1988). Special education reform: School psychology revolution. *School Psychology Review, 17*, 459–75.

Reynolds, M. C., Wang, M. G., and Walberg, H. J. (1987). The necessary restructuring of special and regular education. *Exceptional Children, 53*, 391–8.

Rosenfield, S. (1985). Teacher acceptance of behavioural principles. *Teacher Education and Special Education, 8*, 153–8.

Rosenholtz, S. J. (1985, May). Effective schools: Interpreting the evidence. *American Journal of Education, 93*, 352–87.

Rosenshine, B. V. (1983). Teaching functions in instructional programmes. *Elementary School Journal, 83*, 335–52.

Salvia, J., and Ysseldyke, J. E. (1985). *Assessment in special and remedial education*, 3rd edn. Boston: Houghton Mifflin.

Sarason, S. B., and Doris, J. (1979). *Educational handicap, public policy and social history: A broadened perspective on mental retardation*. New York: Free Press.

Schön, D. (1983). *The reflective practitioner: How professionals think in action*. New York: Basic Books.

—— (1987). *Educating the reflective practitioner: Toward a new design for teaching and learning in the professions*. San Francisco: Jossey Bass.

Slavin, R. E. (1988). Synthesis of research on grouping in elementary and secondary schools. *Educational Leadership*, September, 67–76.

Smith, C. J. (1982). Helping colleagues cope: A consultant role for the remedial teacher. *Remedial Education, 17*(2), 75–9.

Solity, J., and Raybould, E. (1988). *A teacher's guide to special needs: A positive response to the 1981 Education Act*. Milton Keynes: Open University Press.

Speece, D. L., and Mandel, C. J. (1980). Resource room support services for regular teachers. *Learning Disabilities Quarterly, 3*, 49–54.

Thomas, G. (1985). Room management in mainstream education. *Educational Research, 27*(3), 186–93.

—— (1986). Integrating personnel in order to integrate children. *Support for Learning, 1*(1), 19–26.

—— (1987). Extra people in the primary classroom. *Educational Research, 29*(3), 173–81.

Thurlow, M., Graden, J., Greener, J., and Ysseldyke, J. (1983). LD and non-LD students' opportunities to learn. *Learning Disabilities Quarterly, 6*(2), 172–83.

Tindal, G. A., and Taylor-Pendergast, S. J. (1989). A taxonomy for objectively analysing the consultation process. *Remedial and Special Education, 10*, 6–16.

Vogel, S. A. (1990). Gender differences in intelligence, language, visual–motor abilities, and academic achievement in students with learning disabilities: A review of the literature. *Journal of Learning Disabilities, 23*(1), 44–52.

Wang, M. C., and Baker, E. T. (1985–6). Mainstreaming programmes: Design features and effects. *Journal of Special Education, 19*, 503, 525.

Wang, M. C., Gennari, P., and Waxman, H. C. (1985). The adaptive learning environments model: Design, implementation and effects. In M. C. Wang and H. J. Walberg (eds), *Adapting instruction to individual differences*. Berkeley, CA: McCutchan Pub. Co. (pp. 191–235).

West, J. F., and Cannon, G. S. (1988). Essential collaborative consultation competencies for regular and special educators. *Journal of Learning Disabilities, 21*(1), 56–63.

West, J. F., and Idol, L. (1987). School consultation, Part I. An interdisciplinary perspective on theory, models and research. *Journal of Learning Disabilities, 20*(7), 388–408.

Wilson, A. J., and Silverman, H. (1990). Teachers' assumptions and beliefs about exceptionality. In N. Jones (ed.), *Special Education Needs Review* (Vol. III) (pp. 69–81). Lewes: Falmer Press.

—— (1991). Teachers' assumptions and beliefs about the delivery of services to exceptional children. *Teacher Education and Special Education, 14*(3), 198–206.

Winkley, D. (1985). The school's view of parents. In C. Cullingford (ed.), *Parents, teachers and schools*. London: Robert Royce Ltd.

Wolfendale, S. (1988). *Primary schools and special needs*. London: Cassells.

Wolfgang, A. (1984). *People-watching across cultures* (booklet). Toronto: Department of Applied Psychology, OISE.

Yarger, S. J. (1990). The legacy of the teacher centre. In B. Joyce (ed.), *Changing school culture through staff development*. Alexandria: Association for Supervision and Curriculum Development.

Ysseldyke, J. E. (1983). Current practices in making psychoeducational decisions about learning disabled students. *Journal of Learning Disabilities, 16*, 226–33.

Zischka, P. C., and Fox, R. (1985). Consultation as a function of school social work. *Social Work in Education, 7*(2), 69–79.

Name index

Subject index

accountability 105
achievement 10, 75, 90, 91, 92, 100; monitored 88; Record of 29; tests of 19, 20, 27, 51
action plans *see* plans of action
active listening 44
adaptive instruction 92
administrators *see* school administrators
ALEM (Assisted Learning Model) 106
ancillary staff 4
assessment 9, 26, 34, 92, 97; competence in 28; consulting 47–9, 57; curriculum-based 51; cycle 46, 47, 48, 52, 53; diagnostic 24, 79, 89; dynamic 51; effective 35, 53; formal scientific, criteria of 54; further 84; instructional 27, 33, 87; programme-based 24; psychological 8, 19, 27; purpose of 25; referral for 19; results 53; tools 51; *see also* preventive approach; referrals; restorative approach
at-risk pupils 1, 7, 14, 17, 30; characteristics 88–9; needs 4, 5, 107
attitudes of staff 4, 41, 79, 96, 104; current/prevailing 5, 12; negative or prejudicial 12

barriers 95–6
behaviour 52; classroom 74; group management of 63, 96; interpersonal 88; performance and 20, 53, 75; problems of pupils 4, 22, 59–60, 75
beliefs 4, 5, 8–12, 63, 79
biases 57
body language 64

'Brad the Brat' 68, 115–16
brainstorming of ideas 61, 83, 85
Britain *see* Great Britain
'buddy systems' 90

Canada 3n, 7, 29, 84, 92, 94; primary schools 88; *see also* Ontario
change agents 5, 94
characteristics: at-risk pupils 88–9; psychological and intellectual 20; special needs pupils 88–9
clients 32–3, 42; assurance to 44–5; expectations of collaboration 40; not being caught off guard by requests of 42; professional responsibility to balance needs of 52; relationship between consultant and 62–71, 76–8; role of 64; *see also* resistance
coaching 2, 93
cognitive and social training 73
communication 13; parent–teacher 7, 29, 73
comprehensive schools 7, 29, 30, 87, 88–90
conflicts of interest 62, 69
connotations: emotional 60; potentially harmful 75
consolidation 34, 47
consultant's role 32, 41, 53, 64, 99; facilitative rather than expert 77; modelling 3, 33; multicultural 103; needs/offers 40–2; subject-matter expertise 29; understanding of 46
consultation 70, 71; classroom, overview of 94–107; collaborative, in context 1–15; confidential 77–8; difficult situations 62–7; skills 4–5,